Conference Proceedings of the Society for Experimental Mechanics Series

Series Editor

Kristin B. Zimmerman, Ph.D.
Society for Experimental Mechanics
Bethel, CT, USA

More information about this series at http://www.springer.com/series/8922

Anders Brandt • Raj Singhal
Editors

Shock & Vibration, Aircraft/Aerospace, Energy Harvesting, Acoustics & Optics, Volume 9

Proceedings of the 34th IMAC, A Conference and Exposition on Structural Dynamics 2016

Springer

Editors
Anders Brandt
Department of Technology and Innovation
University of Southern Denmark
Copenhagen, Denmark

Raj Singhal
Canadian Space Agency
Montreal, Canada

ISSN 2191-5644 ISSN 2191-5652 (electronic)
Conference Proceedings of the Society for Experimental Mechanics Series
ISBN 978-3-319-30086-3 ISBN 978-3-319-30087-0 (eBook)
DOI 10.1007/978-3-319-30087-0

Library of Congress Control Number: 2016935715

Printed on acid-free paper

This Springer imprint is published by Springer Nature
The registered company is Springer International Publishing AG Switzerland

Preface

Shock & Vibration, Aircraft/Aerospace, Energy Harvesting, Acoustics & Optics represent one of ten volumes of technical papers presented at the 34th IMAC: Conference & Exposition on Structural Dynamics, organized by the Society for Experimental Mechanics and held in Orlando, Florida, January 25–28, 2016. The full proceedings also include volumes on nonlinear dynamics; dynamics of civil structures; model validation and uncertainty quantification; dynamics of coupled structures; sensors and instrumentation; special topics in structural dynamics; structural health monitoring, damage detection and mechatronics; rotating machinery, hybrid test methods, vibro-acoustics and laser vibrometry; and topics in modal analysis and testing.

Each collection presents early findings from experimental and computational investigations on an important area within structural dynamics. The topics represent papers on practical issues improving energy harvesting measurements, shock calibration and shock environment synthesis, and applications for aircraft/aerospace structures.

The organizers would like to thank the authors, presenters, session organizers, and session chairs for their participation in this track.

Copenhagen, Denmark — Anders Brandt
Montreal, QC, Canada — Raj Singhal

Contents

1 **Improved Cutting Force Measurements in Milling Using Inverse Filtering** 1
Martin Magnevall and Tomas Beno

2 **Use of a Depth Camera as a Contactless Displacement Field Sensor** 13
Jean Michel Franco, Johannio Marulanda, and Peter Thomson

3 **Uncertainty of Digital Image Correlation with Vibrating Deformable Targets** 21
Alfredo Cigada, Navid Hasheminejad, and Emanuele Zappa

4 **Physical Vibration Simulation of an Acoustic Environment with Six Shakers on an Industrial Structure** 29
Randall L. Mayes and Daniel P. Rohe

5 **Developing Conservative Mechanical Shock Specifications** 43
Matthew Baker, Kelsey Neal, Katrina Sweetland, Garrison Stevens, Dustin Harvey, and Stuart Taylor

6 **Force Limited Vibration Using the Apparent Mass Method** 53
Paul Marchand, Raj Singhal, and Mark O'Grady

7 **Harmonic Force Excitation Analysis of a Small-Body Asteroid/Satellite System** 67
Joshua Johnson, William H. Semke, Shankar Nag Ramaseri Chandra, and Ronald Fevig

8 **A Study on the Dynamic Interaction of Shock Response Fixtures and Test Payload** 77
Jesus M. Reyes and Peter Avitabile

9 **Modal Analyses and Experimental Verifications of Joined-Wing Configurations** 87
Berkan Alanbay, Melin Şahin, and Güvenç Canbaloğlu

10 **Modal Testing of James Webb Space Telescope (JWST) Optical Telescope Element (OTE)** 103
Douglas J. Osterholt, David Cloutier, Timothy Marinone, and Reem Hejal

11 **Quantification of Dynamic Differences Between Boundary Conditions for Environment Specification Improvement** 117
Julie M. Harvie and Randy Mayes

12 **Modal Updating of Tail of a Military Helicopter** 133
Kurtuluş Ersoy, Mert Atasoy, and Cem Genç

13 **Modeling of High Frequency Shock Tests** 145
Washington J. DeLima, Melanie N. Ambrose, and Richard Jones

14 **A Novel Method to Correlate a Rocket Launcher Finite Element Model Using Experimental Modal Test Measurements and Identification Algorithms** 153
Ronald N. Couch, Eliott J. Radcliffe, and Rickey A. Caldwell

15 Numerical Studies on the Reduced Order Modeling of Frictionless Joint Contact Interfaces 167
M. Breitfuss and H.J. Holl

16 Structural Design with Joints for Maximum Dissipation 179
M. Stender, A. Papangelo, M. Allen, M. Brake, C. Schwingshackl, and M. Tiedemann

17 A Hybrid Piezoelectric and Electrostatic Vibration Energy Harvester 189
H. Madinei, H. Haddad Khodaparast, S. Adhikari, and M.I. Friswell

18 Design of Scaled-Down Composite I-Beams for Dynamic Characterization in Subcomponent Testing of a Wind Turbine Blade 197
Mohamad Eydani Asl, Christopher Niezrecki, James Sherwood, and Peter Avitabile

Chapter 1
Improved Cutting Force Measurements in Milling Using Inverse Filtering

Martin Magnevall and Tomas Beno

Abstract Accurate estimates of cutting forces in metal cutting are important in the evaluation of e.g. different cutting tool geometries and concepts. However, dynamic influences from the measurement system affect the measurement result and may make the obtained cutting force data erroneous and misleading. This paper presents a method to construct an inverse filter which compensates for the dynamic influences from the measurement system. Using the suggested approach, unwanted dynamic effects from the measurement system can be counteracted. By applying the inverse filter it is possible to retain information related to the cutting forces at higher frequencies than possible with unfiltered data. The advantage of using the proposed method is illustrated by comparing simulated, inverse- and low-pass filtered cutting forces to unfiltered forces at different cutting speeds. The results indicate that inverse filtering can increase the usable frequency range of the force dynamometer and thereby provide more accurate and reliable results compared to both low-pass and unfiltered force measurements.

Keywords Metal cutting • Cutting force • Dynamometer • Inverse filter • Deconvolution

1.1 Introduction

Cutting forces are one of the most important quantity in the metal machining process. The cutting forces govern power and torque requirements in the machine tool and drive heat generation which catalyzes tool wear and determines the magnitude and direction of residual stresses in the machined component. Cutting forces also cause deflection of the cutting tool, machine tool and work piece and may have a negative influence on the machined components quality. The cutting forces are therefore important parameters in the evaluation of different cutting tool geometries and concepts. However, dynamic influences from the measurement system affect the result and make it difficult to obtain accurate cutting force data. A commonly used method to remove unwanted dynamic effects from the measured cutting forces is low-pass filtering. Low-pass filtering removes all information above a specified cut-off frequency and may therefore also remove important information related to the true cutting forces contained in frequencies above the cut-off frequency. This is especially evident in milling with transient cutting conditions when the rise times are short and the cutting forces thereby have high frequency content. Therefore, it is difficult to get reliable estimates of the amplitudes and rise times by low-pass filtering transient cutting forces, especially at high cutting speeds.

Accurate estimates of both rise times and force amplitudes are important, e.g., when evaluating and comparing different tools and insert geometries. An alternative approach, to low-pass filtering, that improves the cutting force estimates, i.e. increases the effective frequency range of the force dynamometer, is therefore of great interest.

For example, Tlusty et al. used accelerometers to compensate for the inertia and structural damping of the dynamometer, thereby increasing the effective frequency range [1]. This method has been proven to work under certain conditions, but encounters difficulties around resonance frequencies when the system inertia or damping is large. The method also has problems handling systems with more than one dominating mode [2]. Park and Altintas used a Kalman filtering technique to compensate for unwanted dynamics and process and measurement noise of a spindle integrated force sensor, [3, 4]. The same method has also been applied to cutting force measurements in micro end milling, [5]. Jensen et al., [2], developed a

M. Magnevall (✉)
AB Sandvik Coromant, SE-811 81 Sandviken, Sweden
e-mail: martin.magnevall@sandvik.com

T. Beno
University West, SE-461 86 Trollhättan, Sweden
e-mail: tomas.beno@hv.se

A. Brandt, R. Singhal (eds.), *Shock & Vibration, Aircraft/Aerospace, Energy Harvesting, Acoustics & Optics, Volume 9*, Conference Proceedings of the Society for Experimental Mechanics Series, DOI 10.1007/978-3-319-30087-0_1

method to obtain an invertible filter of a non-minimum phase frequency response function (FRF) of a force dynamometer. The method was tested on experimental data with promising results. The basic principle of the method is to divide the system's transfer function into one stable and one unstable causal IIR-filter. The unstable causal IIR-filter is transformed into a non-causal stable FIR-filter. Then, by using these two filters in series a stable inverse filter is obtained. Depending on the location of the non-minimum-phase zeros of the system's transfer function, the length of the non-causal FIR-filter can become large, introducing long time delays. However, if the method is applied on large data sequences, this will not cause a problem.

In the case when the force dynamometer can be considered linear and the cross-frequency responses between the directions x, y and z are negligible, a minimum-phase inverse filter can be constructed and used to counteract the dynamometer dynamics and thereby increase the usable bandwidth of the dynamometer. A procedure for creating a minimum-phase inverse filter is described in this paper and applied in both simulations and on experimental data from milling under various cutting conditions.

1.2 Inverse Filtering

Consider the FRF $H(\omega)$ between the applied force $F_R(\omega)$ and the force output from the dynamometer $F_D(\omega)$, Eq. (1.1).

$$H(\omega) = \frac{F_D(\omega)}{F_R(\omega)} \tag{1.1}$$

In an ideal case, i.e. when there are no dynamic influences from the dynamometer, the magnitude of $H(\omega)$ is equal to unity and the phase equal to zero for all frequencies. However, due to e.g., the mass and shape of the work piece, the stiffness and damping of the force dynamometer and the distance between the actual cutting position and the positions of the force transducers in the dynamometer, the magnitude and phase of the frequency response between applied force and force output will deviate from the ideal values and thus the measured force will differ from the applied force. By applying an inverse filter, describing $H^{-1}(\omega)$, on the recorded force signal these unwanted dynamic effects can be counteracted. If $H(\omega)$ is stable and minimum-phase (all zeros of the system lies within the unit circle in the z-domain) then the system is directly invertible. Usually, mechanical systems are stable and mixed-phase (zeros both inside and outside the unit circle). If the system has zeros outside the unit circle it cannot be directly inverted, since the result will then have unstable poles and the filter output exponentially tends toward infinity. However, a mixed or maximum-phase FRF can be transformed into a minimum-phase FRF while still keeping the amplitude characteristics but changing the phase. An invertible filter describing the characteristics of the minimum-phase FRF can then be estimated as described in this section.

1.2.1 Transformation into Minimum-Phase

A mixed- or maximum-phase transfer function can be transformed into a minimum-phase transfer function by, e.g., using the Hilbert transform [6] or cepstrum [7, 8]. In this paper, real cepstrum is used to transform the FRF of the force dynamometer into minimum-phase.

Let $h(n)$ be a real sequence with $H(\omega)$ as its Fourier transform. Its real and complex cepstrum $\hat{c}(n)$ and $\hat{h}(n)$ are defined as:

$$\begin{aligned} \hat{C}(\omega) &= (H(\omega)) = \log|H(\omega)| \\ \hat{c}(n) &= \mathcal{F}^{-1}(\hat{C}(\omega)) \\ \hat{H}(\omega) &= \log((H(\omega))) \\ \hat{h}(n) &= \mathcal{F}^{-1}(\hat{H}(\omega)) \end{aligned} \tag{1.2}$$

where $\log|H(\omega)|$ refers to the natural logarithm of $|H(\omega)|$ and $\mathcal{F}^{-1}$ denotes the inverse Fourier transform. Some useful relations between minimum-phase and maximum-phase sequences and their complex cepstrums are, [8]:

- If $h(n)$ is a minimum-phase sequence, $\hat{h}(n)$ will be a casual sequence.
- If $h(n)$ is a maximum-phase sequence, $\hat{h}(n)$ will be an anti-causal sequence.

Let the minimum-phase counterpart to $h(n)$ be denoted by $h_{min}(n)$ and its complex cepstrum denoted by $\hat{h}_{min}(n)$. The Kramers–Kronig relations for a causal sequence states that the entire sequence can be described by its even part. The relationship between the even part of $\hat{h}(n)$ and its Fourier transform is given in [9]:

$$\hat{h}_e(n) = \mathcal{F}^{-1}((H(\omega))) = \hat{c}\,(n) \tag{1.3}$$

Since $\hat{h}(n)$ is a causal sequence when $h(n)$ is minimum-phase, $\hat{h}_{min}(n)$ can be estimated as:

$$\hat{h}_{min}(n) = \begin{cases} 2\hat{c}\,(n) & n > 0 \\ \hat{c}\,(n) & n = 0 \\ 0 & n < 0 \end{cases} \tag{1.4}$$

According to Equation (1.2), the minimum-phase transfer function $H_{min}(\omega)$ can be obtained by:

$$H_{min}(\omega) = e^{\mathcal{F}\left(\hat{h}_{min}(n)\right)} \tag{1.5}$$

1.2.2 Fitting an Invertible Digital Filter to the Minimum Phase FRF

The Steiglitz-McBride iteration algorithm is used to find the filter coefficients describing the given impulse response of the estimated minimum phase FRF, $H_{min}(\omega)$, [10]. This method is based on non-parametric frequency response characteristics and describes the identified FRF using a polynomial model. The identification is performed using in-house written Python code, alternatively the MATLAB function invfreqz.m can be used to estimate the coefficents [11]. The identified model is a discrete representation of the minimum-phase transfer function of the force dynamometer, $h_{min}(n)$, and can be represented as:

$$H_{min}(z) = \frac{B(z)}{A(z)} = \frac{b(1) + b(2)z^{-1} + \ldots + b(nb+1)z^{-nb}}{a(1) + a(2)z^{-1} + \ldots + a(na+1)z^{-na}} \tag{1.6}$$

The result is an invertible IIR-filter described by the coefficients b and a, where nb and na are the total number of coefficients in the numerator and denominator, respectively. nb and na are selected by visual inspection ensuring a satisfactory fit between the measured and estimated FRFs.

1.3 Simulations

To verify the proposed method and identify usable frequency ranges for the inverse filters, simulations using a model of the force dynamometer obtained from experimental data are performed. The model is based on the FRF matrix of the force dynamometer, with the work piece, mounted in the machine tool. The FRF matrix was estimated using impulse excitation in x- and y-directions, results are shown in Figs 1.1 and 1.2. The cross-frequency response in the region 0 to 3000 Hz is small in both directions. Also, the coherence functions are close to unity up to approximately 3000 Hz, implying linear relationships between inputs and outputs. These results indicate that a linear model, with negligible cross frequency response, of the force dynamometer is valid for frequencies up to approximately 3000 Hz.

Based on the FRF measurements of the dynamometer, inverse filters in both x and y-directions are estimated and evaluated with respect to both amplitude and phase correction. Comparisons between the measured FRFs and the minimum-phase FRFs are shown in Figs 1.3 and 1.4. To clearly show the amplitude and phase characteristics of the inverse filters, the combined FRFs are also displayed; these are calculated as:

1. Inverse Fourier transform the measured mixed-phase FRF, $H(\omega)$, the result is the impulse response.
2. Apply the inverse filter to the obtained impulse response; the result is a unit impulse.
3. The combined FRF is the Fourier transform of the signal obtained from the inverse filter.

If the inverse filter behaves perfectly, the magnitude of the combined FRF should be unity and the phase zero for all frequencies.

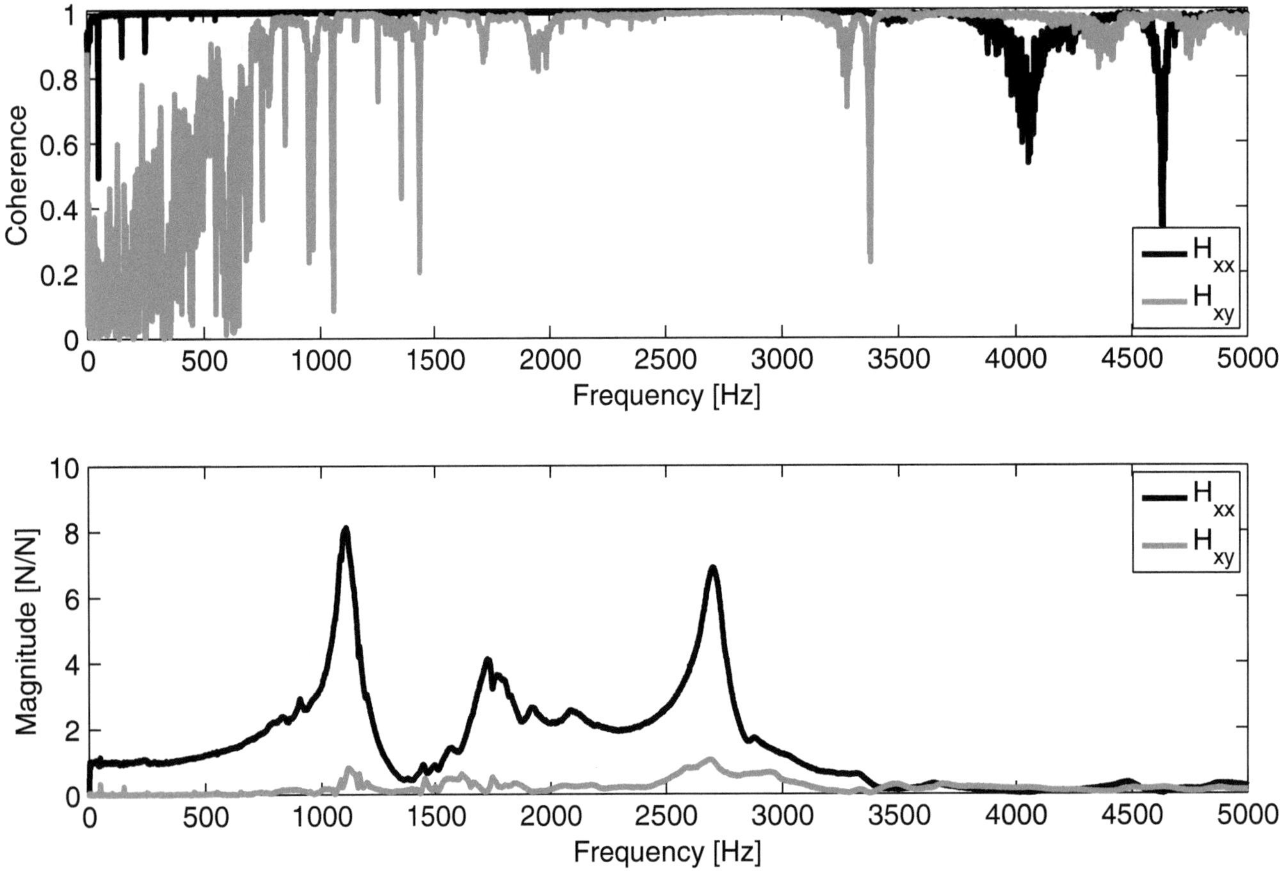

Fig. 1.1 Measured FRFs of the force dynamometer from impulse excitation. The dynamometer is excited in the *x*-direction and responses collected in *x*- and *y*-directions

The magnitudes of $H_{xx}(\omega)$ and $H_{yy}(\omega)$ matches well with their respective minimum-phase FRFs $H_{xx(min)}(\omega)$ and $H_{yy(min)}(\omega)$. The combined FRFs indicate a good amplitude correction over the entire frequency range, even at dominant modes. Due to the minimum-phase transformation, the phase responses of measured and minimum-phase FRFs differ, which is clearly visible in the phase responses of the combined FRFs. However, the phase responses of the combined FRFs are dominated by linear trends which can be related to constant delays in the time domain and will not affect the amplitude responses of the inverse filtered signals. Thus, the phase responses affecting the inverse filtered signals are obtained by removing the linear phase trends from the combined FRFs. As seen in Figs 1.3 and 1.4, the phases of the combined FRFs have deviated approximately 7 degrees from zero at 2500 Hz. Therefore, it is expected that frequencies above 2500 Hz will be out of range for the inverse filter. To avoid influences from the resonances at 2550 Hz in the *x*-direction and 2700 Hz in the *y*-direction, the frequency limit of the inverse filters was set to 2400 Hz in both directions; higher frequencies are removed by low-pass filtering. Low-pass filtering also removes any high-frequency noise present in the inverse filtered signal. High frequency noise is common when performing inverse filtering due to the nature of the filter. The reason is that mechanical systems normally act as low-pass filters, attenuating high frequencies. When these systems are inverted they will instead act as high-pass filters and therefore respond badly to high frequency noise. The low-pass filter design used in the simulations is a Butterworth filter of order 3. To remove any phase distortion caused by the low-pass filter, zero-phase filtering is performed by filtering the data in both forward and reverse directions.

To test the behavior of the inverse filters, simulations are carried out for the cutting speeds and feed rates listed in Table 1.1. Down milling and 50 % radial immersion is used to excite the dynamometer with a high frequency transient signal and thereby clear effects from the dynamometer dynamics appear. The simulations are carried out using both *x* and *y* as feed directions. Mechanistic modeled cutting forces are used as input (see Table 1.2), these are filtered through ETFEs of the measured mixed-phase transfer functions, $h_{xx}(n)$, $h_{xy}(n)$ and $h_{yy}(n)$, $h_{yx}(n)$. Each output is then inverse filtered and compared to the reference cutting force (simulated input force), Figs 1.5 and 1.6. The simulated output forces have contributions both from the point FRFs and the cross FRFs. Due to the phase distortion caused by the minimum-phase transformation,

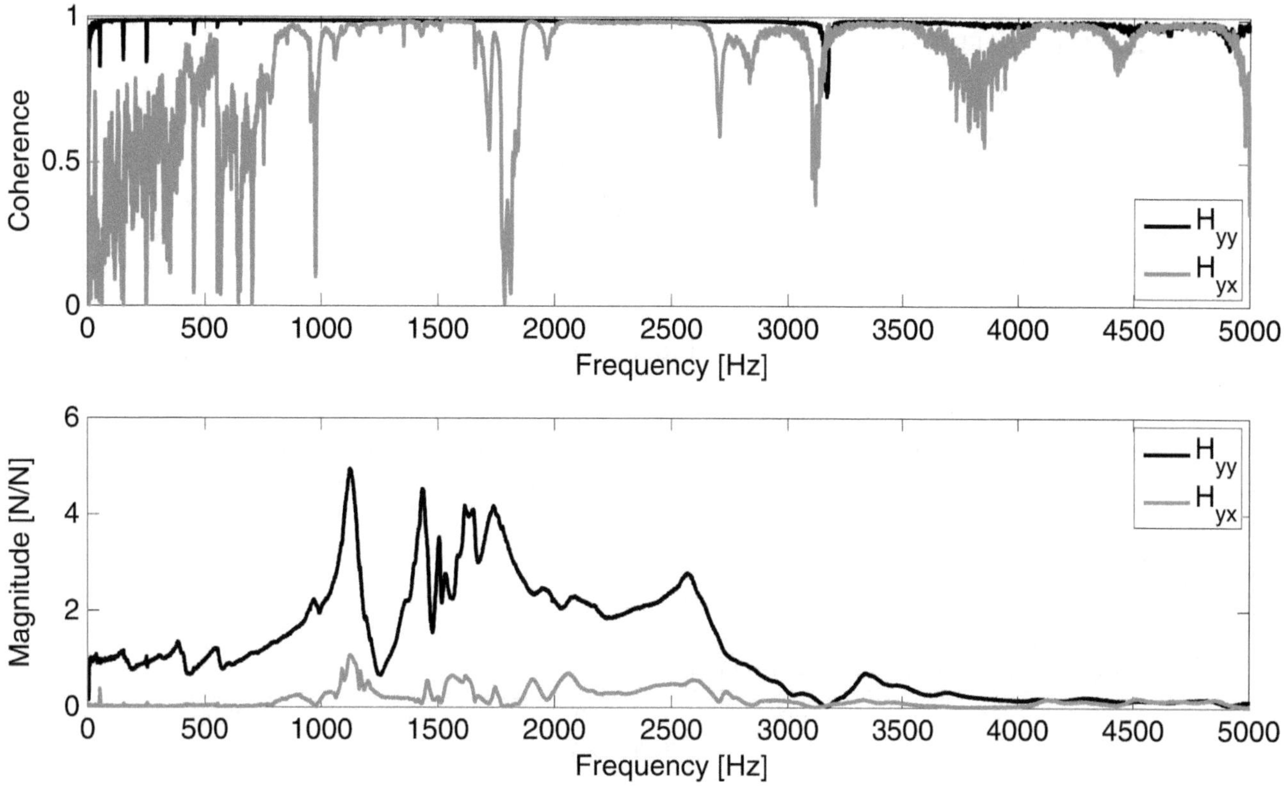

Fig. 1.2 Measured FRFs of the force dynamometer from impulse excitation. The dynamometer is excited in the *y*-direction and responses collected in *y*- and *x*-directions

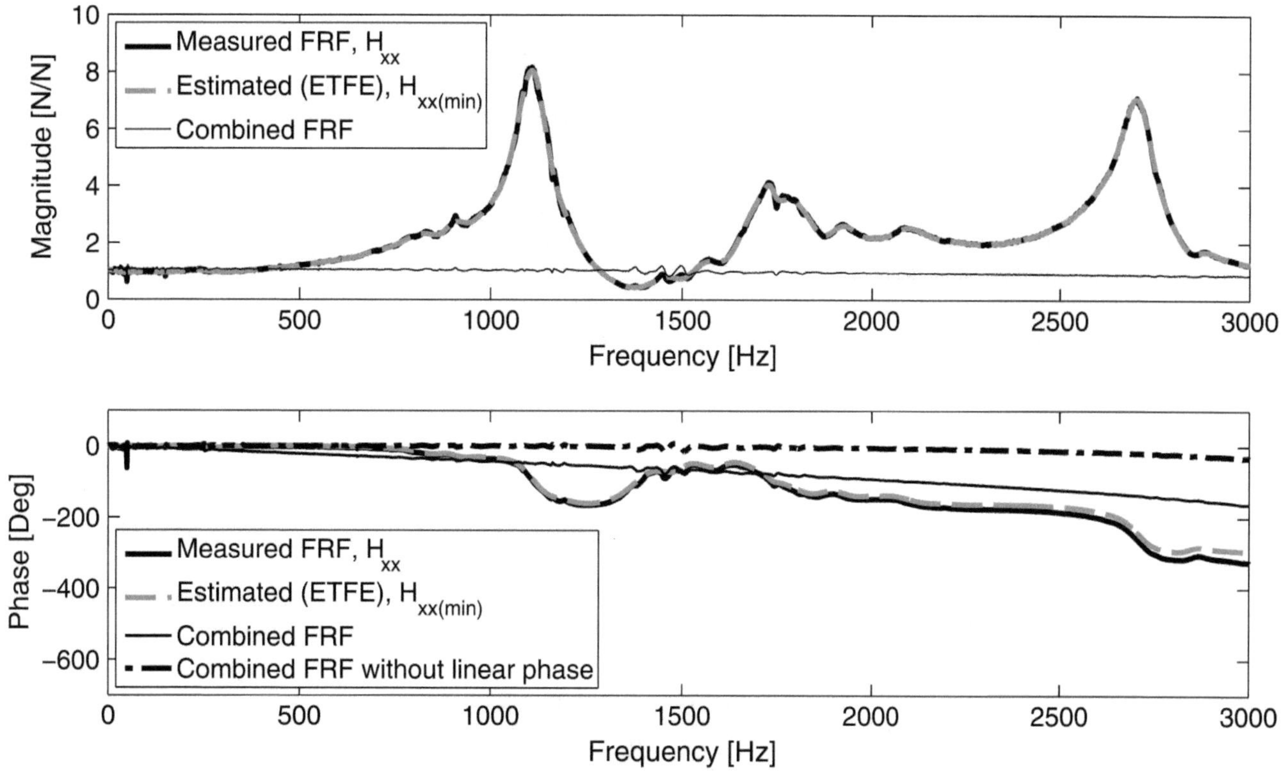

Fig. 1.3 Magnitude and phase responses of $H_{xx}(\omega)$, $H_{xx(min)}(\omega)$ and the combined FRF, $nb = 200$ and $na = 55$. The dynamometer's dominant modes are: 1110 Hz, 1725 Hz, 2700 Hz

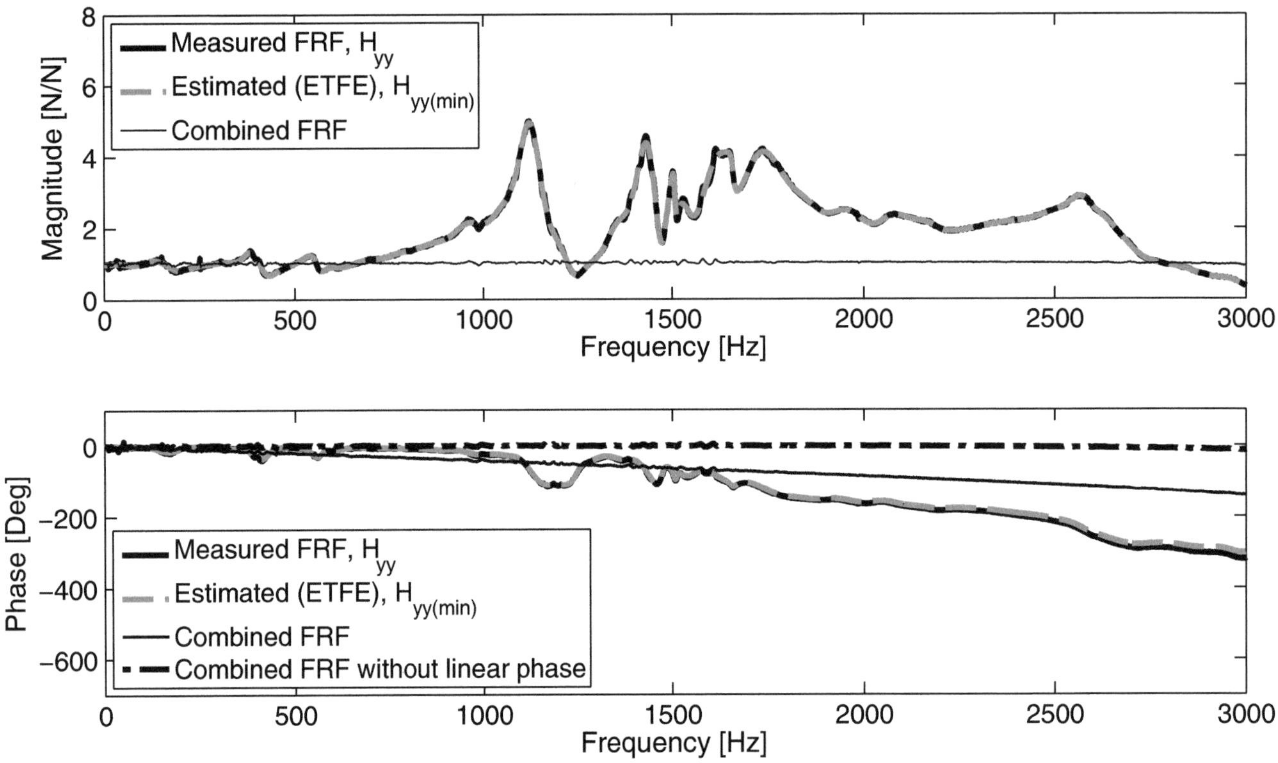

Fig. 1.4 Magnitude and phase responses of $H_{yy}(\omega)$, $H_{yy(min)}(\omega)$ and the combined FRF, $nb = 200$ and $na = 55$. The dynamometer's dominant modes are: 1120 Hz, 1430 Hz, 1630 Hz, 1750 Hz and 2550 Hz

Table 1.1 Equipment and cutting data used in the cutting force measurements

Machine tool	MORI SEIKI NV5000
Cutting tool	Sandvik R331.35-050A20EM100
Insert	Sandvik N331.1A-08 45 08H-NL H10
Work piece material	AISI 7075
Force dynamometer	Kistler 9255B
Charge amplifier	Kistler 5011
Cutter diameter, D_{cap}	50 [mm]
Radial cutting depth, a_e	10; 25; 40 [mm]
Axial cutting depth, a_p	3 [mm]
Number of teeth, z_c	1
Cutting speeds, v_c	200; 400; 800; 1200; 1490; 2000 [m/min]
Tooth passing frequencies, f_t	21.2; 42.4; 84.8; 127.3; 158.1; 212.2 [Hz]
Feed rates, f_z	0.05; 0.1; 0.15; 0.2; 0.25 [mm/tooth]

superposition of e.g. $h_{xx(min)}(n)$ and $h_{xy(min)}(n)$ does not apply. Therefore only $h_{xx(min)}(n)$ or $h_{yy(min)}(n)$ are used in the inverse filters. Thus, for the results to be accurate, the effect of the cross FRFs have to be negligible.

The similarity between the signals is estimated by calculating the ratio of the energy in the difference between reference and inverse filtered force, $E(f_R(t) - f_I(t))$ and the energy in the reference signal, $E(f_R(t))$, according to:

$$\eta = \min\left(1 + \frac{R_{f_I f_I}(0) - \bar{f}_I^2 - 2\left(R_{f_R f_I}(\tau) - \bar{f}_R \bar{f}_I\right)}{R_{f_R f_R}(0) - \bar{f}_R^2}\right) \tag{1.7}$$

where $R_{f_R f_R}(0)$ and $R_{f_I f_I}(0)$ are the autocorrelations at zero time delay between reference and inverse filtered forces, respectively. $R_{f_R f_I}(\tau)$ refers to the cross correlation between reference and inverse filtered forces, where τ is the time delay

Table 1.2 Estimated cutting force coefficients from milling tests

v_c		K_{tc}	K_{te}	K_{rc}	K_{re}	K_{ac}	K_{ae}	
200	[m/min]	732.26	26.38	163.71	15.65	−52.07	18.90	[N/mm^2]
400	[m/min]	693.05	18.37	128.67	9.66	−36.80	14.771	[N/mm^2]
800	[m/min]	638.00	17.92	99.43	11.33	−5.47	12.69	[N/mm^2]
1200	[m/min]	649.03	21.78	42.14	14.67	−23.76	4.61	[N/mm^2]
1490	[m/min]	617.35	19.71	24.12	12.77	−87.01	16.72	[N/mm^2]
2000	[m/min]	597.66	17.40	24.44	11.61	−105.45	15.61	[N/mm^2]

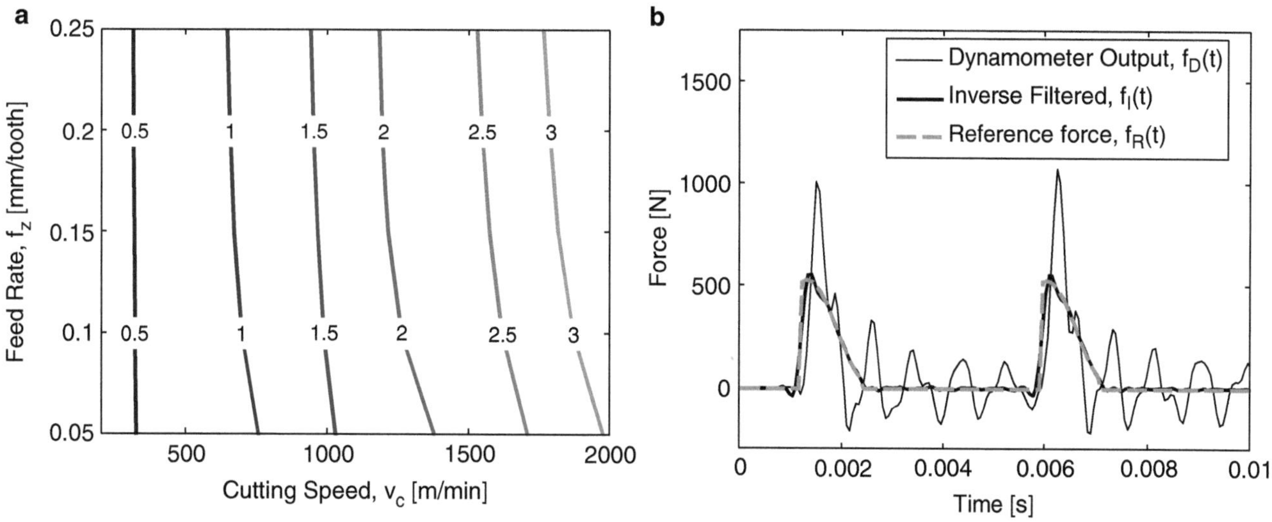

Fig. 1.5 Simulation results with feed in *x*-direction; (**a**) The isolines of η [%], Eq. (1.7). (**b**) Comparison between dynamometer output, inverse filtered and reference input forces (Cutting speed, $v_c = 2000$ [m/min]; Feed rate, $f_z = 0.25$ [mm/tooth])

between the two signals, [12]. $\bar{f}_R$ and $\bar{f}_I$ are the average values of the reference and inverse filtered forces, respectively. The best fit between the signals is found at the time delay, τ, where the energy in the difference between the two signals has a minimum. If $\eta = 0$ the two signals are identical.

The simulation results show that the inverse filters are able to counteract the dynamic influences and reconstruct the reference cutting forces within a small error margin for all cutting speeds and feed rates listed in Table 1.1. As seen in Figs 1.5a and 1.6a the difference between reference and inverse filtered forces increases as the cutting speed increases. This is expected since the cutting speed affect ramp up times and thereby the frequency content in the reference force signal. Higher cutting speed leads to higher frequencies in the reference force. Thus, the frequency range of the inverse filter may not be enough to fully describe the transient behavior in the force signal at high cutting speeds. Comparisons between reference and inverse filtered forces using maximum feed rate and cutting speed are shown in Figs 1.5b and 1.6b. Additionally, simulations were performed without considering the effect of the cross FRFs on the dynamometer outputs. Neglecting the cross FRFs did not show any significant changes in the results confirming that the cross FRFs are negligible.

1.4 Experimental Tests

The proposed method was evaluated in experimental cutting tests using different cutting speeds, feed rates and radial immersion. To be able to test the method over a large span of cutting speeds, the tests were performed in aluminum. Forces in both x- and y-directions were recorded and inverse filtered. The results are compared with unfiltered, low-pass filtered and simulated cutting forces. The simulated cutting forces are mechanistic modeled and the cutting coefficients are estimated from milling tests in the work piece used in the experimental tests, see Table 1.2, [13]. The test setup is illustrated in Fig. 1.7. Equipment and cutting data used are listed in Table 1.1.

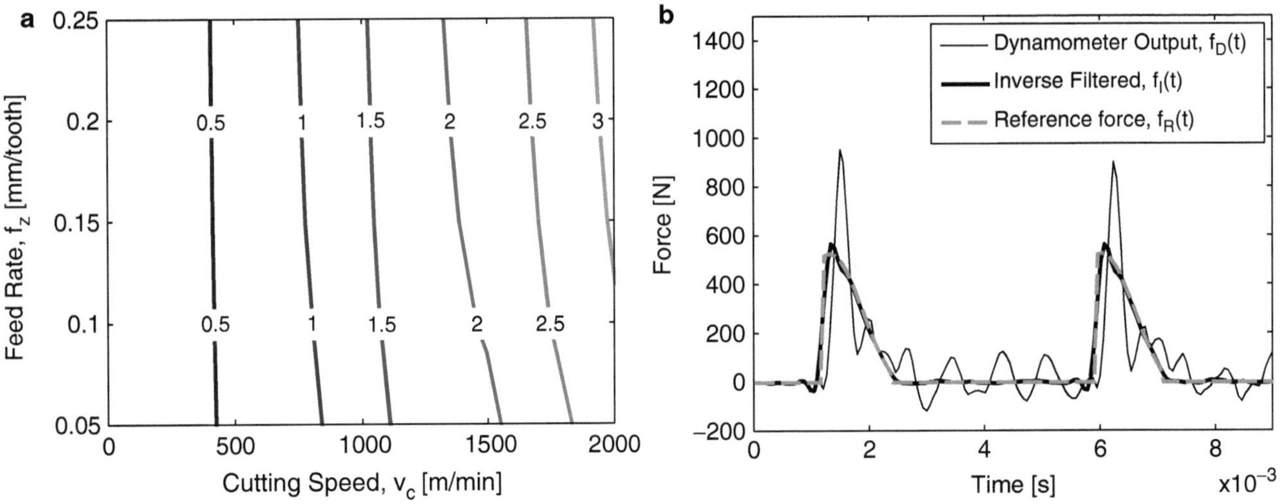

Fig. 1.6 Simulation results with feed in *y*-direction; (**a**) The isolines of η [%], Eq. (1.7). (**b**) Comparison between dynamometer output, inverse filtered and reference input forces (Cutting speed, $v_c = 2000$ [m/min]; Feed rate, $f_z = 0.25$ [mm/tooth])

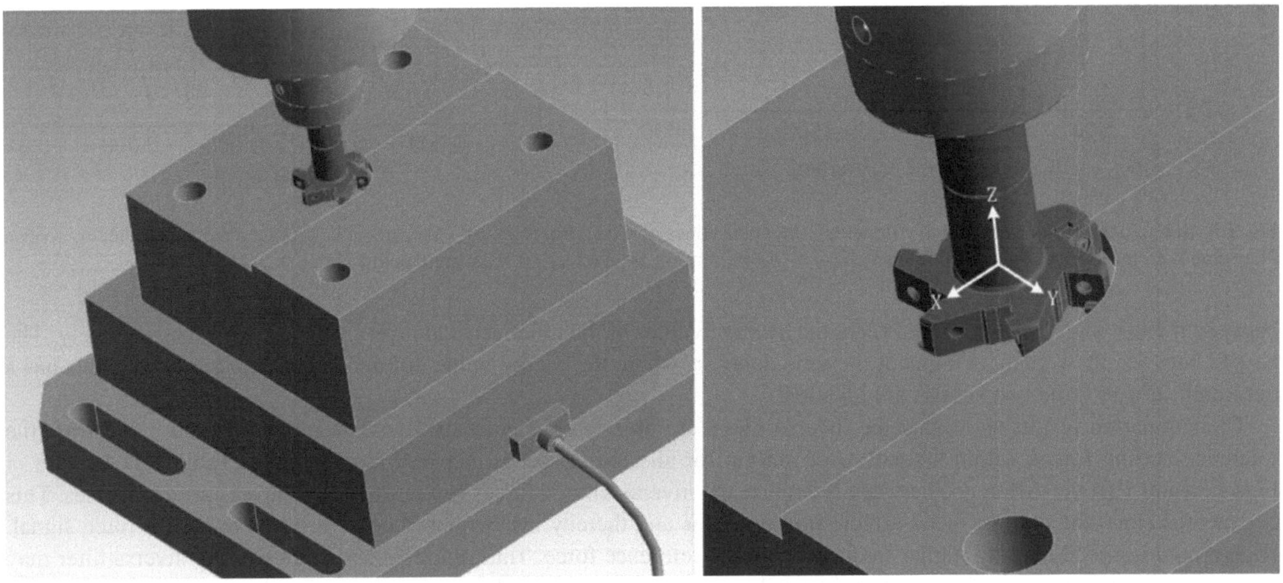

Fig. 1.7 The test setup showing force dynamometer, workpiece and cutter used in the measurements

The cut-off frequency used for direct low-pass filtering of the cutting forces was set by visually inspecting the FRFs in Figs 1.1 and 1.2. As seen, the frequency response of the force dynamometer is relatively flat up to 600 Hz. Information contained in the region above 600 Hz is expected to be effected by the force dynamometer dynamics and generate an erroneous output. Thus, the cut-off frequency used for direct low-pass filtering was set to 600 Hz. The cut-off frequency used for low-pass filtering the inverse filtered signals is the same as used in the simulations, 2400 Hz. The filter design and filtering techniques used in the simulations are also used in the experimental tests. Figs 1.8 and 1.9 show results from tests with different radial immersion and cutting speeds.

The results show that inverse filtered cutting forces are able to predict both amplitude and ramp-up in a reliable manner. Compared with simulated forces the difference in amplitude is small for all cutting conditions tested. Both amplitude and ramp-up prediction is better for inverse filtered compared to low-pass filtered forces. The difference is especially clear for transient cutting conditions and high cutting speeds. Another issue using low-pass filtering is that changes in the estimated cutting forces with respect to cut-off frequency make it difficult to tell if and when the obtained results are accurate.

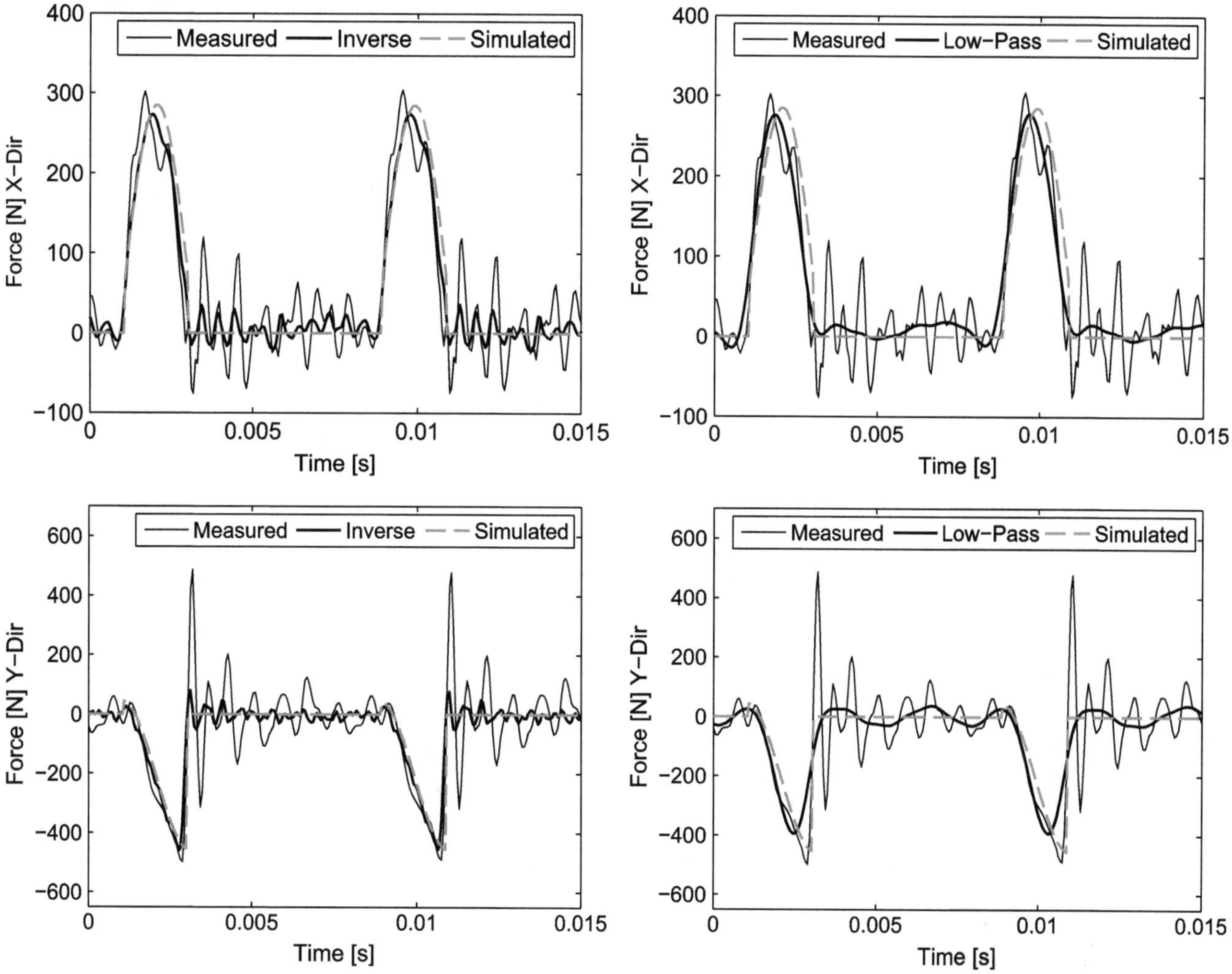

Fig. 1.8 Comparison of unfiltered, simulated, inverse filtered and low-pass filtered cutting forces. Down milling; Cutting speed, $v_c = 1200$ [m/min]; Tooth passing frequency, $f_t = 127.3$ [Hz]; Feed rate, $f_z = 0.2$ [mm/tooth]; Radial immersion $a_e = 25$ [mm]

Also, using low-pass filtering the cut-off frequency should be set to the highest possible value in order to preserve as much information as possible. This requires measurements of the dynamometers frequency response for each specific set-up. Thus, to properly select a cut-off frequency for low-pass filtering, the same number of frequency response measurements are needed as are needed to construct an inverse filter. Since the frequency range, in this case, was extended by 400 % using inverse filtering compared to low-pass filtering, the results become more reliable and less sensitive to changes in cutting conditions.

1.5 Conclusions

A method to create an inverse filter for improved cutting force measurements based on minimum-phase FRFs has been presented. The method is based on the assumptions that the force dynamometer can be described by a linear model and that the cross FRFs of the system are negligible. These assumptions were verified by simulations, combining both measured mixed-phase and minimum-phase FRFs of the force dynamometer, confirming the inverse filters validity within the boundaries of the cutting conditions used in the measurements. The method successfully counteracted the dynamometer dynamics in experimental tests with different feed rates, cutting directions and cutting speeds. The results show that a more reliable estimation of the cutting forces can be obtained using the proposed method compared to traditional low-pass filtering, especially under transient cutting conditions and high cutting speeds. Since the dynamic corrections are performed

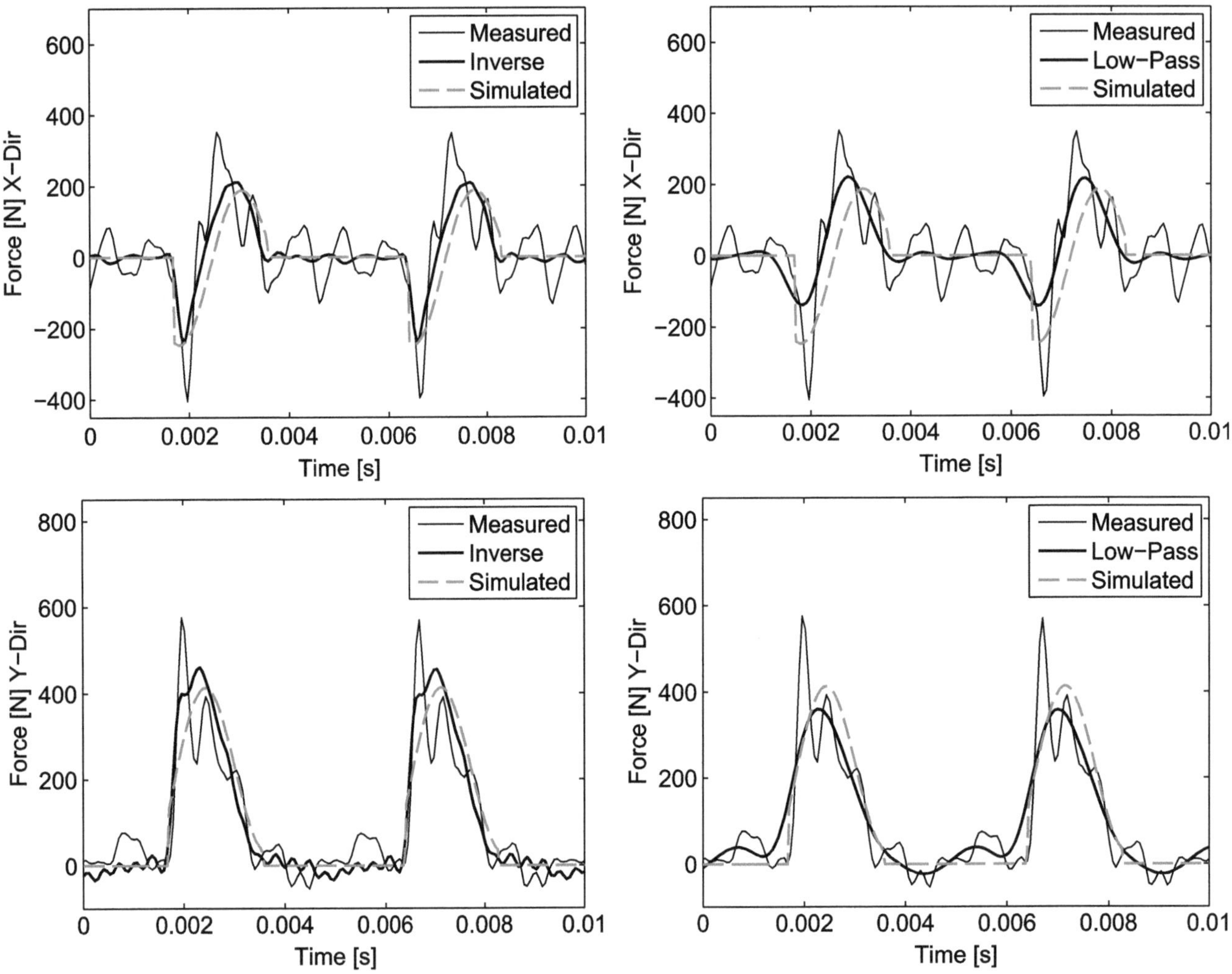

Fig. 1.9 Comparison of unfiltered, simulated, inverse filtered and low-pass filtered cutting forces. Down milling; Cutting speed, $v_c = 2000$ [m/min]; Tooth passing frequency, $f_t = 212.2$ [Hz]; Feed rate, $f_z = 0.2$ [mm/tooth]; Radial immersion $a_e = 40$ [mm]

in the time domain, compensation can be performed on both stationary and non-stationary signals, which allows the method to be used under both constant and varying cutting conditions. The method can be used to study detailed force responses such as transient entering and/or exiting forces, which is important, e.g., for cutting tool design.

References

1. Tlusty, J., Jang D., Tarng, Y.: Measurements of milling force over a wide frequency range (1987)
2. Jensen, S.A., Shin, Y.C., Davies, P.: Inverse filtering of unwanted system dynamics in cutting force measurement. In: American Society of Mechanical Engineers, Dynamic Systems and Control Division (Publication) DSC, vol. 58, pp. 167–174 (1996)
3. Park, S.S., Altintas, Y.: Dynamic compensation of spindle integrated force sensors with Kalman filter. J. Dyn. Syst. Meas. Control **126**(3), 443–452 (2004)
4. Altintas, Y., Park, S.: Dynamic compensation of spindle-integrated force sensors. CIRP Ann. - Manuf. Tech. **53**(1), 305–308 (2004)
5. Park, S., Malekian, M.: Mechanistic modeling and accurate measurement of micro end milling forces. CIRP Ann. - Manuf. Tech. **58**(1), 49–52 (2009)
6. Hawksford M.J.: Minimum-phase signal processing for loudspeaker systems (1996)
7. Pei, S.-C., Lin, H.-S.: Minimum-phase FIR filter design using real cepstrum. IEEE Trans. Circ. Syst. II-Express Briefs **53**(10), 1113–1117 (2006)
8. Oppenheim, A.V., Schafer, R.W.: Discrete-Time Signal Processing, 3rd edn. Pearson Education, Upper Saddle River (2010)
9. Proakis, J.G., Manolakis, D.G.: Digital Signal Processing: Principles, Algorithms and Applications, 3rd edn. Prentice-Hall, Upper Saddle River (1996)

10. Hayes, M.H.: Statistical Digital Signal Processing and Modeling. Wiley, New York (1996)
11. Magnevall, M., Lundblad, M., Ahlin, K., Broman, G.: High frequency measurements of cutting forces in milling by inverse filtering. Mach. Sci. Technol. 16(4), 487–500 (2012)
12. Bendat, J.S., Piersol, A.G.: Engineering Applications of Correlation and Spectral Analysis, 2nd edn. Wiley, New York (1993)
13. Altintas, Y.: Manufacturing Automation: Metal Cutting Mechanics, Machine Tool Vibrations, and CNC Design, pp. 35–42. Cambridge University Press, Cambridge (2000)

Chapter 2
Use of a Depth Camera as a Contactless Displacement Field Sensor

Jean Michel Franco, Johannio Marulanda, and Peter Thomson

Abstract During experimental tests, optical displacement measures can provide reliable data about the behavior of structural elements without altering key parameters, such as damping, stiffness, or mass, with low cost and high spatial density of measurements. Motion capture Systems are used in different application from medicine to cinematography, involving different types of image processing techniques, but its application to measure the response of civil structures is costly and of limited value in terms of real implementations. Range/Depth Cameras, on the other hand, can provide a 3-D imaging Solution to capture motion and displacements at an affordable cost. These cameras are widely available and used in the videogames industry. This paper presents the first steps for the implementation of a large-displacement measurement methodology and its application.

Keywords Artificial vision • Instrumentation • Displacement measures • Depth camera

2.1 Introduction

Kinect for Xbox360™ [1] is essentially a set of sensors which comprises a triaxial accelerometer, an RGB camera and an infrared camera, initially developed for detecting human features in three dimensions, with a primary application in the field of video games Through a pattern generated by an infrared laser a Range/Depth camera is achieved for three-dimensional scene detection where lighting stops playing an important role as it is in other artificial vision systems, making it a sensor with good performance [2–4] and remarkably low cost. (<150 US $) [5, 6].

The methodology is proposed based on improvements over the Kinect for Xbox360™ raw data using 3D interpolations and a 3D correspondence technique [7] for the measurement of the actual displacement field at a certain time. Test were made using two acquisition methodologies, one based on a continuous 3D reconstruction using Kinect Fusion that is included on the Microsoft Kinect Framework; other extracting raw Kinect depth using Matlab through the Image Acquisition Toolbox that supports Kinect for Xbox360™ devices. Once an acquisition is made, it is performed a 3d interpolation for a normalization of the scattered data in aims to provide normalized data and a posterior processing with a 3D correspondence technique to improve results.

Due to the measuring characteristics and low cost, has become a multipurpose sensor, in different areas from surveys of complex three-dimensional scenes [8], applications focused on improving and reducing costs in augmented reality systems [5], to characterization of turbulent flows using multiple Kinect's [9]. There are studies of the use of the Kinect for the realization of whole plant phenotypes [10], evaluation of postures in the human body and on-line medical evaluation using Kinect generated point clouds from a clinical environment [3, 6, 11]. There are applications for tracking objects in 3D space [16], machine vision applications in robotics for automated three-dimensional survey [12], also there are multiple calibration approaches [13–15] and even improvements to sensor characteristics [16] as are further comparisons with instruments such as laser scanners commonly used for three-dimensional surveys [13] demonstrating the versatility of this sensor.

J.M. Franco (✉) • J. Marulanda • P. Thomson
Escuela de Ingeniería Civil y Geomática, Universidad del Valle, Cali, Colombia

Grupo de Investigación en Ingeniería Sísmica, Eólica, Geotécnica y Estructural (G-7), Cali, Colombia
e-mail: jean.franco@correounivalle.edu.co

A. Brandt, R. Singhal (eds.), *Shock & Vibration, Aircraft/Aerospace, Energy Harvesting, Acoustics & Optics, Volume 9*,
Conference Proceedings of the Society for Experimental Mechanics Series, DOI 10.1007/978-3-319-30087-0_2

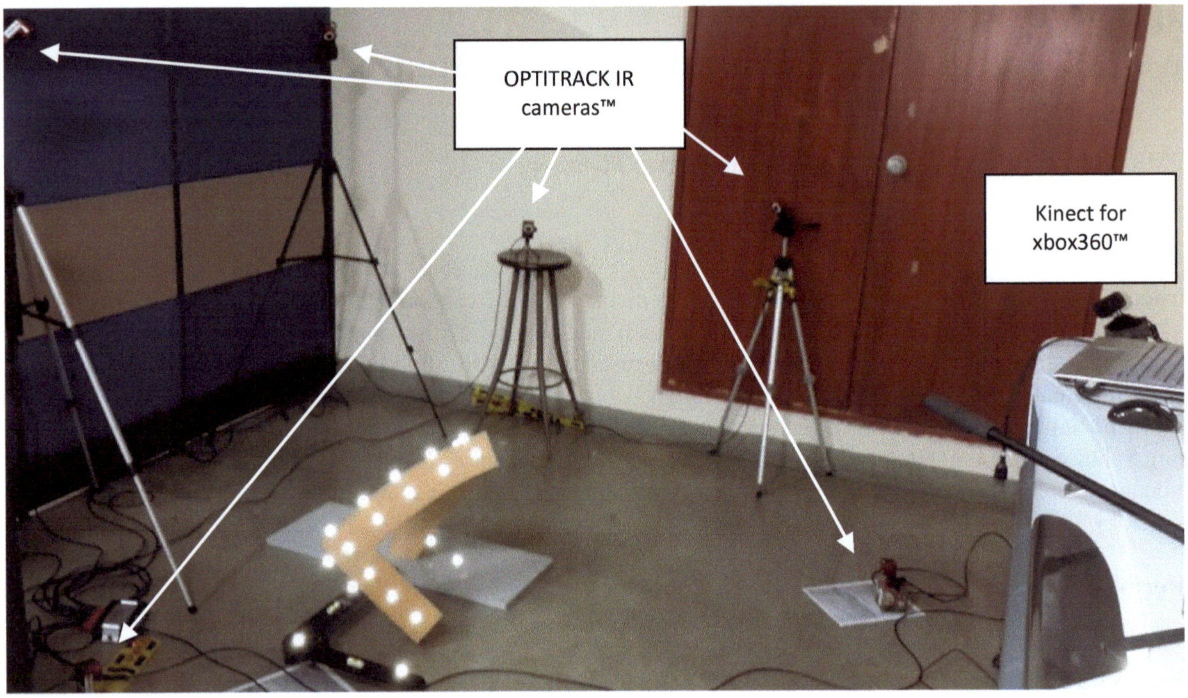

Fig. 2.1 Experimental setup for static displacement approach

2.2 Static Displacement Approach

A foam piece with three straight perpendicular segments, each one in a different coordinate direction, is deformed in a complex arbitrary displacement form as an initial setup. On the surface of the piece, reflective markers used on the OPTITRACK motion capture system are located, composed of six Near Infrared Cameras to obtain a precise 3D location of each marker on the surface of the foam (Fig. 2.1).

The OPTITRACK camera array location is calibrated following the instructions provided by the developer and the positions of each marker are captured. Using Kinect Fusion software the geometry of the foam is measured, moving the Kinect until a full survey of the piece is completed and a three-dimensional point cloud is saved. The process is repeated for another arbitrary displacement form as a final setup of the piece (Fig. 2.2).

Using a CAD software, marker positions where used to compare the Euclidean displacements between the initial and finals position of the piece. A total of 21 points were evaluated on four faces of the piece. It is found that the differences between displacements found using OPTITRACK motion capture system and the Kinect are well related (Table 2.1).

Using the Coherent Point Drift algorithm [7] to establish the correspondence between point clouds it is found a good correspondence visually, as seen on Fig. 2.3. The CPD algorithm fails to achieve a good assessment of the displacement at the six positions showed on Fig. 2.3, but provides a good guess of the Euclidean distances (Table 2.2).

2.3 Dynamic Displacement Approach

A Dynamic shaker APS 400 ELECTRO-SEIS® is instrumented using a UNIMEASURE LX-PA linear position transducer and a Kinect for Xbox360™ (Fig. 2.4) to measure a frequency sweep between 0.1 and 2.5 Hz using an average amplitude of 5 cm. Kinect Depth raw data is acquired using the Image Acquisition Toolbox for Matlab and the displacement data is acquired using a NI DAQ PAD60-15 data acquisition system.

Pinhole camera calibration parameters are applied to the Kinect depth raw data. The shaker positions are extracted from the same point over time on the Kinect depth data. A comparison between the Kinect extracted positions over the measured positions using the show a correlation of 79.2 % (Fig. 2.5). A spectrogram of the Kinect retrieved displacement signal is shown in Fig. 2.6.

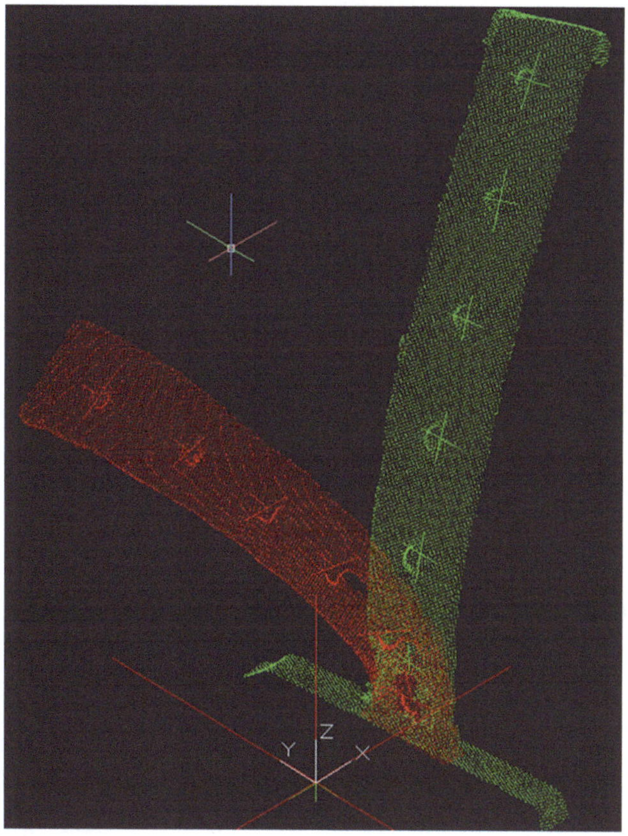

Fig. 2.2 Initial (*green*) vs. final (*red*) for one face

Table 2.1 Kinect vs. Optitrack marker displacements

		Displacement Kinect (m)	Displacement OPTITRACK (m)	Difference (mm)	% Difference
Face 1	1	0.0306	0.0305	−0.1	0.3
	2	0.0724	0.0765	4.1	−5.7
	3	0.1503	0.1531	2.8	−1.9
	4	0.2496	0.2500	0.4	−0.2
	5	0.364	0.3574	−6.6	1.8
	6	0.4882	0.4813	−6.9	1.4
Face 2	7	**0.0089**	**0.0058**	−3.1	**34.8**
	8	0.0605	0.0562	−4.3	7.1
	9	0.1459	0.1396	−6.3	4.3
	10	0.2560	0.2453	−10.7	4.2
	11	0.3733	0.3625	−10.8	2.9
	12	0.4967	0.4884	−8.3	1.7
Face 3	12	0.4967	0.4884	−8.3	1.7
	13	0.5444	0.5363	−8.1	1.5
	14	0.5947	0.5897	−5	0.8
	15	0.6566	0.6513	−5.3	0.8
	16	0.7154	0.7133	−2.1	0.3
Face 4	17	0.5747	0.5742	−0.5	0.1
	18	**0.6110**	**0.6112**	**0.2**	**0.0**
	19	0.6560	0.6599	3.9	−0.6
	20	0.7070	0.7102	3.2	−0.5
	21	0.7599	0.7684	8.5	−1.1
					Mean
					3.3

(The indicated bold values are for the maximun and minimun Difference %)

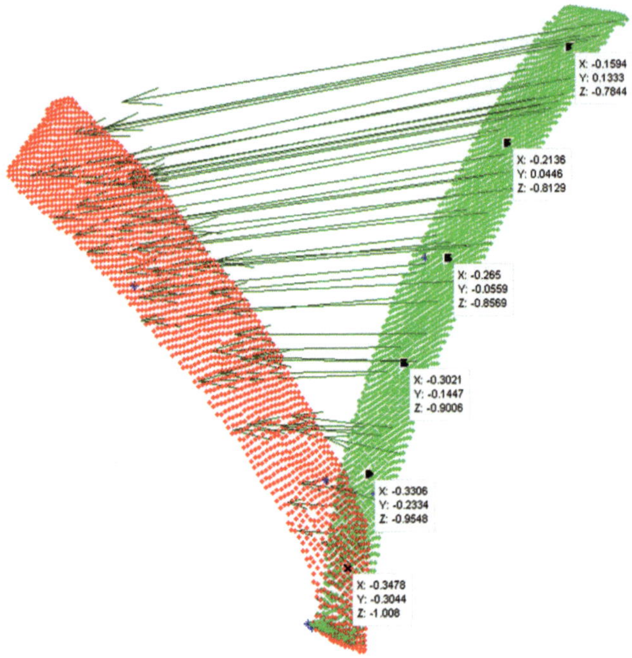

Fig. 2.3 3-Dimensional correspondences for Face 1

Table 2.2 Kinect CPD vs Optitrack marker displacements

		Displacement Kinect CPD (m)	Displacement Optitrack (m)	Difference (mm)	% Difference
Face 1	1	0.0376	0.0305	−7.1	18.9
	2	0.063	0.0765	13.5	−21.4
	3	0.1574	0.1531	−4.3	2.7
	4	0.2597	0.25	−9.7	3.7
	5	0.375	0.3574	−17.6	4.7
	6	0.4671	0.4813	14.2	−3.0
					Mean
					9.1

Fig. 2.4 Experimental setup for dynamic displacement approach

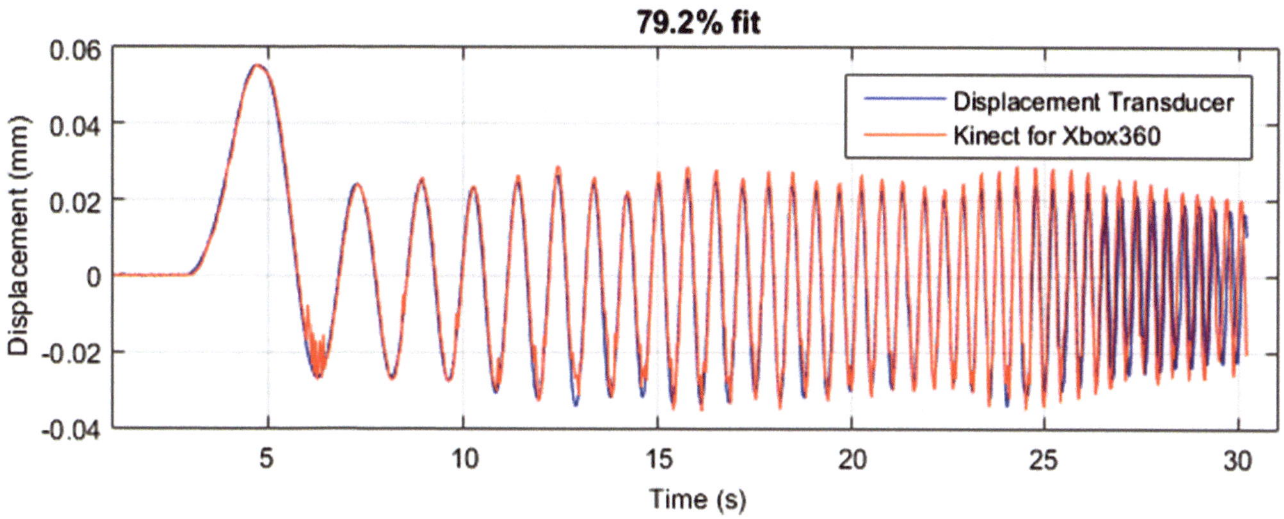

Fig. 2.5 ENCODER vs. Kinect retrieved displacements

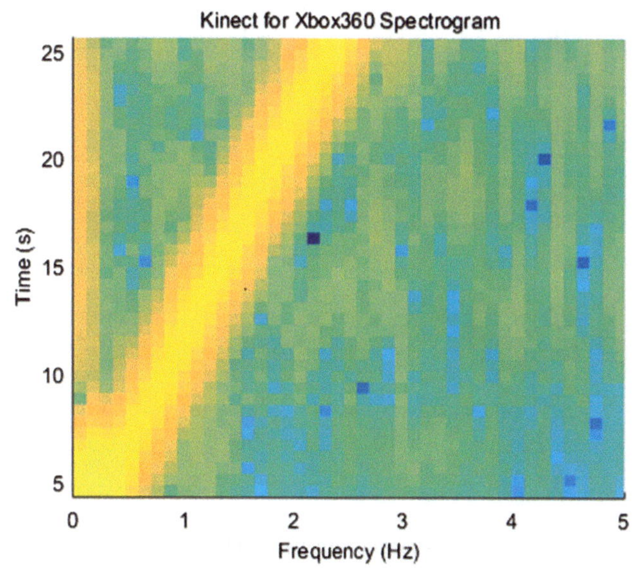

Fig. 2.6 Spectrogram for the Kinect retrieved displacements

Using the same methodology, the response of a benchmark structure of 3 DOF with a first natural frequency of 2.67 Hz (Fig. 2.7) is measured during free vibration and the displacements are retrieved from the Kinect for Xbox360™ as shown on Fig. 2.8. A Power Spectral density of the retrieved displacements is shown in Fig. 2.9 where a predominant frequency of 2.69 Hz is identified

Fig. 2.7 Benchmark structure

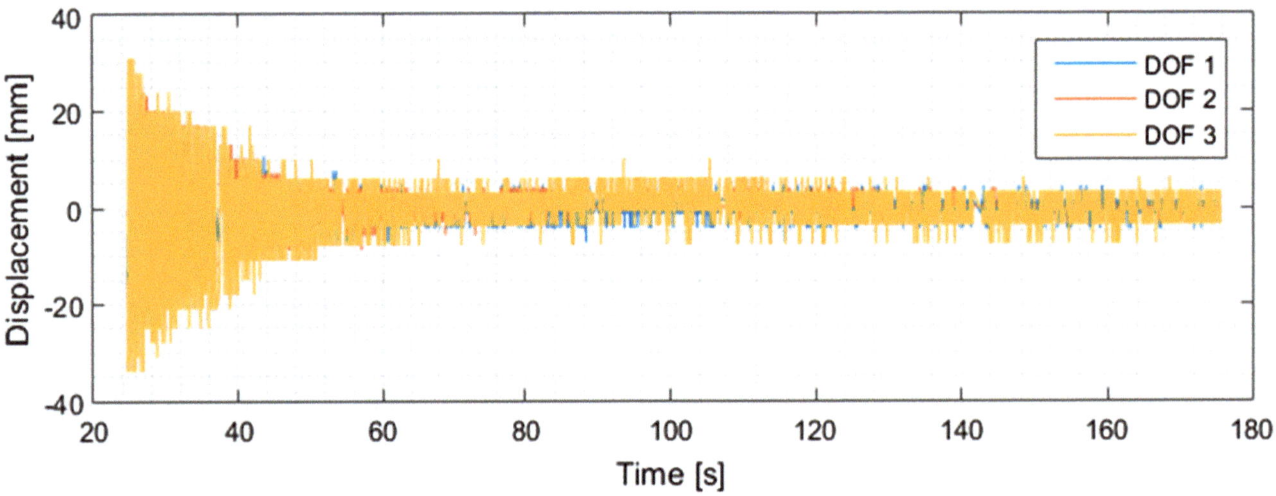

Fig. 2.8 Benchmark structure retrieved displacements

Fig. 2.9 Kinect displacement power spectral density

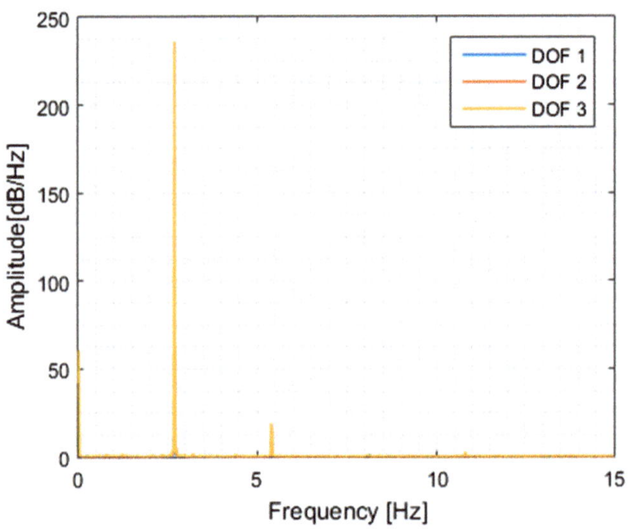

2.4 Conclusions and Perspectives

Kinect Fusion acquisition methodology showed a poor performance on low magnitude displacements in comparison to a motion capture system, but has the advantage of being able to perform a 3-dimensional survey and can provide a good guess of the displacements on elements with complex geometry. Kinect depth raw data calibrated using a pinhole model parameters showed high accuracy assessing the displacements of a dynamical shaker and the dynamic measures shows a good agreement up to 2.5 Hz when large displacements are present. On a benchmark structure it is found a good correlation on the identified first natural frequency. It is expected to improve the displacement measurements using the proposed methodology and a Kinect for Windows v2 sensor with better performance and a Time of Flight technique to obtain depth data.

References

1. Microsoft®. Kinect™ for Xbox360™ Available from: http://www.xbox.com/en-us/kinect/ (2011)
2. Chéné, Y., et al.: On the use of depth camera for 3D phenotyping of entire plants. Comput. Electron. Agric. **82**, 122–127 (2012)
3. Jing, T., et al.: Scanning 3D full human bodies using s. IEEE Trans. Vis. Comput. Graph. **18**(4), 643–650 (2012)
4. Nakamura, T.: Real-time 3-D object tracking using Kinect sensor. In: IEEE International Conference on Robotics and Biomimetics (ROBIO), pp. 784–788. (2011)
5. Placitelli, A.P. and Gallo L.: Low-cost augmented reality systems via 3D point cloud sensors. In: Signal-Image Technology and Internet-Based Systems (SITIS), 2011 Seventh International Conference on, 2011
6. Clark, R.A., et al.: Validity of the Microsoft Kinect for assessment of postural control. Gait Posture **36**, 372–377 (2012)
7. Myronenko, A., Song, X.: Point set registration: coherent point drift. IEEE Trans. Pattern Anal. Mach. Intell. **32**(12), 2262–2275 (2010)
8. Varadarajan, K.M., Vincze M. (2011) Surface reconstruction for RGB-D data using real-time depth propagation. In: Computer Vision Workshops (ICCV Workshops), 2011 I.E. International Conference on, 2011
9. Berger, K., et al.: The capturing of turbulent gas flows using multiple Kinects. In: Computer Vision Workshops (ICCV Workshops), 2011 I.E. International Conference on, 2011
10. Chéné, Y., et al.: On the use of depth camera for 3D phenotyping of entire plants. Comput. Electron. Agri. **82**, 122–127 (2012)
11. Placitelli, A.P., Gallo, L.: 3D point cloud sensors for low-cost medical in-situ visualization. In: Bioinformatics and Biomedicine Workshops (BIBMW), 2011 I.E. International Conference on, 2011
12. Nakamura, T.: Real-time 3-D object tracking using Kinect sensor. IEEE International Conference on Robotics and Biomimetics (ROBIO), pp. 784–788. (2011)
13. Chan-Soo, P., et al.: Comparison of plane extraction performance using laser scanner and Kinect. In: Ubiquitous Robots and Ambient Intelligence (URAI), 2011 8th International Conference on, 2011
14. Dutta, T.: Evaluation of the Kinect™ sensor for 3-D kinematic measurement in the workplace. Appl. Ergon. **43**(4), 645–649 (2012)
15. Khoshelham, K.: Accuracy Analysis of Kinect Depth Data. GeoInf. Sci. **38**(5), 6 (2010)
16. Suttasupa, Y., Sudsang, A., Niparnan, N.: Plane detection for Kinect image sequences. In: Robotics and Biomimetics (ROBIO), 2011 I.E. International Conference on, 2011

Chapter 3
Uncertainty of Digital Image Correlation with Vibrating Deformable Targets

Alfredo Cigada, Navid Hasheminejad, and Emanuele Zappa

Abstract Digital Image Correlation (DIC) is a full-field, non-contact optical technique to measure the contour, deformation, vibration and strain on the surface of an object. Studies on this method and its uncertainty are mostly focused on motion in static conditions. In dynamic conditions, as the target is moving with respect to the camera, the images acquired during the experiments have a considerable amount of motion effect (blurring). This motion effect is the main cause of uncertainty in measurement with DIC method. This work experimentally investigates the effect of the shutter time and the motion of the target on the uncertainty of DIC method. Tests were done on an aluminum cantilever beam, with first natural frequency of 7.6 Hz and a random speckled pattern on its surface. An 8 MP camera acquired the decay free vibration of the full length of the beam (0.93 m) with 50 frames per second while a laser triangulation sensor was used as a reference transducer in a section of the beam. Effect of exposure time, sensor gain and vibration amplitudes on measuring the uncertainty is explored. A software was developed to synchronize the laser and camera, while DIC analysis was done with Vic-2D software. The results show that the uncertainty of DIC is proportional to the product of the exposure time and the initial displacement of the beam (i.e. the stripe length). Afterwards, deconvolution method, which is a processing technique to estimate the displacement of an object and the motion effect that exist in an acquired image, was used to improve the results in case of having a deformable target in dynamics conditions.

Keywords Digital image correlation • Uncertainty • Motion effect • Deconvolution • Deformable targets

3.1 Introduction

Digital image correlation is a popular vision-based measurement method to track the motion of an object. DIC was first developed in the 1980s by Sutton et al. [1]. This technique has been used mostly in static applications [2], but recently, its application has been extended to dynamic applications too [3–5]. It is known that in static conditions the measuring uncertainty of a vision-based measurement system is dependent on image resolution, contrast, processing algorithm, noise, etc. but in dynamic condition there are also other factors like exposure time of the camera or relative motion between the camera and the object [6]. Therefore, setup of the camera and specifying its different parameter can be important for a DIC analysis in dynamic conditions. In this study, experiments were done on a deformable object (cantilever beam) to investigate the important parameters of the camera that can affect the uncertainty of DIC method in dynamic conditions.

In dynamic conditions, the motion of the target causes blurring in the acquired images. This blurring is an important source of uncertainty in DIC measurement. To solve this issue, one can decrease the exposure time of the camera. However, it is not possible to arbitrarily decrease the exposure time, because decreasing the exposure time leads to darker images. Therefore, a higher light intensity is required to have images with proper brightness, which is not always possible in real applications [7]. Thus, deconvolution method to create new reference images was suggested to solve this issue for rigid objects [8]. In this method, the motion effect that exists in an acquired image is estimated by deconvolution method. Then this motion is simulated on a reference image. Using this technique, DIC accuracy in dynamic conditions is improved, as both the acquired image and the reference image have the same blurring [8, 9]. In this research, after finding the important parameters of the camera and motion on DIC uncertainty and comparing two different image processing method to estimate the displacement, deconvolution method to generate new reference images was used to improve the accuracy of DIC method for deformable targets in dynamic conditions.

A. Cigada • N. Hasheminejad • E. Zappa (✉)
Department of Mechanical Engineering, Politecnico di Milano, Via La Masa 1, Milan, Italy
e-mail: emanuele.zappa@polimi.it

A. Brandt, R. Singhal (eds.), *Shock & Vibration, Aircraft/Aerospace, Energy Harvesting, Acoustics & Optics, Volume 9*,
Conference Proceedings of the Society for Experimental Mechanics Series, DOI 10.1007/978-3-319-30087-0_3

3.2 Experiments

The experiments are done on a beam made of aluminum. The cross section of the beam is a 10*30 mm rectangular and the length of the beam is 993 mm. First, using stencil technique, a speckle pattern was painted on one side of the beam. The speckle pattern was painted on the 10*993 mm face of the beam by a frame with circle shape holes. The nominal diameter of these circles was chosen 1.2 mm, which corresponds to speckles with a diameter of 4 px in the images for the specific setup condition, of these experiments. This is the optimal size according to the literature [10, 11]. A portion of the painted side of the beam can be seen in Fig. 3.1.

Then one side of the beam was fixed to a grip and an initial displacement was applied to the tip of the cantilever beam. The camera was in front of the beam, capturing images during the motion. The variables of the experiments are the amount of applied initial displacement to the beam and the exposure time of the camera.

Before applying the excitation, a picture of the beam in static conditions was acquired; this image will be used as the reference image during the DIC analysis. The beam has free vibration and stops after a few seconds. Target's motion was recorded by a digital camera with a frame rate of 50 fps. The shutter speed of the camera was changed from 1000 to 2000, 3000, 4000 and 5000 μs. Meanwhile, the gain was 0, 3, 6 and 9 depending on the camera's shutter speed. The vertical displacement of the beam, at a section 800 mm apart from the constraint, was measured simultaneously with a laser triangulation sensor. A software was developed to synchronize the camera and laser sensor. The camera used during the experiment is a GX3300 produced by Allied Vision Technologies with 3296*2472 px resolution and the laser triangulation sensors is an ILD 1400-20(00) produced by Micro Epsilon company with a range of 20 mm.

3.3 Digital Image Correlation

The acquired images during the experiments were analyzed by Vic-2D software. Vic-2D uses DIC algorithms to calculate the displacement of each portion of the images acquired during the test with respect to the reference image acquired in static conditions. The first test of the experiment was analyzed with different subsets (15*15, 21*21 and 27*27 pixels) and it was proven that the results are not considerably affected by this parameter. The rest of the analysis is done using a 15*15 px subset and a step of 3 px.

The uncertainty associated with the estimated displacement of each subset depends on different parameters, including the speckle quality in the subset, the local lighting conditions and the noise. Sigma is an output of Vic-2D, which shows standard

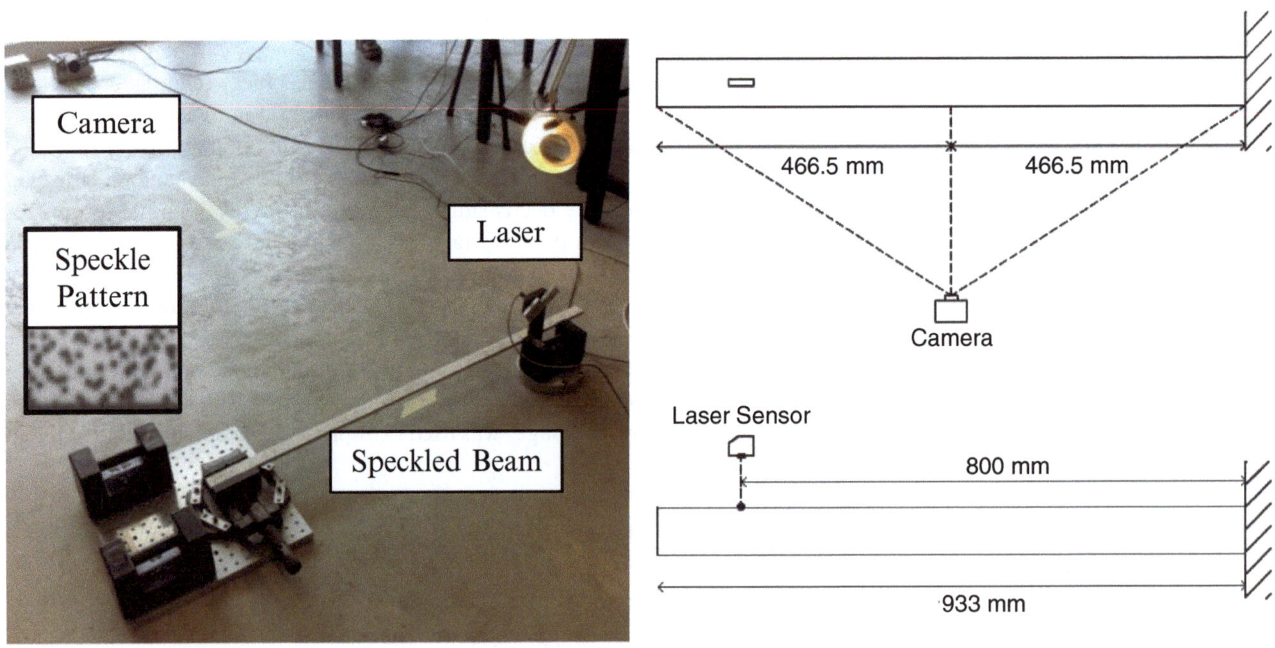

Fig. 3.1 *(Left)* Measurement setup, *(Right)* Schematic configuration of the experiment, front view *(up)*, top view *(down)*

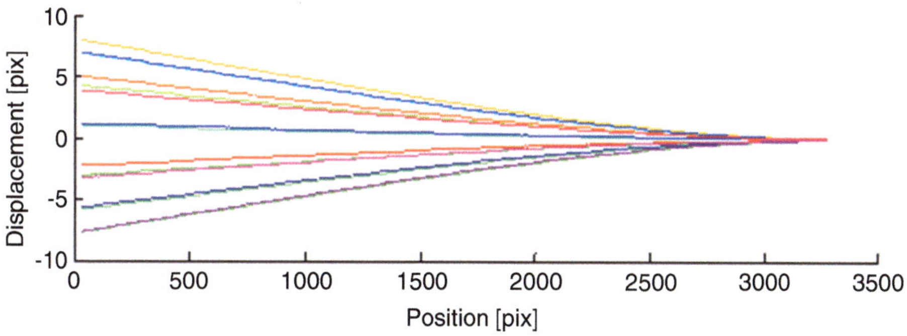

Fig. 3.2 Estimation of beam profile with DIC for 15 images in [Exposure time = 3000 μs, gain = 0 and Initial displacement = 2.8 mm]

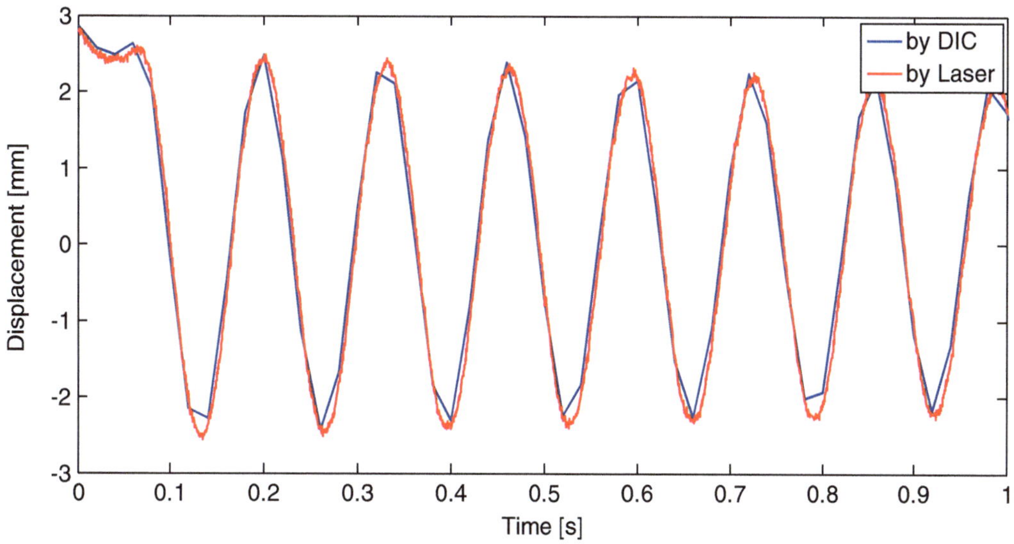

Fig. 3.3 Motion of the point in front of laser sensor, estimated by DIC method and measured by Laser sensor for [Exposure time = 3000 μs, gain = 0 and Initial displacement = 2.8 mm]

uncertainty associated to the estimated displacement of each region of the beam. The values of sigma was used to detect and remove the values of the estimated displacements affected by a low accuracy. After removing the data with low accuracy, the average of the estimated displacements of all the regions in each column of the beam are calculated and considered as the displacement on each section of the beam. Thus, in case the error is high in all regions of a column, using a low sigma limit can remove all of the data on that point. So the value of sigma for the experiments were chosen between 0.01 and 0.1 px to have less than 5 % of missing points. Figure 3.2 represents the profile of the beam estimated by DIC for 15 different images and Fig. 3.3 shows the motion of the beam in front of the laser sensor, estimated by DIC and measured by the laser sensor with 3000 μs exposure time, 0 gain and 2.8 mm initial displacement.

Two parameters are defined to investigate the accuracy of the analysis: NRMSE (Normalized Root Mean Square Error) and peak spectrum error. These parameters show the accuracy of the motion estimated with DIC method in front of the laser sensor. If A(i) is the reference motion measured by the laser sensor in time history and B(i) is the motion in a point in front of laser sensor estimated by DIC method, NRMSE is calculated as bellow:

$$\mathbf{RMSE} = \sqrt{\sum_{i=1}^{N} \frac{[\mathbf{A(i)} - \mathbf{B(i)}]^2}{\mathbf{N}}} \tag{3.1}$$

$$\mathbf{NRMSE} = \frac{\mathbf{RMSE}}{\max(\mathbf{A(i)}) - \min(\mathbf{A(i)})} \tag{3.2}$$

Furthermore, peak spectrum error is the discrepancy between the peak of the spectra of the motion, estimated by DIC method and the laser sensor. Peak spectrum error can be computed from the spectrum diagram as following:

$$\mathbf{Peak\ Spectrum\ Error} = \frac{|\mathbf{A_r} - \mathbf{A_m}|}{\mathbf{A_r}} * 100 \tag{3.3}$$

Where A_r is the peak of the spectrum diagram of the motion, measured by the laser sensor, and A_m is the peak of the spectrum diagram of the motion estimated by DIC method. Figure 3.4 represents the spectra of the two estimated motions after applying a hamming window to reduce the leakage due to decreasing the amplitude generated by the damping.

After calculating NRMSE and peak spectrum error for all the tests, it is evident that error is higher for the tests in which exposure time and initial displacement are higher. To show this relation, the product of exposure time multiplied by the initial displacement of the beam on a point in front of the laser sensor is calculated for each condition and plotted in Fig. 3.5. This parameter can be a representative of the stripe length in the acquired images. Moreover, NRMSE and peak spectrum error for each condition are plotted in Fig. 3.6. It can be seen that the diagrams in Fig. 3.6 have almost the same trend of Fig. 3.5. This same trend proves high values for product of the exposure time and initial displacement means having a high

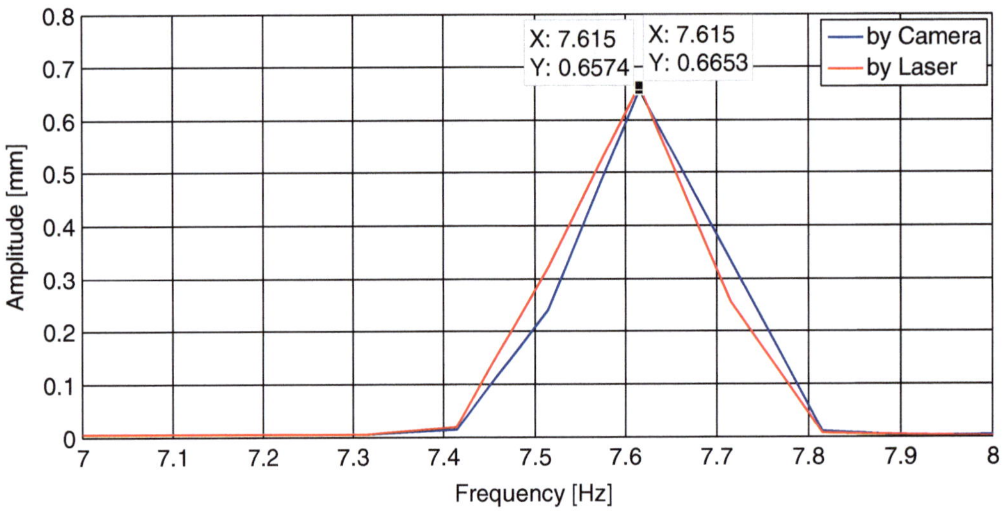

Fig. 3.4 Spectrum of the motion of the point in front of laser obtained by camera and laser after applying a hamming window

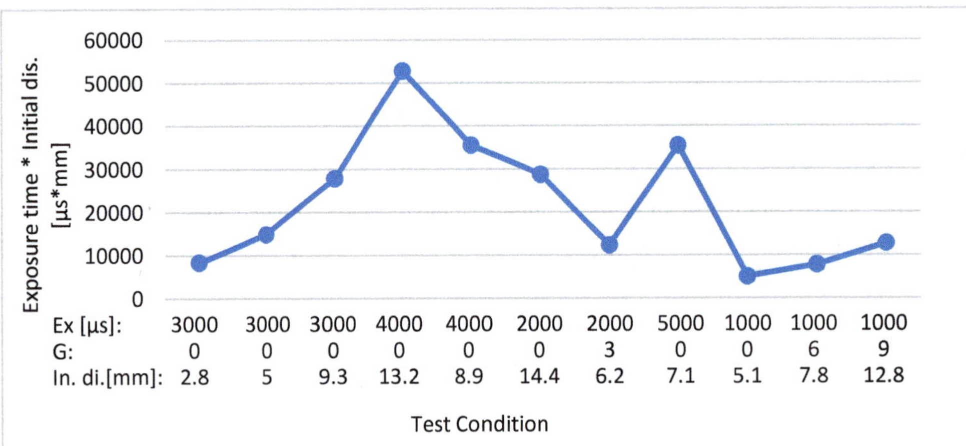

Fig. 3.5 Product of exposure time of the camera multiplied by initial displacement at a point on the beam in front of the laser sensor in different conditions (Ex = Exposure time [µs], G = Sensor gain [dB], In. di. = Initial displacement applied to beam [mm])

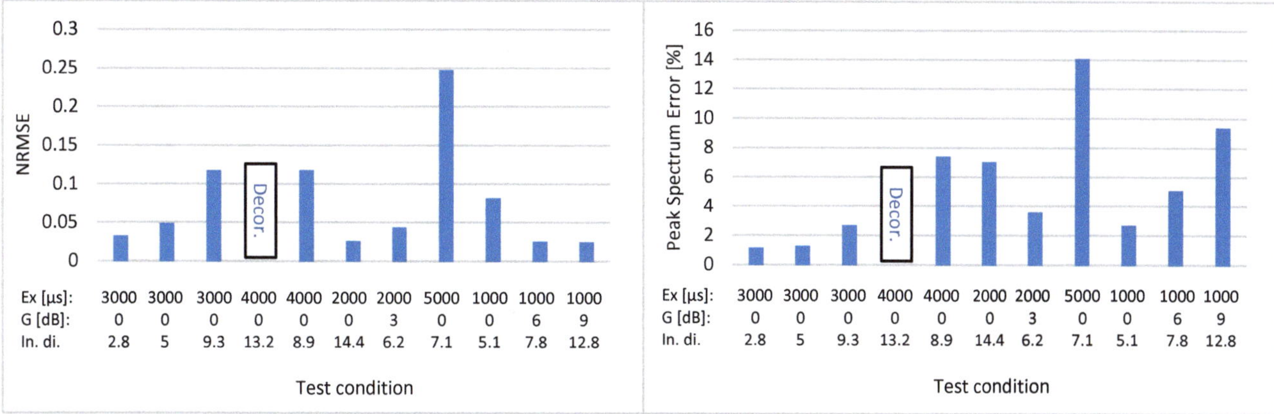

Fig. 3.6 NRMSE (*left*) and peak spectrum error (*right*) of DIC for different conditions (Ex = Exposure time [µs], G = Sensor gain [dB], In. di. = Initial displacement applied to the beam [mm], Decor. = Decorrelation)

motion effect in the images, which causes high error in estimation of displacement with DIC method. There is even one condition (exposure time 4000 µs, gain 0 dB and initial displacement applied to beam 13.2 mm) in which DIC method was not applicable and deceleration occurs due to high motion effect in the images. However, there are some irregularities that show other conditions like the brightness of the images or saturation in them can be important as well. For instance, for the test with exposure time of 1000 µs, gain of 0 dB and initial displacement applied to beam equal to 5.1 mm, the errors are expected to be lower than the same condition and initial displacement equal to 7.8 or 12.8. But as it can be seen in Fig. 3.6, even though peak spectrum error is increasing, NRMSE is declining. This can be due to change of the gain and therefore brightness of the images in different tests.

3.4 Deconvolution Analysis

One method to estimate the displacement and motion effect in acquired images is to perform a deconvolution analysis [8]. Zappa et al. [9] used deconvolution method to determine the motion effect (w) and the net displacement (a) of an object in dynamic conditions. They proved that using this net displacement can be useful to track the motion of a rigid object in motion. In this section the beam is assumed as a rigid object and this method is used to estimate the motion of the beam. It is clear that the motion of the beam at different points of its length is different. Therefore, to calculate the vertical motion of the beam at each point, the length of the beam was divided to portions with 16 pixels width. The motion of each portion was considered equal to the displacement of its center. After estimating the displacement in each image, NRMSE for this method can be calculated as explained before in formula (3.2).

Figure 3.7 shows the product of the exposure time and initial displacement of the point in front of the laser sensor, i.e. the stripe length, (left) and comparison of NRMSE for DIC and deconvolution method (right) for different conditions. For the tests in which the stripe length is high, the images are blurred, hence NRMSE of normal DIC is higher. Therefore, using deconvolution method can be useful to increase the accuracy of the estimated displacement, even if it does not allow estimating the strain field, while DIC also allows estimating it.

3.5 Deconvolution Analysis to Improve DIC Uncertainty

It was proven in the previous chapters that DIC analysis is accurate for tests in which images are not blurred too much. However, when the shutter time is increased or the speed of the beam is higher, the motion effect reduces the accuracy of the results. Even in some cases, the blurring of the images is so high, that after applying sigma limit on the data, displacements of some points are removed because of low accuracy. Deconvolution method, explained in the previous chapter, has higher accuracy with respect to DIC, but it is useful for rigid objects, as it cannot calculate the strain on the surface of the objects. Thus, to improve the accuracy of DIC for a deformable object the deconvolution analysis to generate new reference images

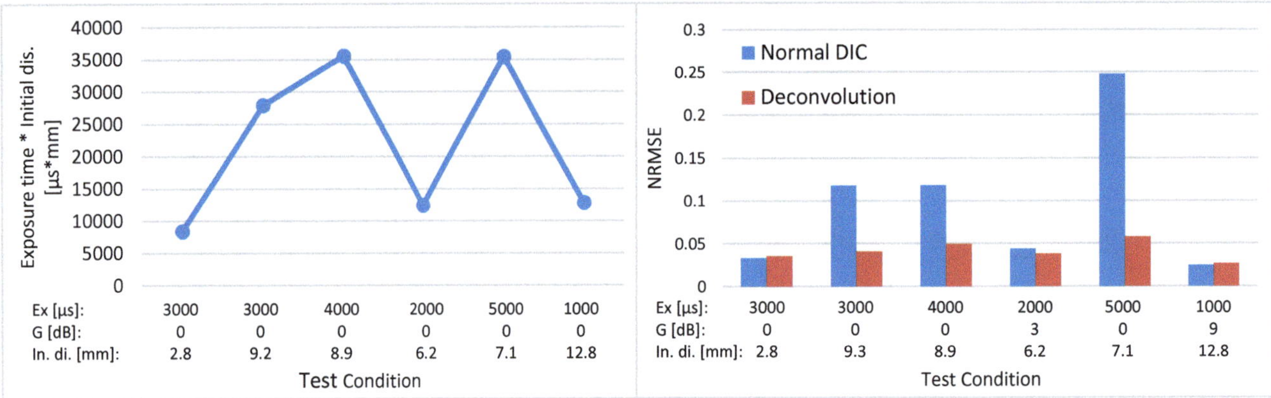

Fig. 3.7 Multiplication of exposure time and initial displacement (*Left*) and NRMSE (*Right*) for six different settings

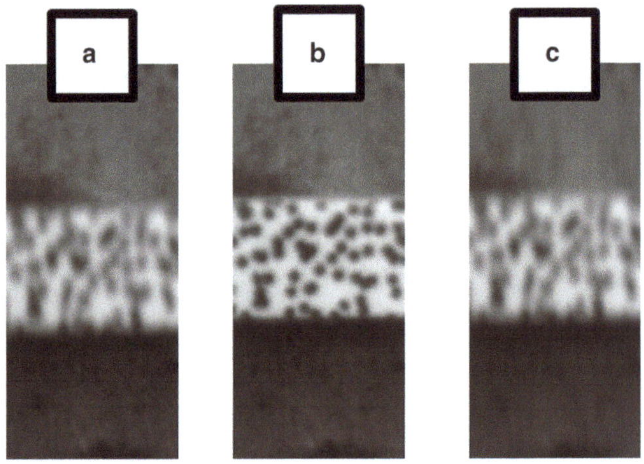

Fig. 3.8 (**a**) acquired image during the test (**b**) original reference image (**c**) generated reference image with w = 5.06 for [Exposure time = 5000 µs, gain = 0 and Initial displacement = 7.1 mm]

can be helpful. With deconvolution analysis proposed in [8], the value of motion effect w is estimated for each image of the test. Afterward, this motion effect is replicated on the reference image to generate a new image. In the case of rigid target, it was proved in [8] that using this new generated image as a reference in DIC, the uncertainty decreases. This is because the new reference image has the same amount of blurring as the image acquired during the test, therefore the correlation can obtain better results. In this method, there is a different reference image for each acquired image during the test.

While using this technique in the case of deformable target, it should be considered that the speed of the beam changes along its length. Therefore, the motion effect is different along the beam and one reference image cannot be generated for the whole beam. Hence in this work, the length of the beam is divided into many small portions, and each portion is assumed to have a constant instantaneous speed over this area. This way, the motion effect for each section is predicted and a properly blurred reference image is generated for each section of the beam.

In this research, one portion of the beam in front of the laser sensor was considered. Figure 3.8b shows the portion of the beam in the case of no motion. Using this image and an image from the same region of the beam in motion (Fig. 3.8a) the amount of motion effect (w) was estimated. Then using this motion effect a new reference image with the same blurring that the acquired image has, was generated (Fig. 3.8c).

After generating the new reference images with a proper blurring, each acquired image during the test was compared with its new generated reference image. For instance, this procedure was done for the test with exposure time equal to 5000 µs and the applied initial displacement equal to 7.1 mm. As it can be seen in Fig. 3.9, the motion of the point in front of the laser sensor for this test was plotted with three different methods. The blue curve is the motion of the beam estimated by DIC, using a normal reference image. The green curve shows the motion of the same point of the beam by the laser sensor. And finally, the red curve is motion estimated with the reference images modified by adding a controlled motion blur. The results obtained by this new method are much more accurate and closer to the result obtained by the laser. Also, in normal DIC sometimes the motion effect is so high that DIC cannot estimate the motion, or estimates the motion with very low accuracy.

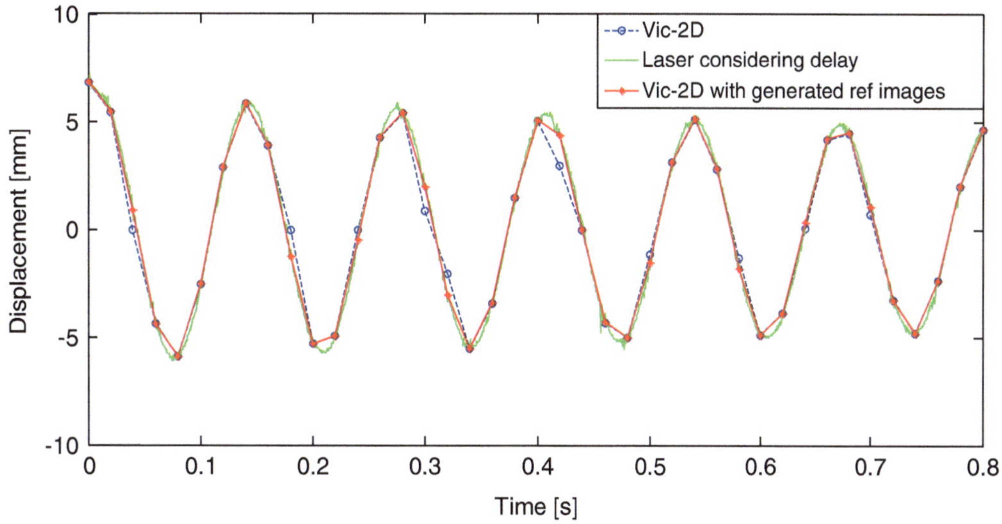

Fig. 3.9 Motion of the beam in front of the laser sensor calculated by Vic-2D with original reference image and generated reference images, and measured by a laser sensor as a reference, for [Exposure time = 5000 μs, gain = 0 and Initial displacement = 7.1 mm]

Fig. 3.10 NRMSE of normal DIC and DIC with Deconvolution to generate new reference images

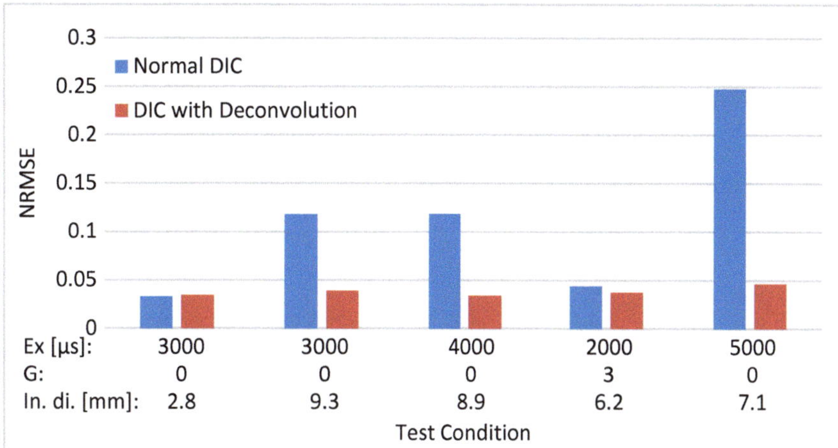

In these cases, there will be a missing point on the motion estimation of the beam. The new method can solve this issue as well.

To conclude, generating new reference images with deconvolution can be useful when the blurring effect is considerable in the images acquired during the tests. There are two important factors that can cause blurring effect. One is the exposure time of the camera and the other one is the speed of the beam. Figure 3.10 shows how the use of deconvolution to generate new reference images improves the results for different conditions.

3.6 Conclusion

Experiments were conducted on an aluminum cantilever beam to investigate the important parameters of the camera and motion on the uncertainty of digital image correlation and deconvolution method in dynamic conditions. Two parameters (NRMSE and peak spectrum error) were defined to compare the accuracy of the results obtained in different conditions by different methods. Finally using deconvolution method, motion effect of the acquired images was simulated in the reference images to reduce the uncertainty caused by motion effect.

References

1. Sutton, M., Wolters, W., Peters, W., Ranson, W., McNeill, S.: Determination of displacements using an improved digital correlation method. Image Vis. Comput. **1**, 133–139 (1983)
2. Pan, B., Qian, K., Xie, H., Asundi, A.: Two-dimensional digital image correlation for in-plane displacement and strain measurement: a review. Meas. Sci. Technol. **20**, 062001 (2009)
3. Wang, W., Mottershead, J.E., Siebert, T., Pipino, A.: Frequency response functions of shape features from full-field vibration measurements using digital image correlation. Mech. Syst. Signal Process. **28**, 333–347 (2012)
4. Siebert, T., Wood, R., Splitthof, K.: High speed image correlation for vibration analysis. J. Phys. Conf. Ser. **181**, 012064 (2009)
5. Wang, W., Mottershead, J.E., Mares, C.: Vibration mode shape recognition using image processing. J. Sound Vib. **326**, 909–938 (2009)
6. Busca, G., Ghislanzoni, G., Zappa, E.: Indexes for performance evaluation of cameras applied to dynamic measurements. Measurement **51**, 182–196 (2014)
7. Caetano, E., Silva, S., Bateira, J.: A vision system for vibration monitoring of civil engineering structures. Exp. Tech. **35**, 74–82 (2011)
8. Zappa, E., Matinmanesh, A., Mazzoleni, P.: Evaluation and improvement of digital image correlation uncertainty in dynamic conditions. Opt. Lasers Eng. **59**, 82–92 (2014)
9. Zappa, E., Mazzoleni, P., Matinmanesh, A.: Uncertainty assessment of digital image correlation method in dynamic applications. Opt. Lasers Eng. **56**, 140–151 (2014)
10. Sutton, M.A., Orteu, J., Schreier, H.W.: Image Correlation for Shape, Motion and Deformation Measurements, vol. 10, pp. 978–970. Springer, New York (2009)
11. Zhou, P., Goodson, K.E.: Subpixel displacement and deformation gradient measurement using digital image/speckle correlation (DISC). Opt. Eng. **40**, 1613–1620 (2001)

Chapter 4
Physical Vibration Simulation of an Acoustic Environment with Six Shakers on an Industrial Structure

Randall L. Mayes and Daniel P. Rohe

Abstract A previous study in the UK demonstrated that vibration response on a scaled-down model of a missile structure in a wind tunnel could be replicated in a laboratory setting with multiple shakers using an approach dubbed as impedance matching. Here we demonstrate on a full scale industrial structure that the random vibration induced from a laboratory acoustic environment can be nearly replicated at 37 internal accelerometers using six shakers. The voltage input to the shaker amplifiers is calculated using a regularized inverse of the square of the amplitude of the frequency response function matrix and the power spectral density responses of the 37 internal accelerometers. No cross power spectral density responses are utilized. The structure has hundreds of modes and the simulation is performed out to 4000 Hz.

Keywords MIMO shaker control • Multi-shaker simulation

Nomenclature

ω	Frequency in radians per second
$+$	Superscript indicating the Moore-Penrose pseudo-inverse of a matrix
c	Tikhonov regularization constant
d	Subscript for desired value
DFAT	Direct field acoustic test
dof	Degree of freedom
FRF	Frequency response function
$\mathbf{H}$	Frequency response function matrix
i	Subscript for initial value
IMMAT	Impedance matched multi-axis testing
MIMO	Multi-input multi-output
PSD	Power spectral density
RMS	Root mean square
$\mathbf{S}$	Absolute value of FRF matrix squared element by element
v	Voltage
x	Physical displacement dof

4.1 Introduction and Motivation

Most engineering laboratories understand that ground based testing of systems on shaker tables has serious shortcomings when compared with the true environment of a payload. A few of these limitations are:

Sandia National Laboratories is a multi-program laboratory managed and operated by Sandia Corporation, a wholly owned subsidiary of Lockheed Martin Corporation, for the U.S. Department of Energy National Nuclear Security Administration under Contract DE-AC04-94AL85000.

R.L. Mayes (✉) • D.P. Rohe
Structural Dynamics, NDE and Model Validation Department, Sandia National Laboratories, Albuquerque, NM 87185, USA
e-mail: rlmayes@sandia.gov; dprohe@sandia.gov

A. Brandt, R. Singhal (eds.), *Shock & Vibration, Aircraft/Aerospace, Energy Harvesting, Acoustics & Optics, Volume 9*,
Conference Proceedings of the Society for Experimental Mechanics Series, DOI 10.1007/978-3-319-30087-0_4

1. The shaker table can only excite the payload in one single axis at a time, so at least three tests have to be performed to capture the three dimensional aspects of the field vibration;
2. The impedance of the shaker table is radically different from the parent structure supporting the payload;
3. Forces are often coming into the payload at multiple points in the true environment, and these effects cannot be simulated with a shaker table;
4. These shortcomings often result in severe overtests for many frequency bands as well as undertests [1] in other bands.

The problems are amplified if one is attempting to simulate vibration that is partially or completely caused by aerodynamic forces on the payload. Theoretically, one does not have to excite a payload with exactly the same loadings as it sees in the operational environment. If enough shakers can be placed at proper locations on the structure and controlled to excite it with the same modal forces, it should see exactly the same vibrational response. There are an infinite number and combination of locations that shakers could be attached to excite the same modal forces as the operational environment. The work in this paper was actually proposed four years ago, but just recently came to fruition. Daborne, et al. presented an approach to the problem [2] and a review of the problem and a successful testing solution [3] on a 1/3 scale model of an underwing missile. In their demonstration, the operational environment of the scale model missile attached to its carriage fixture was generated in a wind tunnel and measured with 13 accelerometers. Then in the vibration laboratory, the missile was mounted to its underwing carriage on a board and excited with three shakers. They achieved an excellent match to the cross power spectral density matrix for all 13 accelerometers with a multiple input multiple output (MIMO) control system. The method was dubbed Impedance Matched Multi-Axis Testing (IMMAT).

Here we attempt to provide a similar demonstration on an industrial payload supported by a mounting rack. The environment is generated acoustically with eight drives for a large system of speakers in the truth test. Then the payload is suspended on its rack with bungee cords, and six shakers are driven to attempt to reproduce the internal vibration of the truth acoustic environment in a modal test lab. Thirty seven internal accelerometers are the target responses. There are hundreds of modes in the 4000 Hz bandwidth. This work focuses on reproducing the accelerometer power spectral densities and does not require cross spectral densities.

In Sect. 4.2 the test hardware and instrumentation is described. Section 4.3 provides the description of the acoustic truth test. Section 4.4 describes the multi-shaker simulation which attempts to reproduce the vibration response of the acoustic truth test. In Sect. 4.5 the theory is explained for generating the proper shaker amplifier input voltages to perform the simulation. Sections 4.6 and 4.7 address some logistic details for a multi-shaker simulation and in Sect. 4.8 conclusions are provided.

4.2 Test Hardware and Instrumentation

The industrial structure of interest is shown in Fig. 4.1 in one of several test setups. The structure was attached to a supporting rack from which it hung. The structure has hundreds of modes up to the desired bandwidth of 4000 Hz. There are dozens of instrumented components inside the structure. For this work 37 internal accelerometers were selected as those of primary interest. In addition there were several dozen accelerometers that were monitoring response on the external skin and on the rack. Most of the accelerometers were of the integrated circuit piezoelectric type with 10 mv/G sensitivity. All data were acquired using IDEAS software for the acoustic test and the later vibration simulation tests. VTI 1436 modules were utilized to acquire the data and VTI 1434 source modules were used to provide either flat random voltage inputs for frequency response function (FRF) calculations or prescribed user defined voltage time histories to excite the required response. For the vibration test of interest six MB 50 shakers with Yamaha model P7000S amplifiers were used to excite the structure. Bungee cords supported the rack and structure.

In Fig. 4.2 one can see three shakers attached to the supporting rack with their connecting hardware. Steel rods of about 1.6 mm diameter and about 25 mm length are used to minimize transmission of moments and shear forces from the shaker to the test article. Drive point accelerometers are utilized, and they are covered with an aluminum cap to which the force gage is mounted. Aluminum receptacles attach the stinger to the force gage and also to the shaker. Set screws hold the steel rods in the aluminum receptacles.

Fig. 4.1 Industrial structure mounted on support rack in one test configuration

Fig. 4.2 Shaker attachment hardware

4.3 Acoustic "Truth" Test

The acoustic test setup is shown in Fig. 4.3. This serves as our "truth" test, or the actual environment for this research. The test was performed at the Sandia National Laboratories acoustic test facility with the direct-field acoustic test (DFAT) approach. Twelve VT-99 model speaker cabinets and six low frequency model VS-Q speakers surround the test article in eight speaker stacks. Multiple Input Multiple Output (MIMO) control was accurately maintained at the six control microphones. All six microphones had the same flat topped drive specification with 0.1 coherence between microphones. The drive levels were at the limits of speaker/amplifier capability. Dozens of accelerometers were mounted on the skin, rack and yoke and dozens more were on internal components. Twelve other response microphones were also placed around the unit. The industrial test structure was supported on its rack with a large yoke made of steel bars, which allowed the structure to be suspended and rotated into a vertical position to facilitate the placement of speakers radially around it. (The rack is not visible in this view, but the steel bars of the yoke are visible).

4.4 MIMO Simulation Test

The MIMO shaker simulation test was set up in a modal testing laboratory at Sandia National Laboratories. The rack and industrial structure were suspended as shown in Fig. 4.1. Several different shaker arrangements were executed for various purposes not germane to this paper. The MIMO simulation configuration was chosen based on information gained from these configurations.

4.4.1 MIMO Simulation Test Setup

The MIMO simulation configuration had two floor mounted shakers attached to the belly of the industrial structure, and four attached to the rack. Two lateral shakers were attached near the forward and aft ends of the rack and two more vertically oriented shakers were attached at the forward and aft ends of the rack. The vertical shakers had a 21.5° tilt from vertical simply to keep the shaker from interfering with the bungee cord support. (See the vertical shaker in Fig. 4.2). Flat random voltage inputs which produced about 45 N RMS force from 25 to 4000 Hz were utilized to calculate accelerometer response to input voltage FRFs for the six shaker drives. One hundred averages were used with a frequency resolution of 2.5 Hz. These FRFs were used to back calculate input voltages to drive the shakers to attempt to produce the same accelerometer power spectral densities (PSDs) as were measured in the acoustic "truth" test. The simulation was adjusted in steps with three iterations to the final physical simulation. 37 internal accelerometers were chosen as accelerometers of interest for which the shaker simulation attempted to match the acoustic test PSDs.

4.4.2 MIMO Simulation Test Results

A metric which provides a single plot comparison of the overall match of the simulation was the sum of the 37 PSDs. This sum of PSDs for the acoustic test (blue) and the vibration test (red) is shown in Fig. 4.4. For most of the band, the vibration test closely envelopes the acoustic test. The band from 3500 to 4000 Hz does not match near as well as the rest of the frequency range. In Fig. 4.5 is shown a couple of the PSDs from individual accelerometers that gives an indication of the "best" and "worst" gage responses. In the worst gage, there are several bands where the vibration test overshoots the desired PSD. These results are very encouraging, and the vibration simulation with six shakers is certainly much better than could be obtained by setting the unit on a shaker table and attempting to control all 37 gages.

There was a portion of the structure that had some intermittent nonlinearity. In Fig. 4.6 one can see the PSD of an accelerometer in this region. The plot shows the truth test result that we desire to match in blue. The red plot shows one physical vibration simulation measurement and the magenta plot shows another physical vibration simulation where everything in the vibration test was repeated exactly as before. The nonlinearity is intermittent.

Fig. 4.3 Acoustic "Truth" test setup

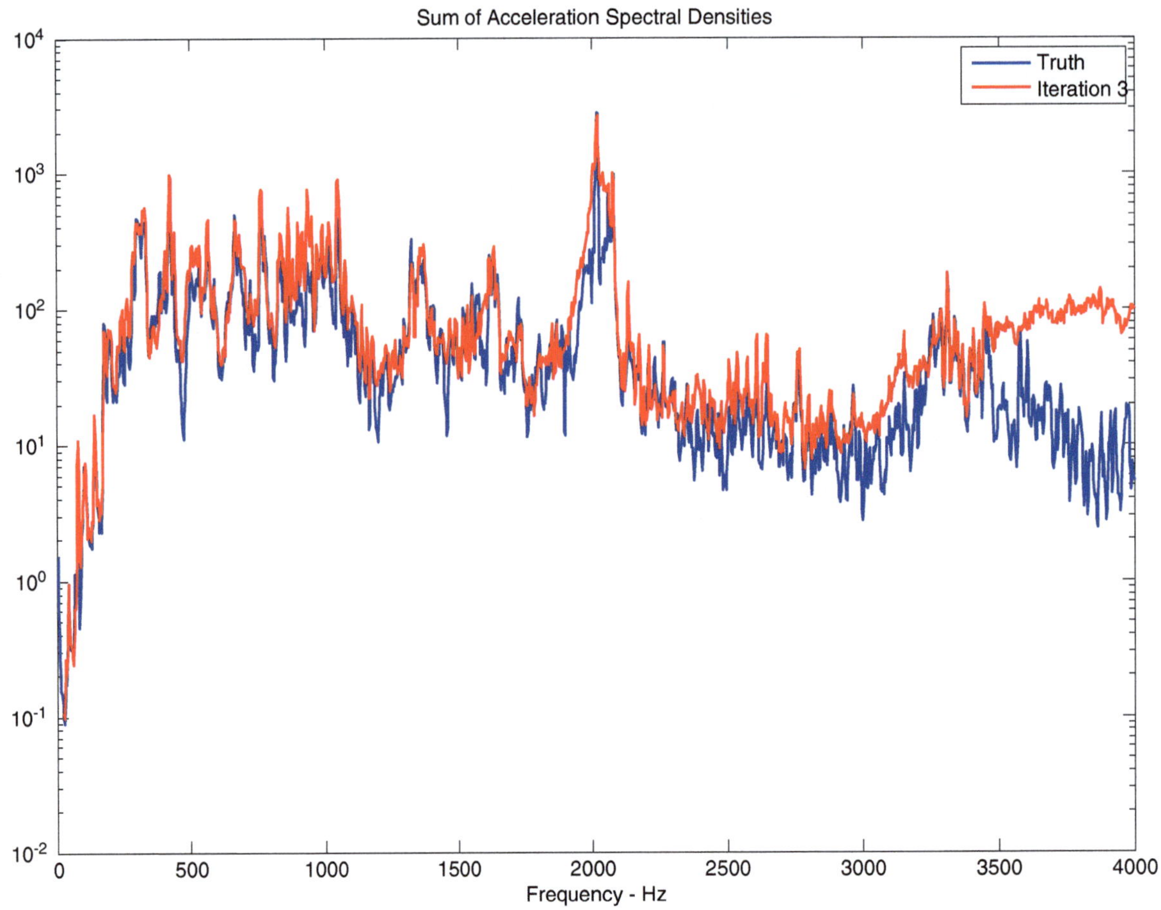

Fig. 4.4 Comparison metric (Sum of PSDs) for 37 accelerometers for the acoustic test (*blue*) and 6 shaker vibration test (*red*)

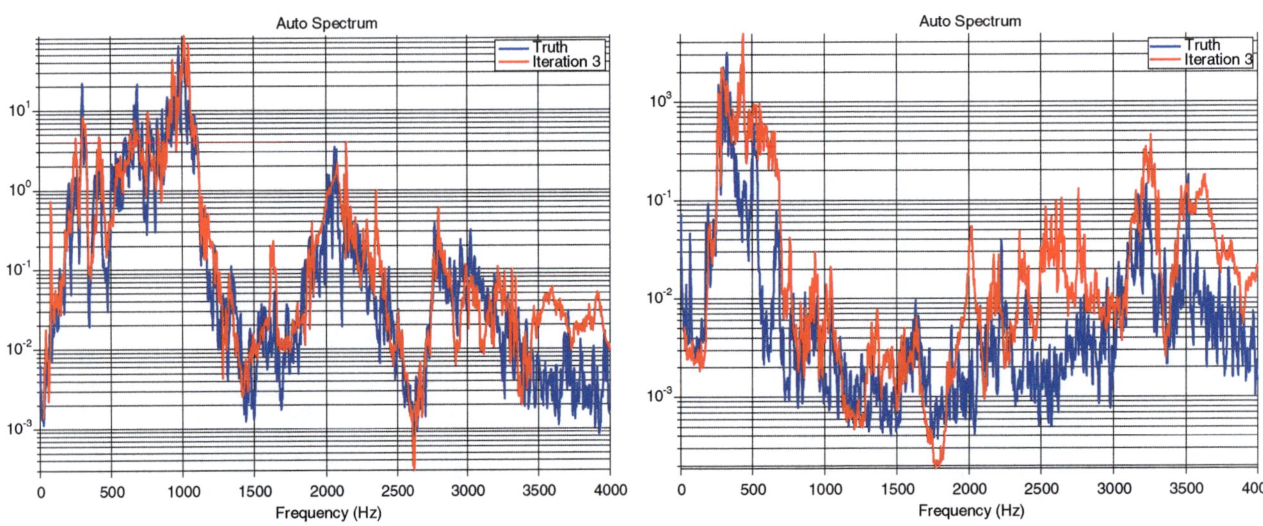

Fig. 4.5 Individual acceleration spectral density—best (*left*) and worst (*Right*) comparisons of acoustic test (*blue*) and vibration test (*red*)

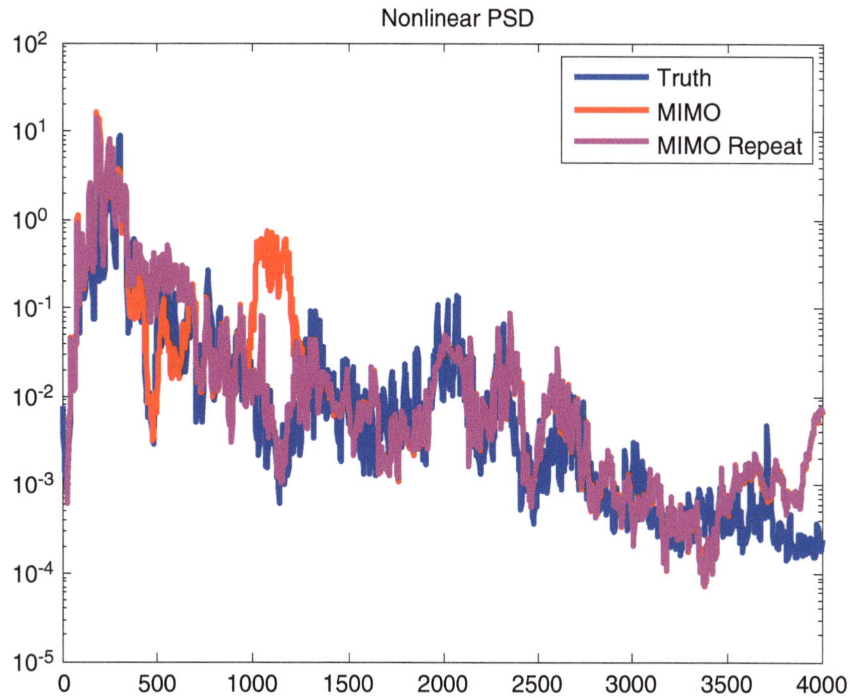

Fig. 4.6 PSD with intermittent nonlinearity—acoustic test (*blue*). Vibration test (*red*)—repeat of vibration test (*magenta*)

4.5 Simulation Theory

For a linear system, the relationship between input voltage to a set of shaker amplifiers and the output response acceleration at many degrees of freedom (dof) can be given as

$$\bar{\ddot{x}}(\omega) = \mathbf{H_{xv}}(\omega)\bar{v}(\omega) \tag{4.1}$$

where ω specifies each frequency line, $\bar{\ddot{x}}$ specifies acceleration at instrumented locations, **H** is matrix of frequency response functions and vector v is the input voltage. Assuming that the calculation is made for each frequency line, let us drop the omega term for convenience. It can be shown for uncorrelated voltage inputs that the autospectra of the responses is the magnitude squared of the FRF matrix times the autospectra of the input voltages, or

$$|\bar{\ddot{x}}|.^{\wedge 2} = |\mathbf{H_{xv}}|.^{\wedge 2} * |\bar{v}|.^{\wedge 2} \tag{4.2}$$

where .^ denotes element-wise exponentiation.

Define matrix **S** as

$$\mathbf{S} = |\mathbf{H_{xv}}|.^{\wedge 2} \tag{4.3}$$

To correct for ill-conditioning of **S**, use the Tikhonov regularization to adjust the Moore-Penrose pseudo-inverse as

$$\mathbf{S}^{+} = \left[\mathbf{S}^{T} * \mathbf{S} + c * \mathbf{I}\right]^{-1} * \mathbf{S}^{T} \tag{4.4}$$

so an estimate of the voltage autospectra is the Tikhonov pseudo-inverse of the squared FRF matrix times the autospecra of the target accelerometers

$$|\bar{v}|.^{\wedge 2} = \mathbf{S}^{+} * |\bar{\ddot{x}}|.^{\wedge 2} \tag{4.5}$$

To handle some mild nonlinearity, after each iteration one calculates the change in voltage needed to create an appropriate change in autospectrum of the accelerations and simply adds that to the last voltage autospectrum (subscript i here).

$$|\bar{v}_d|.^{\wedge 2} = |\bar{v}_i|.^{\wedge 2} + \mathbf{S}^{+} * \left(|\ddot{\bar{x}}_d|.^{\wedge 2} - |\ddot{\bar{x}}_i|.^{\wedge 2}\right) \tag{4.6}$$

4.6 Simulation Implementation

4.6.1 Adjusting for Calculated Negative Autospectrum values

When calculating response from Eqn. (4.5), at some frequency lines the mathematics calculated a negative autospectrum, which is not physically realizable. For those frequencies, the voltage autospectrum was just set to a small positive value. Theoretically, the mathematical calculations show that the pseudo-inverse could achieve a very good match to the desired acceleration autospectra. This can be seen in Fig. 4.7 in the left plot of the sum of the PSDs. However, this requires the negative voltage autospectra which are not achievable. When the negative voltage autospectra are set to a small positive number, the result on the right is achieved, which is an over-prediction. Figure 4.4 indicates that the over-prediction is reduced with the Tikhonov regularization.

4.6.2 Degree of Tikhonov Regularization and Associated Benefit

For the Tikhonov pseudo-inverse, the authors chose the constant c to be 0.01 times the largest element on the diagonal of the $\mathbf{S}^T\mathbf{S}$ matrix. The condition number of the S matrix is shown in Fig. 4.8, with and without this regularization. The regularization results in 2 orders of magnitude improvement of the condition number which helps ensure that the pseudo-inverse does not "blow up" producing unachievable voltage inputs. Regularization is known to degrade the accuracy of the solution slightly with the benefit of not destroying the solution due to an inversion based on noise in the signals.

In Fig. 4.9 is displayed a voltage time history input for each of the six shaker amplifiers based on the pseudo-inverse with and without regularization. The dark blue plot is the required input voltage time history without the Tikhonov regularization and the red plot shows the reduced voltage time history associated with the Tikhonov regularization. The largest difference is a factor of four in RMS voltage, to obtain almost the same acceleration PSD results.

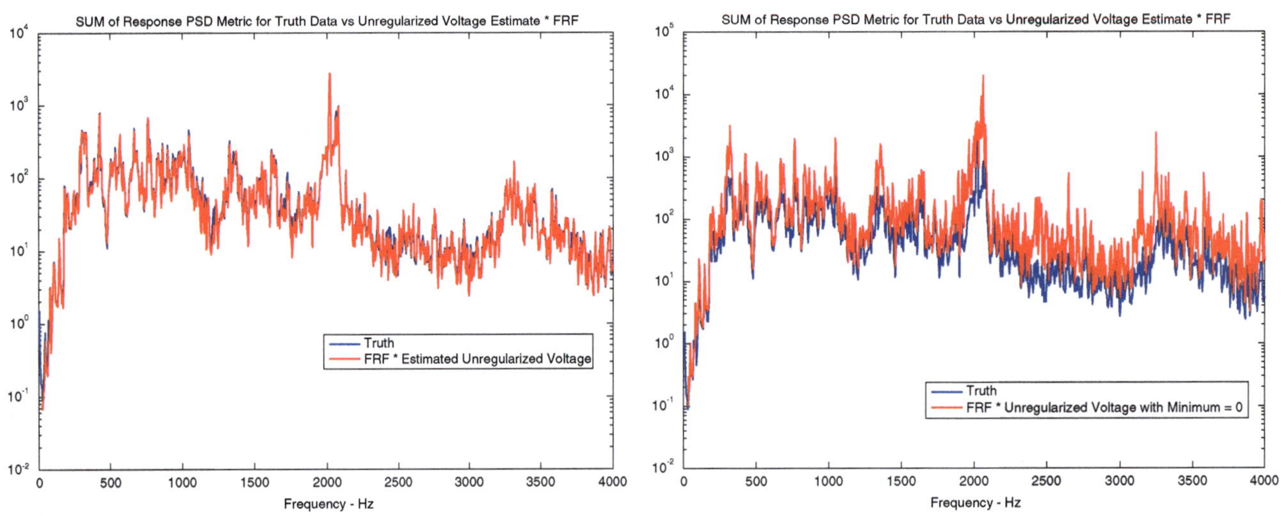

Fig. 4.7 Mathematical prediction of sum of acceleration PSDs. Result with some negative autospectra (*left*)—result with all autospectra positive (*Right*)

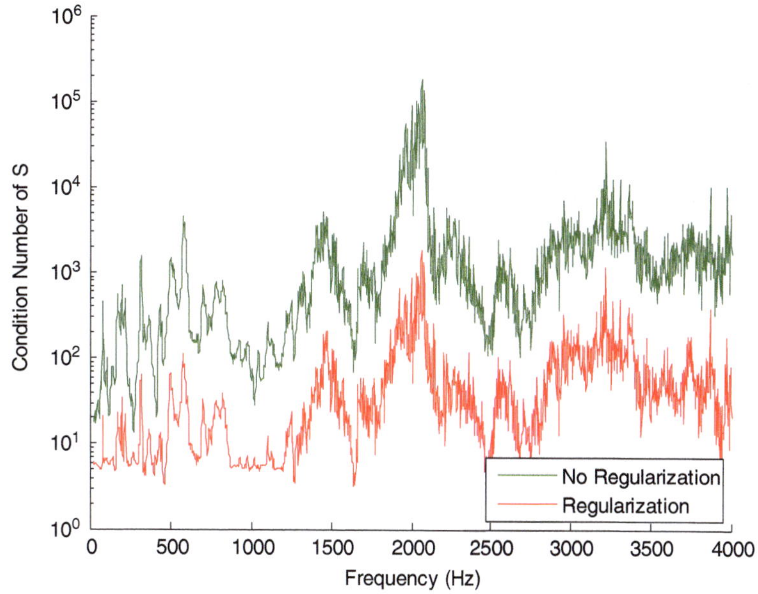

Fig. 4.8 Condition number of the S matrix with and without regularization

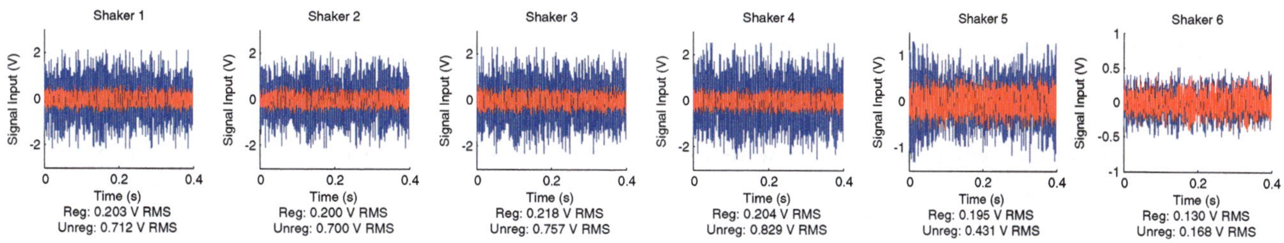

Fig. 4.9 Comparison of voltage input time histories required without regularization (*blue*) and with regularization (*red*)

4.6.3 Mathematical Results with a Decreased Number of Control Accelerometers

In many realistic cases, the number of available target accelerometers may be much less than the 37 used in this research. For a typical industrial system, the shaker voltage might have to be predicted from only a dozen measured responses. The authors randomly picked 12 of the 37 gages to determine the voltage inputs and then compared that result for the sum of all 37 PSDs for another environment in Fig. 4.10. In most frequency bands, the match of the result calculated from only 12 gages looks amazingly good. One region of error is circled.

4.6.4 Calculation of the Voltage Input Time Histories

Voltage time histories had to be generated to produce the shaker forces. After calculating the anticipated voltage autospectra from Eqn. (4.6), the square root of the ratio of the desired voltage autospectrum and initial voltage autospectrum at a particular frequency line was used to scale the initial amplitude of the sine wave of the particular frequency line. The phase of the corresponding sine wave was set randomly. This was performed independently for each source signal, so no two shakers had correlated phases. One frame of data would be constructed from this sum of sine waves with random phase. The next frame would be constructed from the same amplitude sine waves each with a new random phase. Hundred frames were utilized to give many frames whose voltage values were uncorrelated in phase for our assumption that the voltage inputs from one shaker to another were uncorrelated. On subsequent iterations we did not let the voltage at a specific line change more than $+100\ \%$, $-67\ \%$. In hindsight, the first iteration using the Tikhonov regularization could probably be allowed to change to +anything and $-90\ \%$, and further iterations could have tighter bounds.

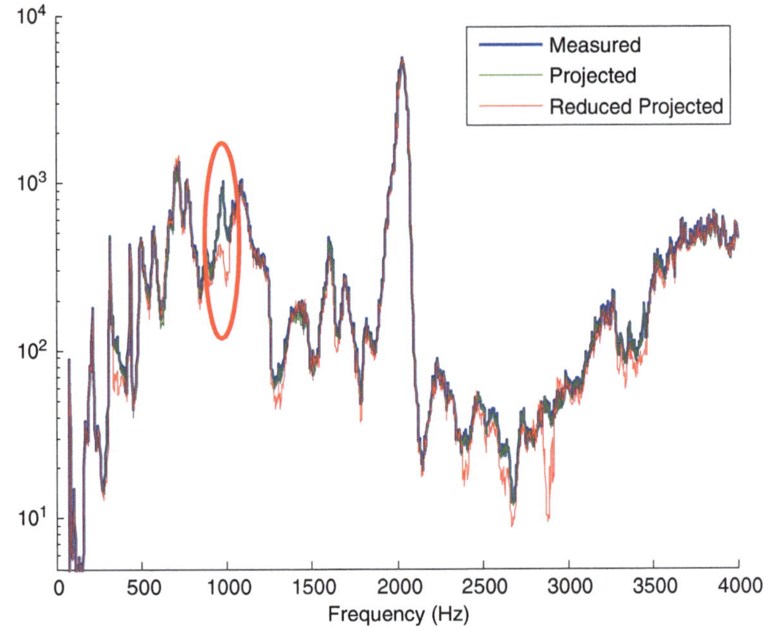

Fig. 4.10 Sum of 37 PSDs from some measured data (*blue*), mathematical prediction from voltage calculated using all 37 responses (*green*), mathematical prediction from voltage calculated using 12 responses (*red*)

4.7 Research to Increase Shaker Force

After the simulation was performed, interested parties wished to know the maximum force we could have put into the test article. Force inputs for the vibration simulation were from 19.8 to 49 N rms for the six shakers, which is about 25–65 % of the shaker rated force capability (225 N). The rms power delivered by the shakers was extremely low, 0.35 W for the shaker which required the most power. The questions that followed were: (1) Could we increase the force input beyond the rated shaker capability; and (2) What was the force capability of the connecting hardware?

4.7.1 MB50 Shaker Capabilities to 4000 Hz

Some testing with flat random voltage inputs was done with the shakers up to 4000 Hz that showed the shakers could deliver three times the rated force capability if the power required was not beyond the capability of the amplifier. The shakers could put out peak forces of 675 N. At massive "hard points" the shaker could put out more force than at non-massive "soft locations". At the soft locations, the velocity was so high that the power capability of the amplifier (force times velocity) was exceeded.

4.7.2 Stinger Force Capability

In considering the force capability of the shaker and connecting hardware, the authors decided that the connecting hardware should provide protection for the shaker and test hardware. This required understanding the force capability of the connecting hardware and developing a failure mechanism that would protect the shaker and test unit. It was decided that having the stinger slip under its setscrews was an inexpensive way to protect the expensive shaker and test article from costly failure. The force gage would also be protected if the stinger slipped below the maximum rated force capability of the force gage and its connecting stud. Static tests were performed on the stinger attached to a receptacle with either one or two size #3 setscrews and torques on the setscrews from 0.228 to 0.8 N-m. 0.8 N-m would begin to deform the threads in the aluminum stinger receptacle to receive the set screw. In Fig. 4.11 is given the static force at which the stinger slipped for several trials of set screw torque for one or two set screws. Having two screws torqued to 0.6–0.7 N-m would allow the maximum 675 N capability of the MB50 shaker to be utilized.

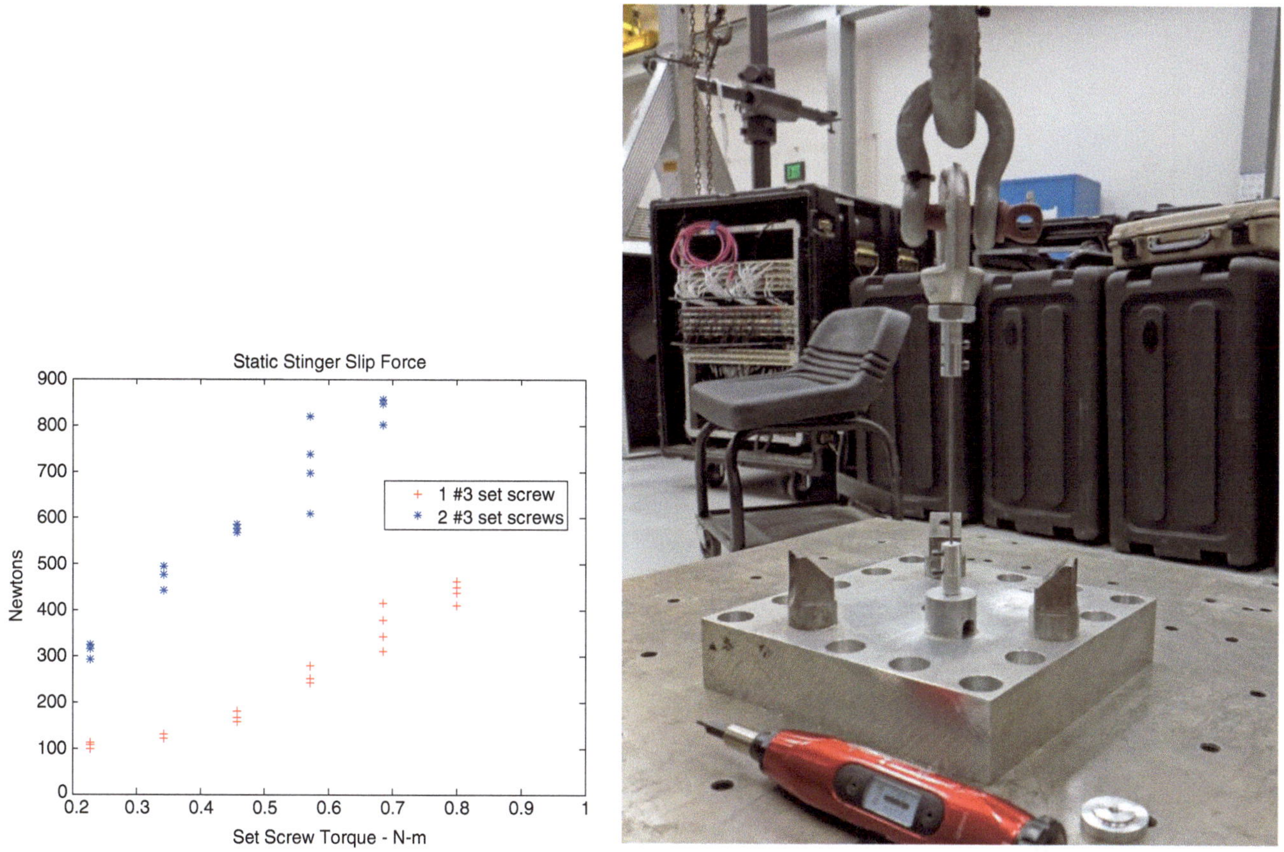

Fig. 4.11 Static stinger slip force vs. set screw torque

4.7.3 Force Capability of Cap

The cap to which the force gage is attached (see Fig. 4.2) was attached to a flat surface with Durakore brand dental cement and pulled to failure while measuring the static force that caused the dental cement to fail. Wide variability in the dental cement failure strength was found for the 200 mm^2 attachment area. If the surface was smooth and not cleaned well, the dental cement could fail as low as 200 N force. With roughed up surfaces well cleaned on both sides, the failure values were above 2800 N, which is considerably above the 675 N capability of the MB50.

4.7.4 Optimal Shaker Placement and Increasing Tikhonov Regularization c *Value to 0.1*

Since the final responses measured indicated that the mathematical calculations for the sum of PSDs metric was almost accurate, some mathematical calculations were performed examining changing the Tikhonov regularization value, c, in Eqn. (4.4). It was found that almost the same accuracy could be obtained by increasing the value of c to 0.1. Also, the calculation was made with the FRFs gathered from six shakers mounted to the belly of the structure as in Fig. 4.1. Figure 4.12 shows the calculated match of the sum of PSDs. With these adjustments, the maximum shaker force required for any of the shakers was about 23 N rms. Using three sigma clipping, and the 675 N shaker peak force capability found in Sect. 7.1, the MB50 shakers could put out about ten times more force than required for this physical simulation, which would equate to roughly a factor of 100 in power. Of course this assumes no significant nonlinearity in the structure.

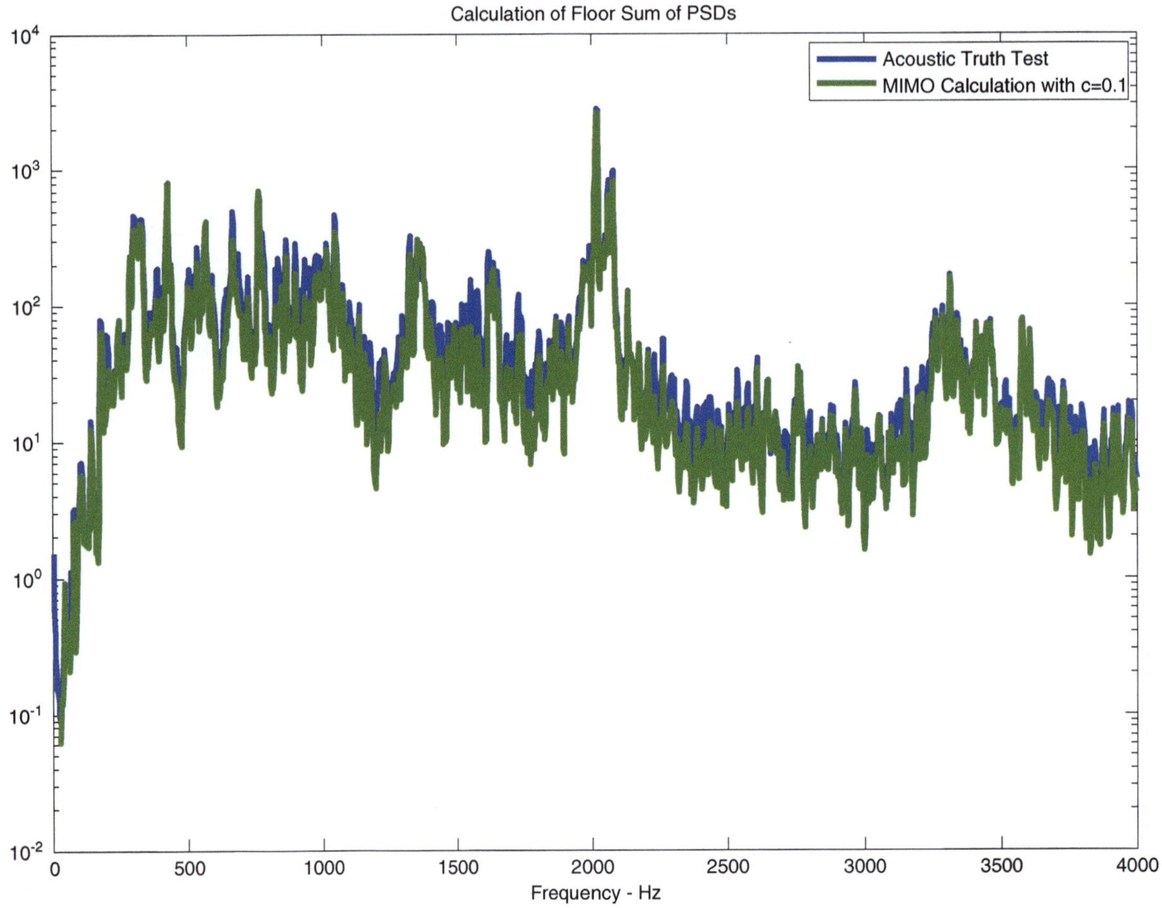

Fig. 4.12 Calculated sum of PSDs match for 6 floor mounted shakers and Tikhonov c value of 0.1

4.8 Conclusions

The six shaker input simulation was able to nearly replicate the results of an acoustic test environment up to about 3500 Hz. The match from 3500 to 4000 Hz was not very good. The element by element square of the absolute value of the FRF matrix was used with the target acceleration autospectra to calculate six voltage autospectra used for the input voltage to the shaker amplifiers in the inverse problem. Tikhonov regularization was utilized to reduce these inputs. Cross spectra were not needed for this method. Reduction of the force requirement was achieved by increasing the Tikhonov constant without a significant reduction in accuracy of matching the 37 internal target acceleration responses. The inverse problem was also calculated for 12 of the original 37 target autospectra with only a small reduction in accuracy of the simulation. Some additional reduction in the required forces appeared available based on the location of the shaker attachment. For the 4000 Hz bandwidth, it was found that the shakers could produce about three times the rated force input when attached to hard points of a structure. At soft structural force connection locations, the maximum force was limited by the amplifiers in the cases we tested. The slip force of the stingers and set screws was characterized to provide a degree of protection for the shakers and the test article.

Acknowledgments The acoustic test was performed by Eric Stasiunas and Ryan Schultz. Anthony Gomez was the primary technologist involved in the data acquisition along with the authors. Jerry Cap was the prime motivator who was able to get the testing programmed and funded. Diane Callow provided a huge portion of the logistic support in terms of writing the test plan and getting the testing approved to meet the operational and safety requirements. Bill Fladung from ATA engineering showed us how to input user defined time histories into the VTI source modules.

References

1. Himelblau, H., Hine, M.J.: Effects of triaxial and uniaxial random excitation on the vibration response and fatigue damage of typical spacecraft hardware. In: Proceedings of the 66th Shock and Vibration Symposium, Arlington, 1995
2. Daborn, P.M., Roberts, C., Ewins, D.J., Ind, P.R.: Next-generation random vibration tests. In: Proceedings of the 32nd International Modal Analysis Conference, paper number 92, Orlando, February 2014
3. Daborn, P.M., Ind, P.R., Ewins, D.J.: Replicating aerodynamic excitation in the laboratory. In: Proceedings of the 31st International Modal Analysis Conference, paper number 42, Garden Grove, 2013

Chapter 5
Developing Conservative Mechanical Shock Specifications

Matthew Baker, Kelsey Neal, Katrina Sweetland, Garrison Stevens, Dustin Harvey, and Stuart Taylor

Abstract Mechanical shock testing and analysis are integral parts of developing new high-value items and ensuring their capability to withstand the environments to which they will be exposed. Most conventional methods for specifying mechanical shock environments provide no mathematically defensible correlation between their parameters and the damage-causing potential of the environment, thus warranting the study of new methods to specify shock environments. In this paper, a variety of parameters that correlate to the damage-causing potential of a shock environment are identified. The parameters investigated are restricted to those that can be constrained during a real-time laboratory test. An analytical study using a sensitivity analysis determined the effect of each parameter on the damage-causing potential of a shock environment. These parameters are used to create shock specifications and investigate the simulated response of a structure. The parameters generated in this study improve mechanical shock testing by providing a strong correlation between the shock environment and the damage-causing potential of a mechanical shock.

Keywords Mechanical shock • Shock test specification • Shaker shock • Shock severity • Sensitivity analysis

5.1 Introduction

A mechanical shock is a sudden acceleration excitation on a structure usually resulting in a significant amount of displacement [1]. Mechanical shocks can be the result of multiple forms of physical excitations, such as jolts, impacts, transportation, and separation. Anticipating mechanical shocks associated with service environments is important because these shocks can damage entire systems or critical components of those systems. The consequences of such damage makes mechanical shock testing critical in the design and development of high-value systems. Without proper mechanical shock testing, there is potential for system failures resulting in devastating consequences.

Mechanical shock testing is widely used for product design and development to ensure the reliability of these products in their service environments [2]. Currently, shock testing uses a variety of methods designed to elicit failure modes of the product. These methods include strategies such as testing in the product's actual service environment, drop testing, and recreating the shock's damage-causing potential in laboratory testing.

Shock testing in a laboratory setting offers some advantages over other forms of testing, including the repeatability of the test [2], which allows for faster design iteration cycles and ultimately faster product development. Laboratory testing also provides a more controlled setting in which mechanical shocks can be applied to the test article. Furthermore, applying mechanical shocks to the test article in a controlled environment, as is the case with laboratory testing, can be safer than alternative testing methods. Recreating the severity of a shock in a laboratory setting requires identification of test specifications, which are derived from the parameters of a mechanical shock. Several methods have been established for defining the test parameters, but the most common method relies on the shock response spectrum (SRS) [3]. The SRS is constructed using the peak acceleration responses from a collection of fictitious single-degree-of-freedom mass-spring-damper systems that have a range of natural frequencies and are excited using base excitation.

M. Baker • G. Stevens
Clemson University, Calhoun Drive, Clemson, SC 29634, USA

K. Neal
Colorado School of Mines, 1500 Illinois Street, Golden, CO 80401, USA

K. Sweetland • D. Harvey • S. Taylor (✉)
Los Alamos National Laboratory, P.O. Box 1663, Los Alamos, NM 87545, USA
e-mail: sgtaylor@lanl.gov

A. Brandt, R. Singhal (eds.), *Shock & Vibration, Aircraft/Aerospace, Energy Harvesting, Acoustics & Optics, Volume 9*,
Conference Proceedings of the Society for Experimental Mechanics Series, DOI 10.1007/978-3-319-30087-0_5

There are several difficulties in the implementation of the SRS. One difficulty is in determining the actual time history with which to excite the structure under evaluation. The SRS does not uniquely identify the shock experienced by the part. This means shocks with different time histories, and parameters in spaces other than the SRS, could produce the same SRS without recreating the damage-causing potential of the structure's service environment. Furthermore, the SRS accounts for neither the damaging potential caused by the interaction between masses on a system nor the damage potential caused by the coupling of translational and rotational movements [4].

This lack of correlation between test inputs and the resulting severity of damage of a shock environment leads to results that have an unknown level of conservatism. The inability to consistently and accurately replicate the severity of a mechanical shock environment warrants the development of new methods. Development of such a method requires the consideration and identification of parameters that are able to be implemented in a laboratory setting.

The study discussed is a simulation based study in which measured mechanical shocks are applied to a finite-element model of an arbitrary test structure. In the context of this study, a simulation based approach offers several advantages over an experimental study. A simulation study allowed for a high volume of simulations which were used to evaluate the distributions of parameters and responses. Collecting this amount of data in an experimental based study would be time consuming and potentially expensive. Furthermore, the mechanical shocks were applied directly to the structure in the finite-element model simulations, and thus hardware constraints were not a factor. Data was collected at many locations on the structure for each finite-element model simulation, while doing so in an experimentally based study would not have been possible due to the amount of instrumentation that would be required. The objectives of this study are to identify possible parameters that correlate to the damage causing potential of a mechanical shock, identify a set of those parameters that can make a specification that uniquely specifies an acceleration time history, and compare the ability of the identified parameters to reproduce the damage causing potential of a shock to the ability of parameters that are currently being used.

The process used to identify parameters of a mechanical shock started with using real transportation shock data to define parameters that could potentially have an effect on the severity of the shock. To determine how each input parameter correlated to the observed measures of severity shock realizations were developed and applied to the structure through finite-element model simulations. Similarly, the transportation shocks were applied directly to the finite-element model in order to observe how these unadulterated shocks, and their associated parameters, correlate to the observed measures of severity. The finite element model was initially evaluated by comparing the results from the model to the results of an experimental modal analysis. This was done to ensure that the finite element model was producing realistic results.

The layout for the report is as follows. The test structure and model development are discussed in Sect. 5.2. Sections 5.3.1 and 5.3.2 describe the process used to identify the parameters that define the measure of severity associated with a mechanical shock. The results of the analytical study and identified parameters are shown and discussed in Sects. 5.3.3 and 5.3.4. Finally, the conclusions drawn from the analytical study are addressed in Sect. 5.4.

5.2 Model Development

For this study the finite element-model was based on an existing structure (Fig. 5.1) as this allowed for confirmation that the finite-element model behaved realistically. The structure is made of approximately 6.5 mm thick aluminum and is composed of multiple components. The base of the structure has a diameter of 165 mm and the stand (the thinnest part) has a diameter of approximately 25 mm. The total height of the structure is approximately 305 mm tall. Composition of these different parts provides discontinuity throughout the structure producing varying measures of severity, which was desired for this experiment. A modal analysis was performed to compare the modes generated from the finite-element model and the modes collected from experimental data. The modal analysis was completed using a roving hammer technique with 3 uniaxial accelerometers and 17 impact points. Each accelerometer was placed on a plane that aligned with the axis of a Cartesian coordinate system in which the "z" axis is parallel to the vertical component of the structure. The impact points were spread around the structure such that multiple response directions on each face would be tested. For the modal analysis, free boundary conditions were simulated by placing the structure on foam. MEScope was used to extract the mode shapes based on the data at the impact and measurement locations.

Finite-element modeling is critical to this study. The finite-element model was developed using Abaqus and consisted of four solid elements: the top and bottom plates, the main cylinder, and the stand. The top and bottom plates are thin cylinders that are connected to the main cylinder. The main cylinder is the largest part, the stand is connected to the top plate. Once the test structure was properly modeled both a modal analysis and the mechanical shocks were simulated.

Fig. 5.1 Test structure

5.2.1 *Test Structure Model*

The parts of the structure were assembled using surface ties, which constrain the two surfaces so there is no relative motion between them at that point, to model the connections between parts. For the modal analysis the boundary conditions along the bottom of the plate were modeled as free in order to match the experimental set-up. The boundary conditions along the bottom of the plate were then set to fixed when applying the mechanical shocks to the model to represent the structure being bolted to the shaker table Fig. 5.2.

5.2.2 *Mesh Refinement*

Mesh refinement studies were conducted for multiple element types, including quadratic hexahedral elements and quadratic tetrahedral elements. Convergence was tested using the frequency for mode 7, the first mode in which the structure did not behave as a rigid body. As can be seen in Figs. 5.3 and 5.4 the frequency was found to converge around 471 Hz for both element types. The study showed the most computationally efficient and accurate mesh would be composed of quadratic hexahedral elements and a 4 mm seed size (approximately 21,000 elements).

5.2.3 *Model Correlation*

The correlation was determined through a direct comparison of the modal frequencies found from the experiment and the finite-element model. A Modal Assurance Criterion (MAC) was used to compare the displacement at each corresponding mode. The numerical results of both comparisons are presented in Table 5.1 while the graphical results of the MAC are shown in Fig. 5.5.

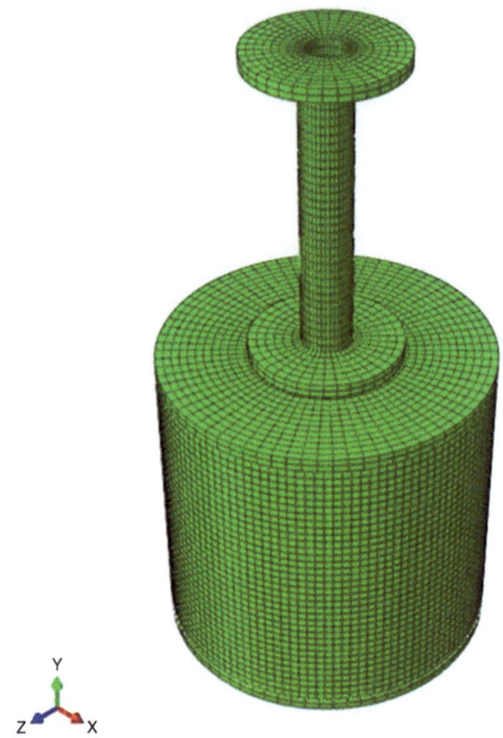

Fig. 5.2 Assembled finite-element model of test structure showing the mesh

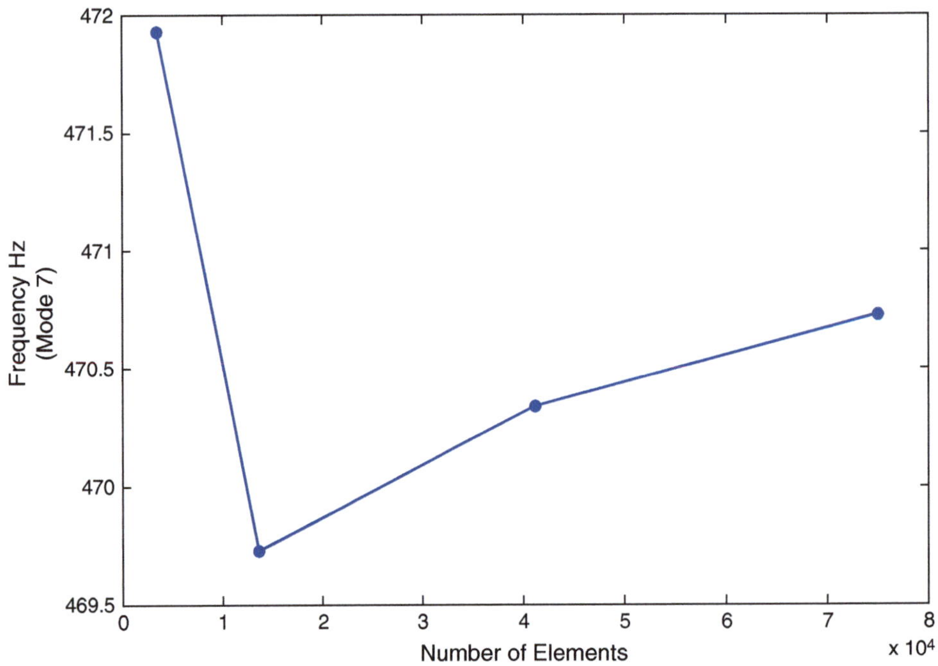

Fig. 5.3 Quadratic tetrahedral mesh refinement study

The finite-element model showed an error in modal frequencies that was acceptable for the purpose of the finite-element model in the study. Mode 3 has a large amount of error in frequency but is of less interest than the other modes because it represents motion in the vertical direction and is not likely to be activated by the mechanical shocks applied in the horizontal direction. This error is likely due to the assumptions made in the finite-element model, such as the surface ties used to simulate the real life bolted connections. The MAC shows strongest correlation between the corresponding modes of the model and the experimental modes. These results show that the model will have a similar response to that of a real life structure, which is sufficient for its purpose in the study.

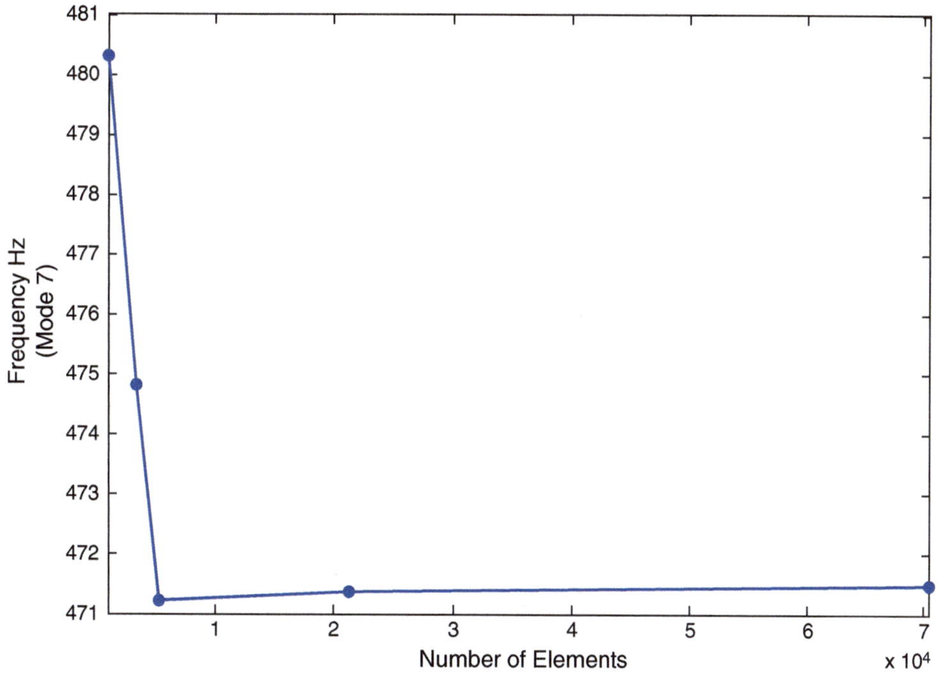

Fig. 5.4 Quadratic hexahedral mesh refinement study

Table 5.1 Model correlation results

Mode number	Experimental frequency (Hz)	Model frequency (Hz)	% Error for frequency	MAC value
1	432	470	8.09	0.795
2	447	470	4.89	0.852
3	1270	1790	29.2	0.979
4	2680	2670	0.34	0.720
5	2750	2670	2.96	0.734

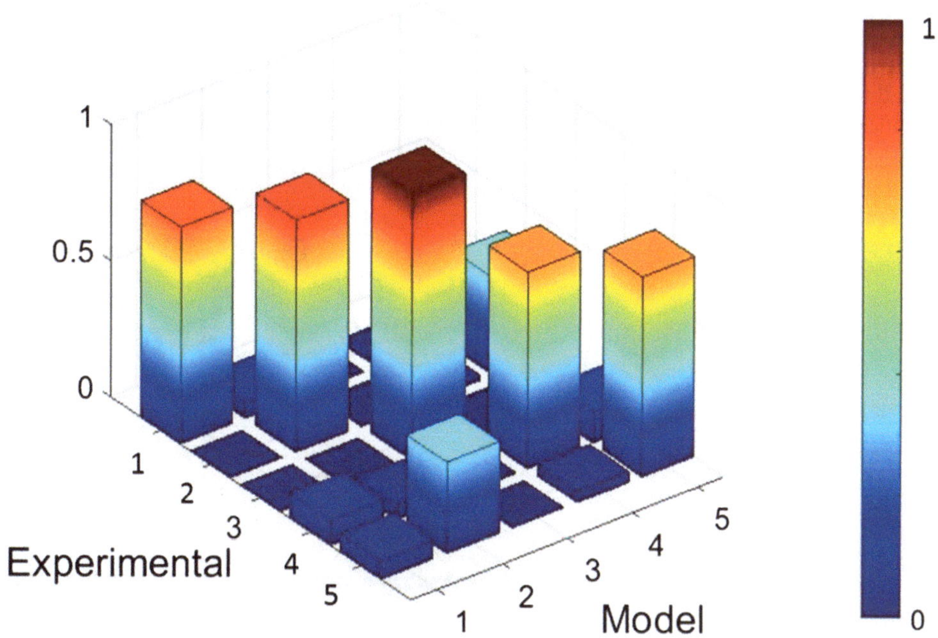

Fig. 5.5 Modal assurance correlation

5.3 Analytical Study

5.3.1 Developing Parameters

The parameters of a shock waveform were chosen based on their potential to affect the severity of the shock. The parameters include the RMS duration, maximum amplitude, root mean square (RMS), and peak signal width. RMS duration was chosen as a parameter because the distribution of the signal energy could affect the maximum strain that the structure experiences. As the RMS duration increases, the energy of the wave becomes more spread out. This increases the time it takes for the wave to deliver an equivalent amount of energy compared to a wave with the same amount of energy but a shorter RMS duration. This could affect the maximum strain and the recovery time of the material. The maximum amplitude was chosen as a parameter because it affects the immediate acceleration applied to the structure. RMS was chosen as a parameter because the amount of acceleration applied to the structure over the duration of a shock could affect the amount of damage experienced by the structure. The peak signal width was chosen because the rate at which the maximum acceleration of a shock is applied to the structure may affect the amount of acceleration or strain experienced.

Temporal moments, M_k, are computed as weighted summations of the time signals squared, Eq. (5.1) computes the temporal moments where k is the order of the moment, t is the time, and $y(t)$ is the system response as a function of time. From the temporal moments and the central time moment, RMS duration can then be calculated [5]. Equations (5.1), (5.2), and (5.3) can be found in "From Shock Response Spectrum to Temporal Moments and Vice-Versa" [5].

$$M_k = \int_{-\infty}^{+\infty} t^k (y(t))^2 \mathrm{dt} \tag{5.1}$$

$$T = \frac{M_1}{M_0} \tag{5.2}$$

RMS duration is based on the central time moment, T, which is the time value between the two areas in Fig. 5.6 such that the energy, (which is equal to the area under the curve) to the left and the right of the point are equal. This means that area A and area B shown in the Fig. 5.6 are equal.

The RMS duration, D in Fig. 5.7, is analogous to standard deviation of a distribution but describes the dispersion of the signal energy in time. Similar to standard deviation, the majority of the energy in a transient signal is within 2 to 3 RMS durations of the central time moment, T [5]. The RMS duration can be found using Eq. (5.3):

$$D = \sqrt{\frac{M_2(T)}{M_0}} \tag{5.3}$$

The peak amplitude that was chosen can be seen in Fig. 5.8. It is the maximum acceleration magnitude over the duration of the mechanical shock. The RMS (Fig. 5.8) is the effective value of a varying amplitude signal, in this case acceleration.

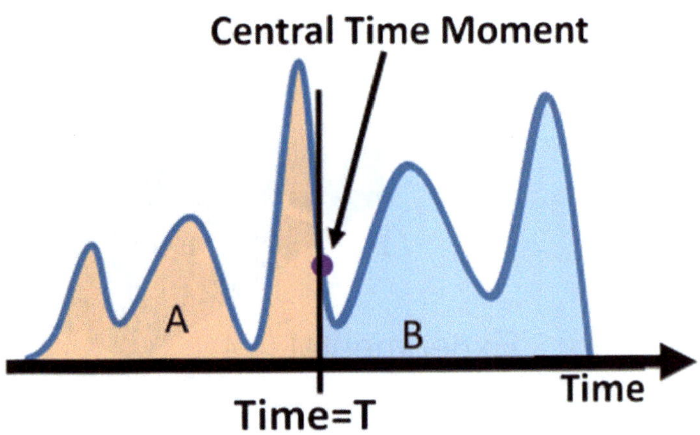

Fig. 5.6 Central time moment, T

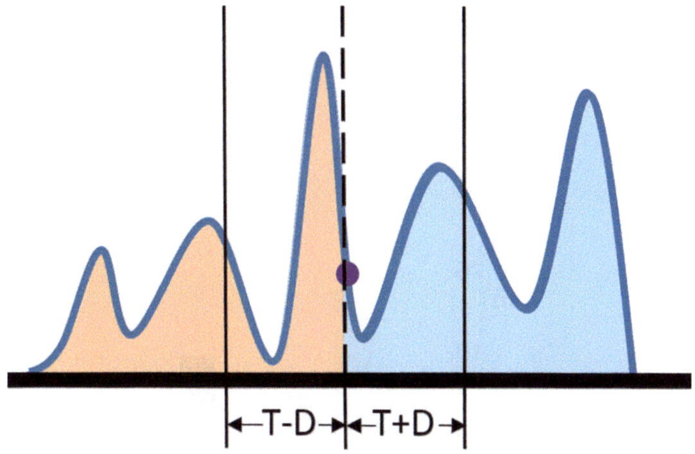

Fig. 5.7 RMS duration, D

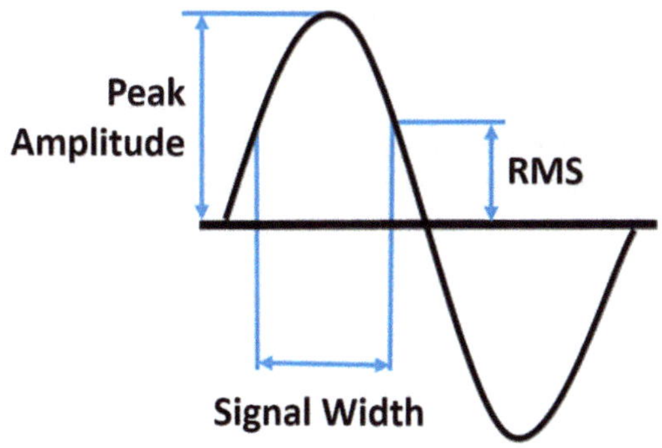

Fig. 5.8 Peak amplitude, RMS, and signal width

Table 5.2 Parameter values

Parameter	Minimum	Mean	Maximum
Peak amplitude	42.8 g's	57.0 g's	71.3 g's
RMS	4.54 g's	6.06 g's	7.58 g's
Signal width	0.002 s	0.003 s	0.004 s
RMS duration	0.061 s	0.085 s	0.109 s

The RMS can be calculated with Eq. (5.4) where n is the number of values and A is the value at each point in time. The peak signal width (Fig. 5.8) is defined as the width of the peak starting and ending at 50 % of its maximum value.

$$RMS = \sqrt{\frac{1}{n}\left(A_1^2 + A_2^2 + \ldots + A_n^3\right)} \tag{5.4}$$

5.3.2 Analysis of Preliminary Parameters

A sensitivity analysis, which is used to investigate the statistical relationships between a factor or a group of factors and one or more independent variables [6], determined the effect each parameter individually or in combination with others had on the severity of the shock. Table 5.2 shows the values for each parameter, which were chosen based on the distribution of shocks extracted from 70 h of road transportation data. The selected maximum and minimum parameter values are the values at ±25 % from the mean. These parameter values were used to synthesize random shocks for the sensitivity analysis.

The measures of severity for brittle failure include maximum strain and maximum acceleration, whereas the measure of severity for fatigue is the total area under the strain curve. These measures of severity were chosen because they represent several types of behavior that could lead to failure as a result of a mechanical shock. The sensitivity analysis utilized a full factorial design of experiments to determine which parameter or combination of parameters had the greatest effect on the previously discussed measures of severity. The results for the sensitivity analysis and parameter correlations can be seen in Fig. 5.9 and Table 5.3.

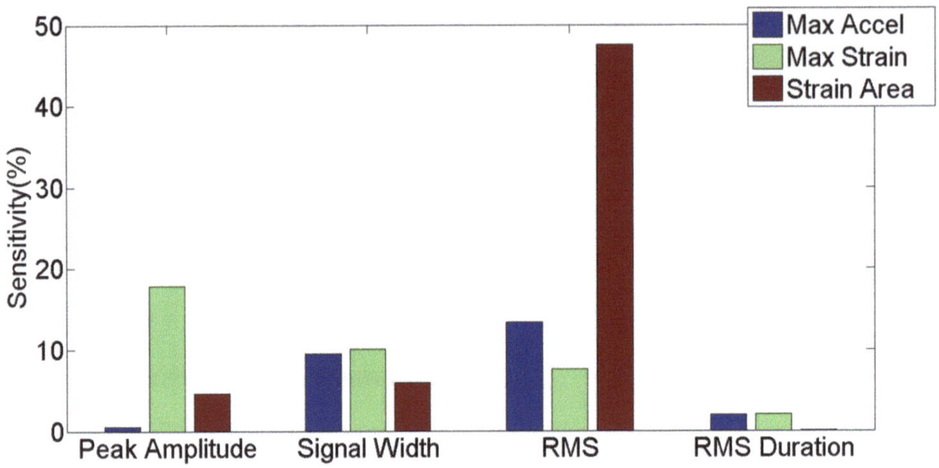

Fig. 5.9 Sensitivity analysis results for preliminary parameters

Table 5.3 Preliminary parameter correlation values

	Maximum acceleration	Maximum strain	Strain area
Peak amplitude	0.741	0.855	0.699
Signal width	0.371	0.379	0.147
RMS	0.534	0.606	0.760
RMS duration	0.401	0.465	0.277

Table 5.4 Filtered parameter correlation values

	Maximum acceleration	Maximum strain	Strain area
Filtered peak amplitude	0.974	0.957	0.977
Filtered RMS	0.968	0.929	0.973

5.3.3 *Final Parameters*

The sensitivity analysis and the correlations indicate that the four preliminary parameters do not have a significant effect on the severity of the shock. Thus, additional parameters needed to be explored. It is expected that the structure will have its greatest response around modal frequencies, since the structure is lightly damped and modeled as a linear system. To investigate this effect, the input shock was filtered using a band-pass filter, and new parameters were calculated from this filtered input. The frequency content of the shock family used in this study only excited the lowest frequency mode of the structure. Therefore, the filter was set to 380 Hz, which is the frequency of the first mode when the structure is fixed. Once filtered, the new peak amplitude and RMS, which will be referred to as "filtered RMS" and "filtered peak amplitude", were calculated and used as shock parameters. The filtered parameters were used to perform a new severity correlation, seen in Table 5.4. The results showed that the filtered parameters have a very strong correlation with severity. As expected, the characteristics of a shock at a structure's resonant frequencies produced the greatest change in the structure's response.

5.3.4 *Filtered Parameters Versus SRS*

SRS is commonly used to develop shock testing specifications. In order to quantify the shock response spectrum, a measure of the average ratio of a single SRS to the mean SRS of the entire data set was used and is referred to as the SRS Ratio in this paper. For this study, the SRS ratio was measured for the transportation shock ensemble and then varied to make synthesized shocks using a sum of decaying sinusoids. These synthesized SRS shocks were applied to the same finite-element model

Table 5.5 Correlation of SRS and filtered parameters for synthesized shocks

	Maximum acceleration	Maximum strain	Strain area
SRS ratio	0.984	0.984	0.917
Filtered peak amplitude	0.974	0.957	0.977
Filtered RMS	0.968	0.929	0.973

Table 5.6 Correlation of SRS and filtered parameters for transportation shocks

	Maximum acceleration	Maximum strain	Strain area
SRS ratio	0.572	0.654	0.698
Filtered peak amplitude	0.943	0.869	0.868
Filtered RMS	0.949	0.894	0.916

used in the study. Both the synthesized and transportation SRS ratios were correlated with the severity of the shock. The correlation between the measures of severity and the SRS ratios of the synthesized shock and transportation shock cases are shown in Tables 5.5 and 5.6 respectively.

5.4 Conclusions

The parameters that were initially investigated in this paper, peak amplitude, signal width, RMS, and RMS duration, did not have a significant effect on the severity of shock. This is likely because the frequency range of the majority of the shock energy was lower than most of the resonant frequencies of the structure. A sensitivity analysis indicated the initial four parameters were not sufficient in reproducing the damage causing potential of a mechanical shock, as RMS was the only parameter to have even moderate effect, about 50 %, on strain area. It was found that the characteristics of a shock at or near the resonant frequencies had the greatest effect on the damage causing potential. The shock inputs were filtered around the structure's first mode, and new parameters were calculated. Both the filtered amplitude and the filtered RMS were found to be strongly correlated with three measures of severity studied. The filtered parameters' correlations with severity were compared to the average SRS level. Although the SRS ratio had slightly higher correlation values for synthesized shocks produced from the same SRS shape at varying levels, the filtered parameters had much stronger correlations for real, measured shock environments. Based on the results from this research, the filtered parameters can be used to make conservative shock test specifications from a family of mechanical shocks for linear, lightly damped systems. Future work in this area would include experimentally verifying the results and expanding this method to accommodate other systems that are not lightly damped and are non-linear. Additionally, shock families that excite more of the systems' modes should be investigated.

Acknowledgements This document has been approved for public release and assigned the number LA-UR-15-28190.

References

1. Harris, C., Piersol, A.: Introduction to the handbook. In: Piersol, A., Paez, T. (eds) Harris' Shock and Vibration Handbook, 6th edn., p. 121. McGraw-Hill, New York (2002)
2. Irvine, T.: An introduction to the shock response spectrum. Rev P Vibrationdata, 1–20 (2002)
3. Goyal, S., Buratynski, E.K., Elko, G.W.: Role of shock response spectrum in electronic product suspension design. Int. J. Microcirc. Electron. Packag. **23**(2), 182–190 (2000)
4. Goyal, S., Papadopoulos, J., Sullivan, P.: Shock protection of portable electronic products: shock response spectrum, damage boundary approach, and beyond. Shock Vib. **4**, 169–191 (1997)
5. Hemez, F.M., Doebling, S.W.: From shock response spectrum to temporal moments and vice-versa. In: 21st International Modal Analysis Conference, Kissimmee (2003)
6. Hemez, F.M., Rutherford, A.C., Maupin, R.D.: Uncertainty analysis of test data shock responses. In: 24th International Modal Analysis Conference, Saint Louis (2006)

Chapter 6
Force Limited Vibration Using the Apparent Mass Method

Paul Marchand, Raj Singhal, and Mark O'Grady

Abstract Force Limited Vibration is a technique that reduces the severity of over-tests during vibration testing by limiting the acceleration input based on the maximum expected in-service interface force. In this paper, the sources of over testing are reviewed and a new method called the Apparent Mass method for deriving the maximum in-service interface force is presented. In this method, the maximum in-service interface force over a given bandwidth is predicted using the full three dimensional linear-elastic theory supplemented by the necessary conditions for a representative vibration test, the assumption that the test article and its in-service mounting structure are lightly damped, and that the in-service forces exciting the mounting structure are broadband. The resulting equation is shown by numerical example to produce significant notching without producing an under-test situation, even for extremely light test articles.

Keywords Interface • Force • Limiting • Vibration • Apparent mass

Nomenclature

A	Acceleration
c	Damping
E	Young's modulus of elasticity
f_{nn}	Natural frequency nn of coupled system
f_o	Notched mode of unit under test
F	Force
$F_{\text{int\_total\_max\_X}}$	Maximum total interface force in the X direction
H	Accelerance matrix
k	Stiffness
K	Stiffness matrix
m	Mass
m_{int}	Mass of nodes at the interface
M	Mass matrix
M_{app}	Apparent Mass matrix
M_{app_total}	Total Apparent Mass in the X direction
$M_{eff\_n,X}$	Effective Mass of mode n in the X direction
$M_{\text{int\_app}}$	Interface Apparent Mass matrix
$S_{\ddot{x}\ddot{x}\text{\_test}}$	Test acceleration PSD in the X direction
$S_{ff\text{\_int\_total\_max\_X}}$	Maximum total interface force PSD in the X direction
u	Displacement
$\ddot{X}$	Acceleration in the X direction
υ	Poisson's ratio
ρ	Density

P. Marchand
Neptec Design Group Ltd., 302 Legget Drive, Suite 202, Ottawa, ON, Canada K2K 1Y5
e-mail: pmarchand@neptec.com

R. Singhal (✉) • M. O'Grady
Canadian Space Agency (David Florida Laboratory), 3701 Carling Avenue, P.O. Box 11490, Station H, Ottawa, ON, Canada K2H 8S2
e-mail: Raj.Singhal@asc-csa.gc.ca

A. Brandt, R. Singhal (eds.), *Shock & Vibration, Aircraft/Aerospace, Energy Harvesting, Acoustics & Optics, Volume 9*,
Conference Proceedings of the Society for Experimental Mechanics Series, DOI 10.1007/978-3-319-30087-0_6

6.1 Introduction

Force Limiting Vibration (FLV) testing is a method of notching the acceleration input during a vibration test to avoid severely over testing the Unit Under Test (UUT). In FLV, the total interface force is measured in situ and the input acceleration is notched automatically in real time to prevent the total interface force from exceeding a prescribed value. Consequently, the method requires the frequency specification of both acceleration and interface force.

This paper aims to provide a method for determining force limits for use in FLV by examining the interface forces predicted by linear Frequency Response Function (FRF) theory between two arbitrary bodies in three dimensions.

6.2 Roots of the Over-Test Problem

The vibration over-test phenomenon has been documented for about 60 years [1] and is the reason why the input acceleration is often notched (i.e. reduced) at and near the main natural frequencies of the UUT during vibration testing. There are two possible causes of an over-test and it is important to understand how they both contribute to an over-test situation.

The first cause is an interface impedance difference between the in-service support structure and the test support structure. In service, the UUT can push back on its interface to reduce the acceleration at its interface. During testing however, the shaker control system simply pushes harder to compensate for any pushback from the UUT in order to meet the programmed test spectrum. The amount of push–back is largest at frequencies where the UUT has modes with high Effective Masses and low damping, and it is these modes that we traditionally wish to notch when using FLV.

The second cause is the enveloping process. The test acceleration can be based on one of two cases. In the first case, the test specification is an envelope of the measured or predicted interface acceleration based on the coupled structure response, i.e. the UUT is attached to its in-service supporting structure. In the second case, the test specification is an envelope of the measured or predicted interface acceleration based on the un-coupled supporting structure response, i.e. the UUT is not attached to its intended supporting structure, and the acceleration at the intended interface is used as a basis for the test specification. An enveloping process is necessary to produce a simplified test spectrum that accounts for possible variations. Regardless of which case is used to derive the test specification, the enveloping process does a good job enveloping the peaks accelerations of the data but significantly over-estimates the acceleration at the valleys. Moreover, because of the vibration absorber effect, some these valleys correspond to the natural frequencies of the UUT in its test configuration.

As an example, consider Figs. 6.1 and 6.2. The lower portions of Figs. 6.1 and 6.2 show the apparent mass of two UUT, each with the same natural frequencies of 100 and 263 Hz but with different masses. The difference between the Test Specification and the Coupled Data in the upper portions of these figures is the amount of over-testing present. Note that the over-test is typically large at the natural frequencies of the UUT because of the vibration absorber effect. The un-coupled data represents the interface acceleration as if the support structure was unaffected by the presence of the UUT, meaning no impedance mismatch just like the test configuration. The systems in Figs. 6.1 and 6.2 were chosen to demonstrate a situation when both the enveloping process and the impedance difference contribute to the over-test (at 100 Hz in Fig. 6.1), when only the impedance difference contributes to the over-test (263 Hz in Fig. 6.1), when the over-test is almost all due to the enveloping process (100 Hz in Fig. 6.2), and when essentially no over-test is present (263 Hz in Fig. 6.2).

In these examples, the UUT first natural frequency of 100 Hz has been chosen to be very different from any of the natural frequencies of the un-coupled supporting structure of 69, 175, 263, and 362 Hz. This can lead to large over-testing from the enveloping process, even if the UUT has a very small mass compared to its supporting structure. Conversely, the UUT second natural frequency of 263 Hz has been chosen to correspond to one of the natural frequencies of the un-coupled supporting structure. In this instance, the cause the over-test is typically dominated by the impedance difference such that the UUT mass compared to its mounting structure becomes an important factor.

6.3 Force Limited Testing

The FLV notching method automatically adjusts the test input to limit the interface force to a level that is representative of the maximum anticipated in-service interface force. By Newton's second law, the sum of the applied forces on any flexible or rigid body is directly proportional to the acceleration of its instantaneous center of mass. Thus, FLV effectively limits the maximum acceleration of the center of mass during testing. This cannot be accomplished by placing an accelerometer at, or

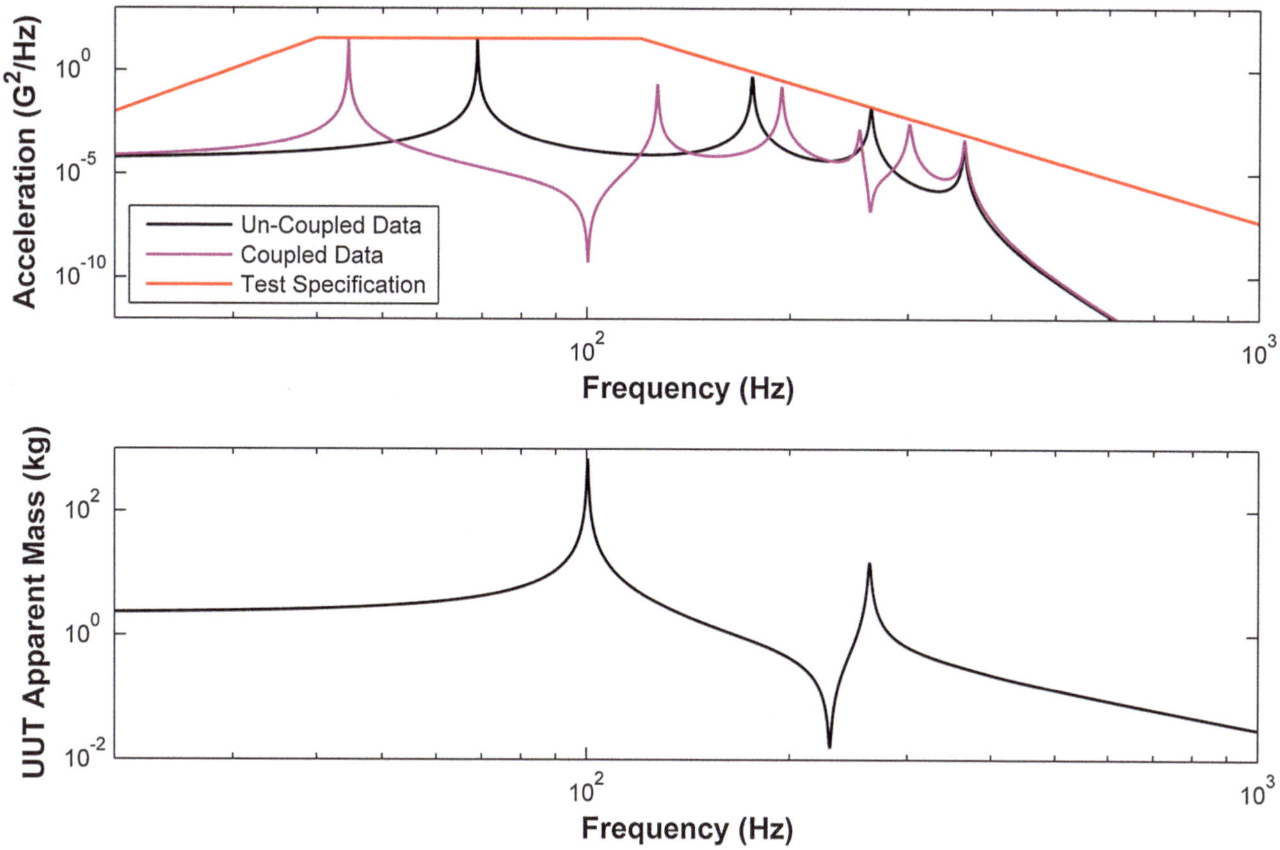

Fig. 6.1 Vibration Over-Testing Example, UUT to Supporting Structure Mass Ratio = 0.5

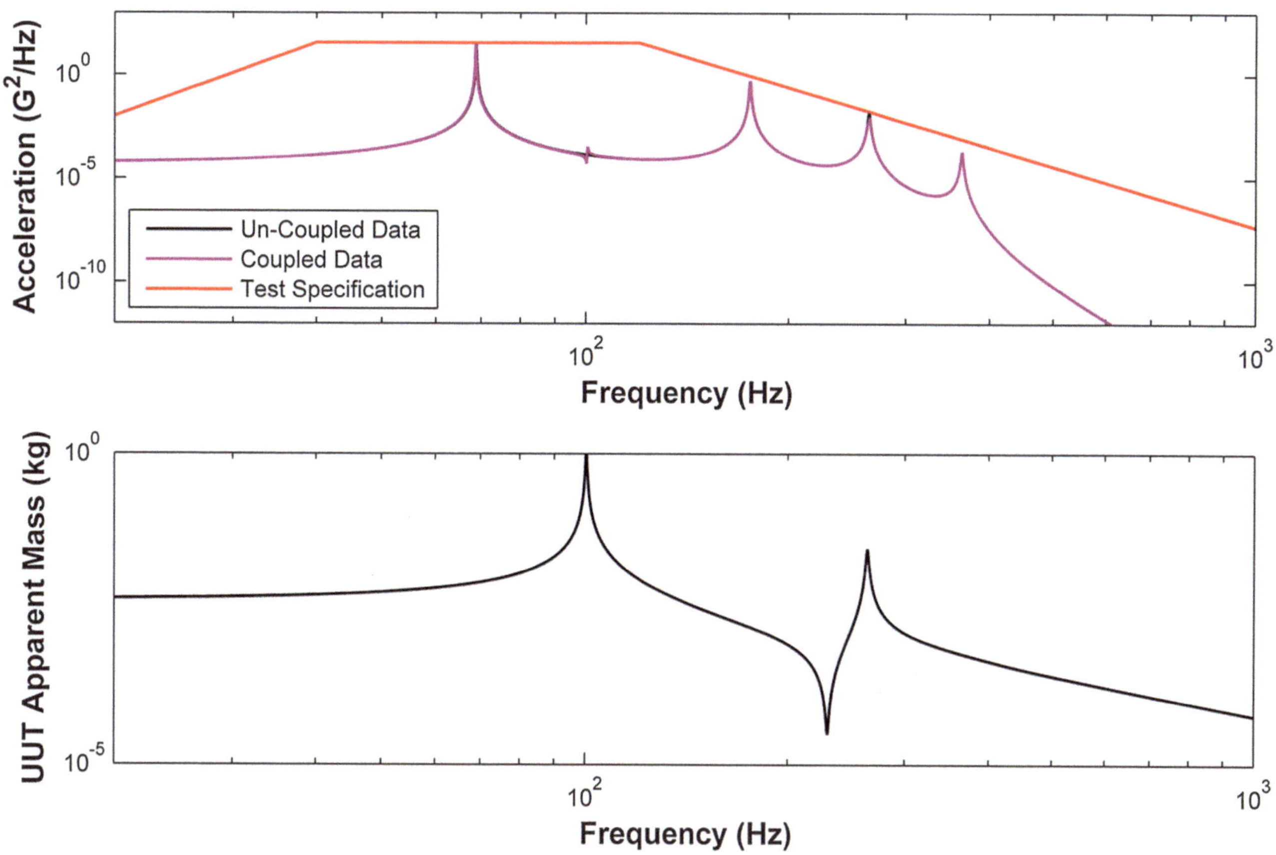

Fig. 6.2 Vibration Over-Testing Example, UUT to Supporting Structure Mass Ratio = 0.001

near, the center of mass because the instantaneous center of mass moves when the structure deforms. Even if the movement is small, the center of mass can have very large accelerations because the acceleration is proportional to both the displacement and the square of the (circular) excitation frequency.

The amount of over testing can be significant. For instance, in-situ acceleration Power Spectral Density (PSD) measurements during launch of the Advanced Composition Explorer spacecraft showed that un-notched random spectral vibration test levels were about 20 dB (~100 times) higher than had occurred during flight near three of its instruments [2]. Similar measurements for a sounding rocket also demonstrated that random spectral vibration test levels were about 20 dB (~100 times) higher during un-notched testing than had occurred during launch at the main natural frequency of the UUT [3]. Smallwood has even suggested that the over-test in acceleration PSD test level can be in practice as much as 30 dB (~1000 times) higher than the in-field level [4].

Force limiting can often provide significant notching even when the force limit is grossly over-estimated. For example, consider reference [1] where the FLV testing on the SVF-2 experiment is described. The PSD force limits used during testing produced a notch of 16 dB, even when the PSD force limit was 10 times greater than the actual force PSD measured in-flight. This is a very important feature of FLV. It means that approximate methods can be used to estimate the maximum interface force and is why there are several successful, but approximate, methods available [1] to estimate the interface force limit and why each method can produce significant reduction in over-testing even if they don't predict the same maximum in-service interface force.

6.4 Linear Dynamic Equations of Motion

The equation of motion of a flexible structure can be derived by considering the force distribution on the structure, Newton's second law applied to differential elements on the structure, the constitutive relationship of the material (i.e. stress-strain curve), and enforcing compatibility such that the material deforms in such a way as to not leave gaps or overlap with itself. The resulting equation is a non-linear differential equation of motion in three dimensions applicable at any instant for any arbitrary force (sine, random, shock, etc.). For example, the resulting equation for a homogeneous isotropic material is given by the Navier elasto-dynamic wave equation [5] as

$$\frac{E}{2(1+\upsilon)}\nabla^2\vec{u}+\left(\frac{\nu E}{(1+\upsilon)(1-2\upsilon)}+\frac{E}{2(1+\upsilon)}\right)\nabla(\nabla\cdot\vec{u})+\vec{F}=\rho\ddot{\vec{u}} \tag{6.1}$$

Note that moments are not part of the general equation of motion, as you cannot apply a moment on a differential element. In general, Eq. (6.1) is difficult to solve, so in practice one typically must resort to approximate methods like Finite Element Analysis (FEA) to solve for the motion of a structure.

Linear FEA transforms the non-linear differential equation Eq. (6.1) into a set of linear discrete differential equations of motion by assuming small strains, small displacements, and linear elastic material behavior. The result is a matrix equation of the form

$$[M]\{\ddot{U}\}+[K]\{U\}=\{F\} \tag{6.2}$$

Damping can be added to the FEA formulation. Like the general differential element upon which it is based, the nodes of a general 3D finite element cannot support rotations or moments at nodes. Node rotations and moments are concepts used to simplify the general 3D element into simpler elements such as plates and beams. Barring finite precision mathematical issues, the FEA formulation will converge to the exact differential equation as the number of nodes is increased. Thus, in principle, the FEA formulation is exact to any desired precision. It is with this idealized FEA formulation, and all associated assumptions, upon which the Apparent Mass method presented in this paper is based on. It is also with this linearized formulation that Frequency Response Function (FRF) theory can be applied.

6.5 Definitions Used for Apparent Masses

FRF theory expresses quantities that relate the steady state outputs in terms of inputs. One of these quantities is the Apparent Mass that expresses the forces in terms of accelerations. There are four different but related types of quantities that will be discussed that can be associated with the term Apparent Mass. For clarity, these will be labelled: Apparent Mass, Interface Apparent Mass, Test (Interface) Apparent Mass, and Total (Test Interface) Apparent Mass. The words in parentheses will be typically omitted for brevity but are included here to show the hierarchy of the concepts.

In FRF theory, the Apparent Mass is defined as the inverse of the Accelerance FRF matrix such that

$$\{F\}_{3Nx1} = \lfloor M_{app} \rfloor_{3Nx3N} \{A\}_{3Nx1} \tag{6.3}$$

Each matrix element of the Apparent Mass is a complex valued frequency-dependent quantity. The Apparent Mass gives the relationship between the forces at ANY node given the acceleration of EVERY node on the body. It is a 3N × 3N square matrix, where N is the number of nodes representing the body and the factor of 3 arises from the three translational Degrees-Of-Freedom (DOF) associated with translational movement in three dimensions.

The Accelerance matrix can be written in terms of p interface points and r non-interface points as

$$\left\{ \begin{array}{c} \{A_{\text{int}}\}_{3px1} \\ \{A_{non-\text{int}}\}_{3rx1} \end{array} \right\}_{3Nx1} = \begin{bmatrix} [H_{11}]_{3px3p} & [H_{12}]_{3px3r} \\ [H_{21}]_{3rx3p} & [H_{22}]_{3rx3r} \end{bmatrix}_{3Nx3N} \left\{ \begin{array}{c} \{F_{\text{int}}\}_{3px1} \\ \{F_{\text{non}-\text{int}}\}_{3rx1} \end{array} \right\}_{3Nx1} \tag{6.4}$$

Assuming only interface forces are present, i.e. $\{F_{\text{non-int}}\} = 0$, leads to the relationship

$$\{A_{\text{int}}\}_{3px1} = [H_{11}]_{3px3p} \{F_{\text{int}}\}_{3px1} \tag{6.5}$$

The Interface Apparent Mass is defined in terms of the inverse of one of the Accelerance sub-matrices as

$$\left[M_{\text{int\_app}}\right]_{3px3p} = [H_{11}]^{-1}_{3px3p} \tag{6.6}$$

such that the interface forces can be described in terms of interface accelerations as

$$\{F_{\text{int}}\}_{3px1} = \lfloor M_{\text{int\_app}} \rfloor_{3px3p} \{A_{\text{int}}\}_{3px1} \tag{6.7}$$

Unlike the Apparent Mass, the Interface Apparent Mass matrix gives the relationship between the forces on INTERFACE nodes given accelerations of the INTERFACE nodes when ONLY INTERFACE FORCES are present. The matrix is a 3p × 3p square matrix where p is the number of nodes on the interface and typically p ≪N. Note that p must consist of at least N = 3 non-collinear points in order to have a stable interface. Although the elements in the Interface Apparent Mass matrix are different than the elements in the Apparent Mass (i.e. the Interface Apparent Mass is not a sub-matrix of the Apparent Mass matrix), the Apparent Mass and the Interface Apparent Mass provide identical answers for the interface force when only interface forces are acting on the body.

The Test (Interface) Apparent Mass is the same as the Interface Apparent mass but with the added two conditions that the interface is perfectly rigid, such that we can set N = 3, and moves in a single translational motion. These restrictions are idealized versions of the interface during testing. For the X direction, the forces at the interface reduce to

$$\left\{ \begin{array}{c} \{F_{\text{int\_}X}\}_{3x1} \\ \{F_{\text{int\_}Y}\}_{3x1} \\ \{F_{\text{int\_}Z}\}_{3x1} \end{array} \right\}_{9x1} = \begin{bmatrix} \left[M_{\text{int\_app\_}XX}\right]_{3x3} & \left[M_{\text{int\_app\_}XY}\right]_{3x3} & \left[M_{\text{int\_app\_}XZ}\right]_{3x3} \\ \left[M_{\text{int\_app\_}YX}\right]_{3x3} & \left[M_{\text{int\_app\_}YY}\right]_{3x3} & \left[M_{\text{int\_app\_}YZ}\right]_{3x3} \\ \left[M_{\text{int\_app\_}ZX}\right]_{3x3} & \left[M_{\text{int\_app\_}ZY}\right]_{3x3} & \left[M_{\text{int\_app\_}ZZ}\right]_{3x3} \end{bmatrix}_{9x9} \left\{ \begin{array}{c} \{1\}_{3x1} \\ \{0\}_{3x1} \\ \{0\}_{3x1} \end{array} \right\}_{9x1} \cdot \ddot{X}_{\text{int\_}X} \tag{6.8}$$

which, for the force in the X direction, reduces further to

$$\{F_{\text{int\_}X}\}_{3x1} = \left(\left[M_{\text{int\_app\_}XX}\right]_{3x3} \{1\}_{3x1} \right)_{3x1,\ \text{test\_app\_}X} \cdot \ddot{X}_{\text{int\_}X} \tag{6.9}$$

The Test Apparent Mass is a 3 × 1 matrix because the rigid interface can be expressed with any 3 non-collinear nodes on the interface.

In FLV, the ratio of the total interface force divided by the rigid interface acceleration is of interest. This quantity is a non-matrix FRF that can be expressed as

$$F_{\text{int\_}X} = \{1\}_{1x3}\left[M_{\text{int\_app\_}XX}\right]_{3x3}\{1\}_{3x1} \cdot \ddot{X}_{\text{int\_}X} \tag{6.10}$$

Total (Test Interface) Apparent Mass in the X direction is a non-matrix quantity that is defined as the value

$$M_{app\_total\_X} = \{1\}_{1x3}\left\lfloor M_{\text{int\_app\_}XX}\right\rfloor_{3x3}\{1\}_{3x1} \tag{6.11}$$

The Total Apparent Mass is the FRF that is measured during testing and will be used by the Apparent Mass FLV method. It gives the relationship between the total interface force given the interface acceleration when only interface forces are present, the interface is perfectly rigid, and the interface is restricted to move in a single translational direction.

The Total Apparent Mass has the special property that it can be expressed as a modal sum of Effective Masses and the Single Degree-of-Freedom (SDOF) force transmissibility functions (T_{SDOF_n}) as

$$M_{app\_total\_X}(f) = m_{\text{int}} + \sum_{n=1}^{r} M_{eff\_n,x} \cdot T_{SDOF\_n}(f) \tag{6.12}$$

The mass of the interface nodes, m_{int}, is often omitted in texts since it can be arbitrarily small, but is included here for completeness of expressing the total force on the interface. The Effective Masses are the mass participation factors, as calculated by FEA programs, multiplied by the total mass. The Effective Masses are a function of the mass matrix, a given mode shape Φ_n, and the rigid body mode shape Ψ when the interface is restricted to move in the single direction considered. For a diagonal mass matrix, the Effective Mass of mode n in direction X can be expressed as [6]

$$M_{eff\_n,X} = \frac{\left(\left(\{\Phi_n\}_{3rx1}\right)^T [M_{UUT}]_{3rx3r}\{\Psi_x\}_{3rx1}\right)^2_{1x1}}{\left(\left(\{\Phi_n\}_{3rx1}\right)^T [M_{UUT}]_{3rx3r}\{\Phi_n\}_{3rx1}\right)_{1x1}} \tag{6.13}$$

The Effective Mass concept can be derived using a single interface node with 6 DOF, 3 translational and 3 rotational [6]. This derivation uses the same three restrictions that we used so far, namely that only interface forces are present, that the interface is perfectly rigid, and that the movement only occurs in one direction at a time. With the 6 DOF of the interface node, one can derive 6 Effective Masses for each mode of the UUT. Although it is tempting to place these in a single 6 × 6 Effective Mass matrix, this practice should be avoided as it gives the impression that you can linearly combine the interface accelerations from different directions to get the correct resulting interface forces. This is not physically correct as it ignores cross-axis force contributions and non-linear rotational effects that arise from combining rotational and translational interface motion. Consequently, the Effective Masses will be written as a non-matrix quantity with the applicable direction indicated with a subscript.

6.6 Interface Forces

Consider the two previously discussed cases for deriving a test specification. In the first case, the test acceleration is based on enveloping the predicted or measured in-service acceleration at the base of the UUT when the UUT is attached (coupled) to its supporting structure. If only interface forces act on the UUT, the exact interface force at each attachment point p predicted by linear theory is given directly by the unit's Interface Apparent Mass as

$$\{F_{\text{int}}(f)\}_{3px1} = \left\lfloor M_{\text{app\_int\_UUT}}(f)\right\rfloor_{3px3p}\{A_{\text{int}}(f)\}_{3px1} \tag{6.14}$$

Applying the conditions that the interface is perfectly rigid and that the interface moves in a single translational direction, the exact total interface force for the X direction can be written in terms of the UUT's Total Apparent Mass as

$$F_{\text{int\_total\_X}}(f) = M_{\text{app\_total\_UUT}}(f) \cdot \ddot{X}_{\text{int}}(f) \tag{6.15}$$

In the second case, the test acceleration is based on enveloping the predicted or measured in-service acceleration of the supporting structure (base) at the intended interface to the UUT but without the presence of the UUT (i.e. un-coupled). If in service only interface forces act on the UUT, the exact interface force at each attachment point p is given by a combination of the supporting structure (base) and the UUT Interface Apparent Mass as

$$\{F_{\text{int}}(f)\}_{3px1} = \left[\left[M_{\text{app\_int\_UUT}}(f)\right]^{-1}_{3px3p} + \left[M_{\text{app\_int\_Base}}(f)\right]^{-1}_{3px3p}\right]^{-1}_{3px3p} \{A_{\text{int\_un-coupled}}(f)\}_{3px1} \tag{6.16}$$

Equation (6.16) is the general 3D elastic version of Norton's electrical circuit theorem. As its derivation is long, it could not be included in this paper. The only additional assumption needed to arrive at Eq. (6.16) was that the supporting structure's FRF obeyed reciprocity. As per the definition of Interface Apparent Mass, the supporting structure's Interface Apparent Mass must be computed at the UUT interface by assuming that only interface forces are present on the base. Interestingly, it is the Interface Apparent Mass of the supporting structure, which is based on having only interface forces acting on it, that is required in order to correctly predict the interface forces when non-interface forces are applied to the supporting structure. This was noticed in the derivation and has been verified for a numerical example.

Applying the conditions that the interface is perfectly rigid and that the interface moves in a single translational direction, the exact total interface force for the X direction can be written in terms of the structures' Interface Apparent Mass as

$$F_{\text{int\_total\_X}}(f) = \{1\}_{1x3p}\left[\left[M_{\text{app\_int\_UUT}}(f)\right]^{-1}_{3px3p} + \left[M_{\text{app\_int\_Base}}(f)\right]^{-1}_{3px3p}\right]^{-1}_{3px3p} \{1\}_{3px1} \cdot \ddot{X}_{\text{int\_un-coupled}}(f) \tag{6.17}$$

Results from Eqs. (6.15) and (6.17) give identical answers for the total interface force, even if one is based on Total Apparent Mass while the other is based on Interface Apparent Mass. The relationship between coupled and un-coupled interface accelerations can be obtained by equating Eqs. (6.14) and (6.16) to each other.

6.7 Conditions for a Representative Test

To properly evaluate the UUT's suitability for its service environment, the vibration test needs to be representative of the in-service environment. Although this statement may seem an obvious criterion for a valid vibration test, there are several assumptions inherent in it that may not be obvious but are nonetheless important and worth noting. A representative test is taken to mean that the response and stresses induced in the UUT during the test will be equal to or greater than, at every frequency and every point, to what it will experience in service. For this to occur, the following five conditions must be true or, at the very least, accepted as approximately true.

Condition 1: In-Service, only interface forces act on the UUT. Any forces, such as acoustic forces or body accelerations, acting directly on the UUT while in service will excite the mode shapes in a combination that cannot be replicated with any amount of fixed based excitation. The forces must therefore be negligible compared to the forces occurring at the interface.

Condition 2: In-service, all points on the interface move with the same direction, magnitude, and phase. In order to efficiently transmit the shaker force to the UUT interface, the structure between where the shaker force is generated and where the interface with the UUT begins (e.g. armature, slip plates, interface plates, etc.) needs to be as rigid as practically possible and be restricted to move only in the same linear direction of the shaker. The in-service configuration must also match this necessary test condition in order for the test response of the unit to be similar to the in-service response. This condition is fulfilled at low frequencies, where the UUT and its base move essentially as a rigid body, but becomes invalid beyond a certain frequency that will be referred to the maximum test frequency. Since the test is no longer representative beyond the maximum test frequency, any form of notching above this frequency becomes a questionable practice.

Condition 3: In service, the interface moves in the same single linear direction as during testing. Most vibration tests are performed with a single shaker. This limits the control of the interface motion to a single, usually linear, direction. In service, however, the interface is typically moving in several directions simultaneously and can rotate. Although the cross-axis movement and interface rotations occurring in-service will excite additional modes that are not excited during testing, we are forced to accept condition 3 as valid and carry on with testing because we cannot do otherwise with a single shaker. One way to look at the situation is that the test simulates the hypothetical ideal in-service situation where there are additional forces exciting the UUT's supporting structure such that the interface moves in a single direction.

Condition 4: The dynamic structural properties of the UUT are the same during testing as in service. The dynamics of the UUT must be the same, or nearly the same, during testing as in-service in order for the test to be representative. This can be an issue if the structural properties are time variant, temperature variant, or if a similar but different unit is tested (e.g. manufacturing tolerances). Condition 4 also entails the assumption that the boundary conditions during testing are representative of the boundary conditions in-service. In practice, this means that seemingly unimportant issues like the amount of bolt torque present can actually modify the structural response and degrade the representativeness of the test.

Condition 5: The test input specification envelops the peaks of the coupled response. For the test to induce equal or greater damage potential as in service, the test acceleration input must envelope the peak accelerations of the coupled in-service structure and not necessarily the un-coupled structure.

6.8 Apparent Mass Method

When the five test conditions are fulfilled, one can consider using either the interface force or the interface motion as the control to correctly reproduce the in-service response of the UUT during testing [7]. Conversely, this means that if the test is not representative, then neither the interface forces nor the interface accelerations are representative.

The Apparent Mass method takes the Apparent Mass FRF given by Eq. (6.3) and applies the necessary test conditions for a representative test to get a formula describing the in-service interface forces. Conditions 1 was the same condition used to define the Interface Apparent Mass. Conditions 1 to through 3 were the same conditions used to define the Test Apparent Mass and the Total Apparent Mass. Since both the Interface Apparent Mass and the Total Apparent Mass are compatible with the necessary test conditions, we can use either Eq. (6.15) or Eq. (6.17) to describe the in-service interface force, as both describe the same interface force. The Apparent Mass method uses the simpler Eq. (6.15) as its key equation to estimate the maximum in-service interface force.

The maximum interface force over any given bandwidth can be found by evaluating Eq. (6.15) at a single frequency. Therefore, the three variables we need to determine are the frequency at which the in-service maximum interface force occurs, the value of the Total Interface Apparent Mass at that frequency, and the value of the in-service acceleration at that frequency.

The Total Interface Apparent Mass can be measured directly at the start of testing using a low level sine or random survey as the ratio of the total interface force to the average interface acceleration, both in the excitation direction. Although we do not usually know the value of the in-service acceleration, we do have the test spectrum, and if the test is representative, then test condition 5 implies that

$$A_{\text{test}}(f) \geq A_{\text{int\_coupled}}(f) \tag{6.18}$$

Since at any given frequency the interface force is proportional to the interface acceleration, using Eq. (6.18) in Eq. (6.15) will always produce equal or higher force limits than if the actual coupled acceleration was used.

The only remaining variable left to determine is the frequency where the maximum in-service interface force occurs. In principle, this frequency can be any frequency. For example, if the supporting structure was excited using a single sinusoidal force of arbitrary frequency, the maximum interface force would occur at that frequency. However, if the coupled structure is lightly damped and is excited by a reasonably constant and broadband excitation, one can argue that the maximum response, and therefore maximum interface force, would occur at one of the coupled system's natural frequencies. Assuredly, if the structure had no damping, this would be the case for any form of broadband excitation because the steady state response at the coupled natural frequencies would go to infinity. Thus, to ensure that the maximum interface force occurs at one of the coupled natural frequencies, we will limit the applicability of the Apparent Mass method to structures that are lightly damped and excited by broadband excitation, at least in the vicinity of f_o, where f_o is the UUT's natural frequency we wish to notch.

Given that the maximum interface force occurs at one of the coupled natural frequencies, we only need to evaluate Eq. (6.15) at the coupled natural frequencies occurring over a bandwidth extending over the Total Apparent Mass response of the mode or modes we wish to notch. For the X direction, this can be written in equation form as

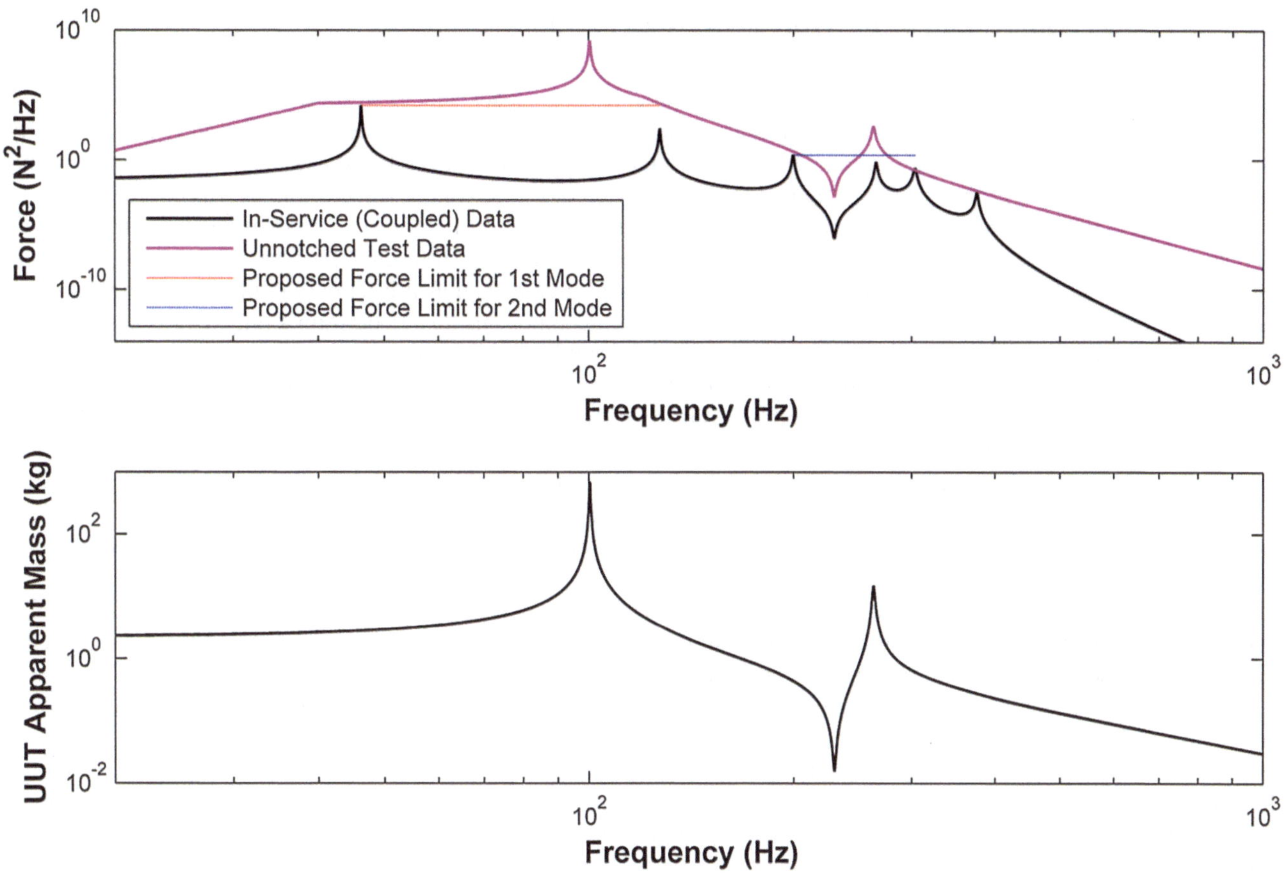

Fig. 6.3 Example of In-Service Interface Forces and Proposed Limits at Different Frequencies

$$F_{\text{int\_total\_max\_X}} = MAX\left[M_{\text{app\_total\_X}}(f_{nn}) \cdot \ddot{X}_{\text{test}}(f_{nn})\right]_{\text{set of } f_{nn} \text{ covering } f_o} \tag{6.19}$$

An example of this procedure is given in Fig. 6.3. Since the first mode of the UUT falls completely between the first and second modes of the coupled structure, the force limit for this mode is taken as the maximum result of Eq. (6.15) evaluated at these two modes and applied over the bandwidth defined by these two modes. The second mode of the UUT is very close to the 4th coupled mode and is better covered by considering the 3rd, 4th, and 5th modes of the coupled structure. The force limit for this mode is taken as the maximum result of Eq. (6.15) evaluated at these three modes and applied over the bandwidth defined by these three modes. In this way, the Apparent Mass method can be applied repeatedly to different modes that we wish to notch. Alternatively, we could have simply used all coupled modes given in the test bandwidth and take the maximum result of Eq. (6.15) evaluated at all these modes to define a single maximum force limit for the entire test.

While the total solution incorporates both the transient and the steady state parts, the solution given by FRF theory is only the steady-state part of the solution. This means that the Apparent Mass concept cannot be applied to transient force excitations such as shock. However, the theory can deal with both periodic excitations and random excitations. Although the ever changing random excitation is characterized more as transient than steady-state, if the random excitation can be described with a time-independent probability distribution (stationary) and if the average of any sample is representative of the average of the entire signal (ergodic), then FRF theory can predict the random responses in a probabilistic sense in terms of the Power Spectral Density (PSD). For random vibration, Eq. (6.19) becomes [6]

$$S_{ff\text{\_int\_total\_max\_X}} = MAX\left[\left|M_{\text{app\_total\_X}}(f_{nn})\right|^2 \cdot S_{\ddot{x}\ddot{x}\text{\_test}}(f_{nn})\right]_{\text{set of } f_{nn} \text{ covering } f_o} \tag{6.20}$$

Similar equations can be derived for the Y and Z directions. The force limits expressed by Eqs. (6.19) and (6.20) are based on the test acceleration. As such, the force limits derived by these equations have the same conservatism, or lack thereof, as the input acceleration.

6.9 Practical Application of the Apparent Mass Method

When the interface loads of a coupled load analysis is available, those loads should be used as the interface force limits. When such analysis is unavailable, the Apparent Mass method can be used to get a theoretically exact interface force limit, given the assumptions used in its derivation. One of the main benefits of the Apparent Mass Method is its ease of use, since only a minimum amount of information is required to correctly predict a maximum value of the interface force.

Since the Total Apparent Mass can be measured directly at the time of testing, the key variables needed when using the Apparent Mass method are the coupled natural frequencies. The coupled natural frequencies can be obtained using a coupled FEA modal analysis. This is an important feature of the apparent mass method as it is simpler to perform a single analysis and report the natural frequencies of the coupled system to all sub-components than to perform a full coupled loads analysis to calculate the maximum interface forces for every sub-component interface over numerous launch load cases.

However, there are times when the fully coupled modal analysis is lacking and one needs to resort to more approximate methods. For theoretically exact results, we need to evaluate the Total Apparent Mass of the UUT at coupled frequencies at either side of f_o. However, it will often be at the lowest coupled natural frequency closest to f_o, where the maximum force occurs. This assumption is included implicitly in the Semi-Empirical Method [1], where the force limit is always larger at frequencies lower than f_o. Using the same assumption, one can simplify the Eq. (6.19) or Eq. (6.20) and base the force limit on the lower frequency. This method is appealing because it is often easier to find or estimate the lower shift in natural frequency when the exact data is unavailable.

This lower mode assumption can be used in conjunction with defining the mode with the largest Effective Mass as f_o in order to derive an equivalent value for the C^2 factor employed in the Semi-Empirical Method [1]. This provides a quick and easy sanity check for the chosen value of C^2 (Figs. 6.4, 6.5, 6.6, and 6.7).

One way to estimate the shifted lower frequency is to employ a simplified FEA model of the support structure. One only needs to model the immediate support structure, such as the panel to which the UUT will be attached to in service, in order to get an estimate of the minimum frequency shift from f_o to $f_{nn\_before\_f0}$. In this method, the nodes at the edges of the simplified support model would be modelled as fixed in all possible degrees of freedom. This is equivalent to assuming that the un-modelled portion of the support structure is perfectly rigid. This always produces smaller frequency shifts than actually present because any added flexibility in the support structure will result in lower values of $f_{nn\_before\_f0}$. The region that one needs to model is arbitrary, with more modelling of the support structure giving more accurate results. As seen if Fig. 6.8, even small frequency shifts can provide significant reduction in interface force.

It is true that the interface force limit given by Eqs. (6.18) and (6.20) may not match the actual in-service maximum interface force, but in such cases the test is not representative and the acceleration will not match either. Even so, since large overestimates of the force limit can provide significant notching, Eqs. (6.19) and (6.20) can still be use to define a maximum value for the force limit. A comfort factor can then be applied e.g. x2 to calculate the force limit used in testing, For example, if the force limits prediction would produce an extreme notch, the force limit may be increased to give at least a 6 dB resonance response, as recommended in NASA 70004C [1].

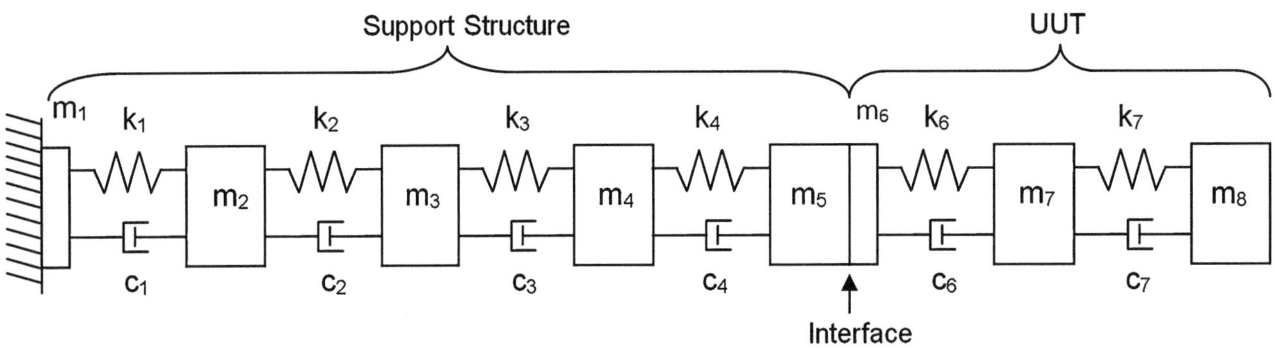

Fig. 6.4 Numerical Model in Coupled Configuration

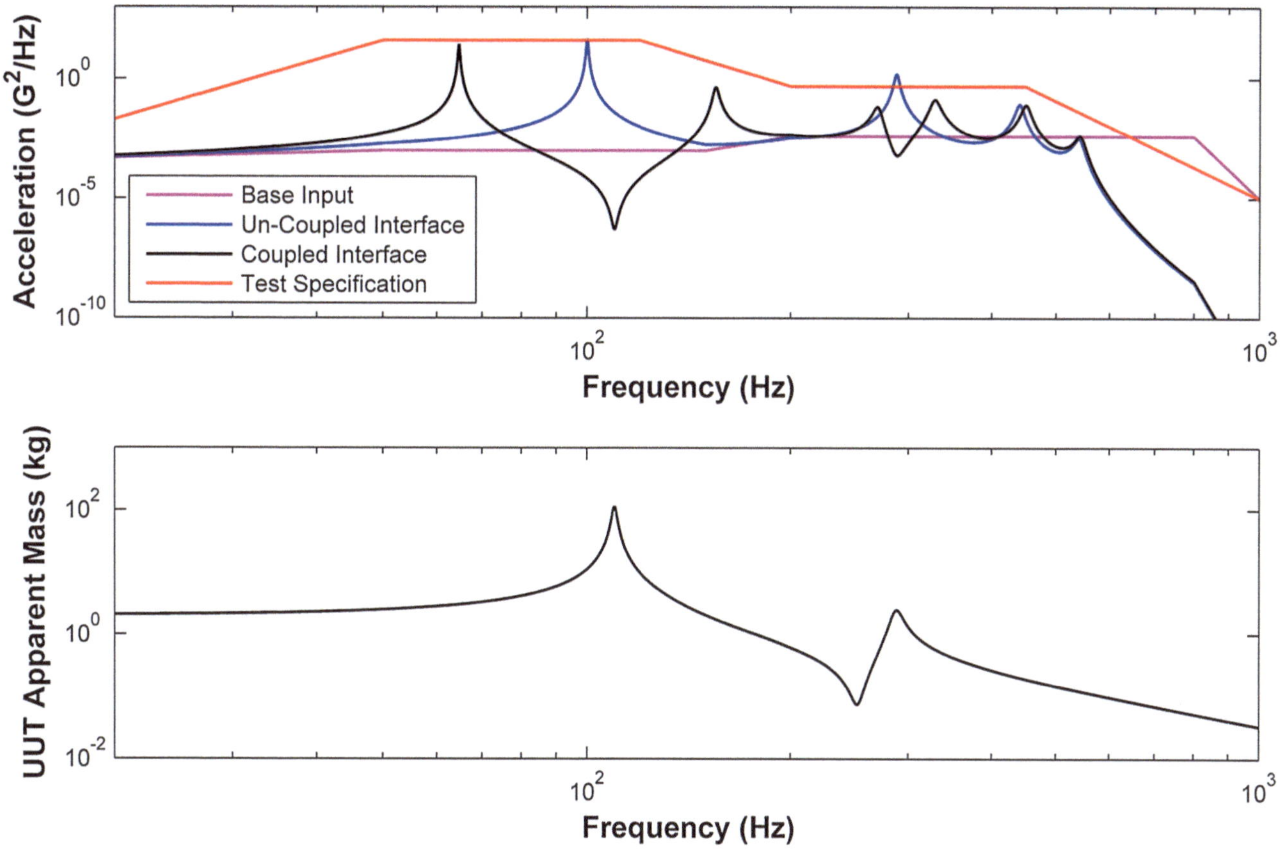

Fig. 6.5 Accelerations and UUT Apparent Mass, Case 1 (Load to Source Mass Ratio is 0.5)

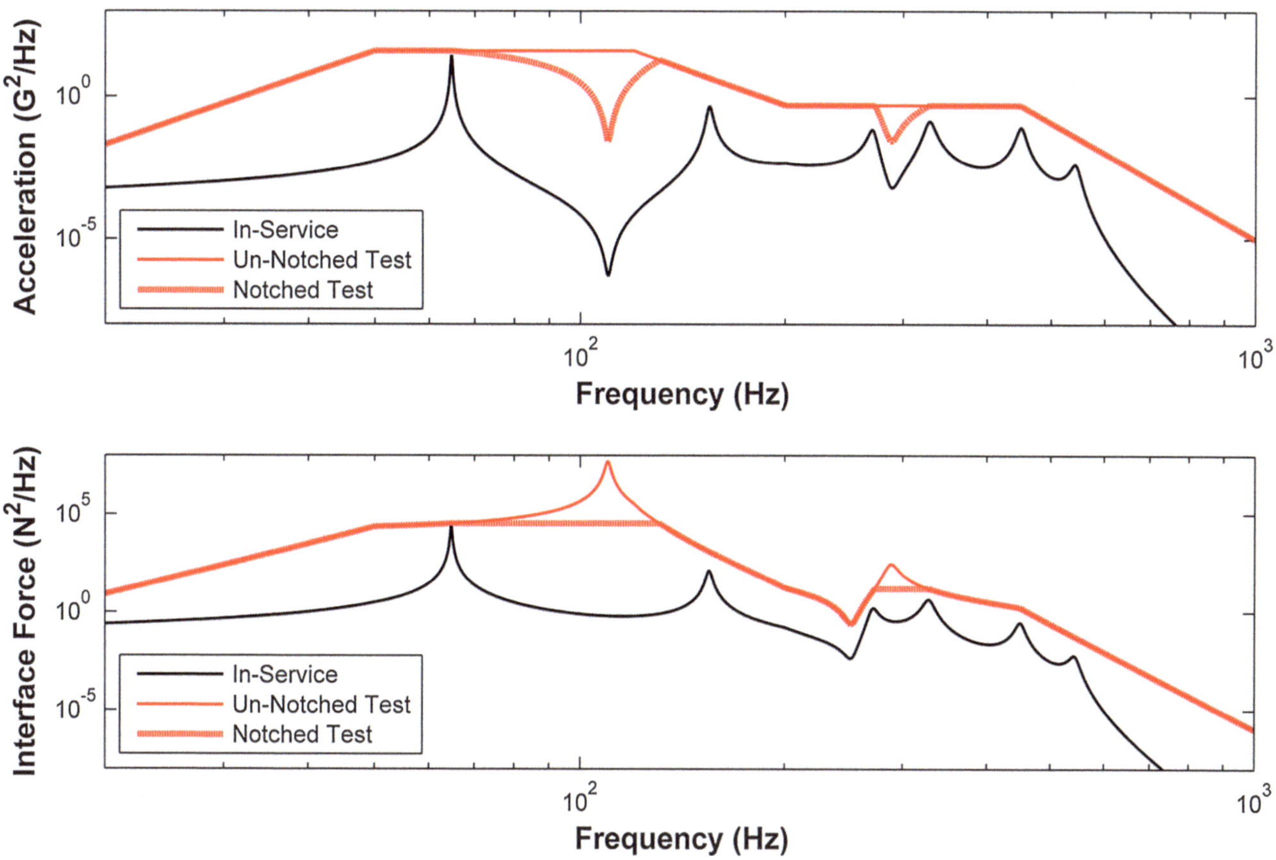

Fig. 6.6 Accelerations and Interface Forces, Case 1 (Load to Source Mass Ratio is 0.5)

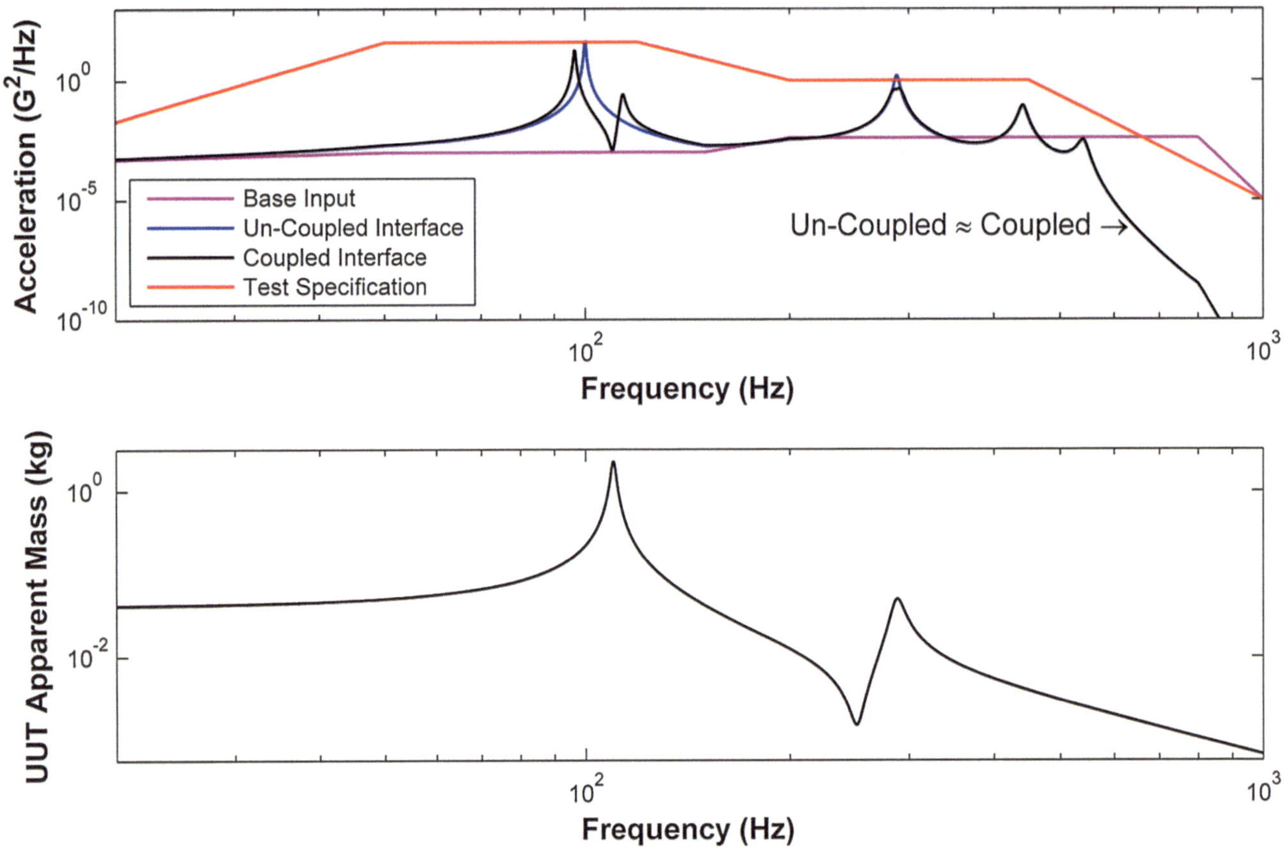

Fig. 6.7 Accelerations and UUT Apparent Mass, Case 2 (Load to Source Mass Ratio is 0.01)

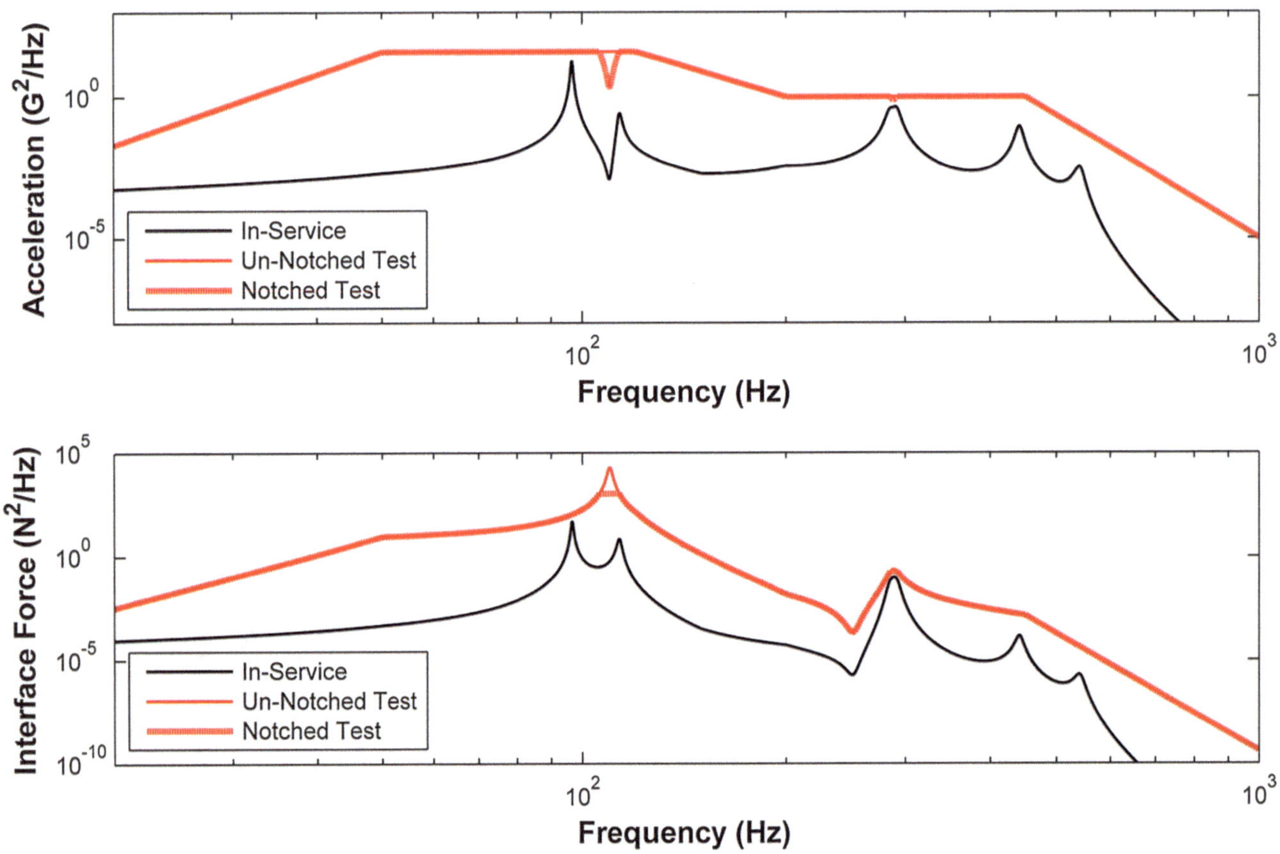

Fig. 6.8 Accelerations and Interface Forces, Case 2 (Load to Source Mass Ratio is 0.01)

6.10 Numerical Examples

To demonstrate the merits of the Apparent Mass method, and to show specifically that one can perform force limiting using the Apparent Mass method when the UUT mass is either significant or very small compared to its support structure, the results of two numerical examples are presented. Both cases use the same configuration shown in Fig. 6.4 but have different values for the masses (m), stiffness (k), and damping (c) as described in Table 6.1. The model could represent an instrument (UUT) mounted on a spacecraft (Support Structure) installed in a rocket. The rocket imparts random vibration at the base of the spacecraft. This base is represented by mass m_1 in Fig. 6.4.

In both cases, the base of the Support Structure is excited with identical acceleration PSD. This base input is broadband as it includes frequencies from 20 to 1000 Hz. From the resulting acceleration at the UUT interface (mass m_5), a test PSD that envelops the coupled structure interface acceleration is derived. The base PSD, along with both test PSDs used in the analyses, are described in Table 6.2 and in the top portion of Figs. 6.5 and 6.7. The final results are shown in Figs. 6.5, 6.6, 6.7, and 6.8.

In both numerical examples, the UUT natural frequencies are at 110 and 287.7 Hz, even though the mass of the UUT is changed from 2 kg in case 1 to 0.04 kg in case 2. The natural frequencies of the un-coupled support structure, meaning when the UUT is not attached, are 100, 288.1, 441, and 541 Hz for both cases. Thus, the natural frequencies of the UUT are very close to some of the un-coupled natural frequencies of the support structure. The support structure has a mass of 4 kg for both cases giving a UUT to supporting structure mass ratio of 0.5 for case 1 and 0.01 for case 2.

The Apparent Mass method was used with Eq. (6.20) in the calculation of the force limits for both numerical examples. Note that in both cases, as required for a representative test, the test input acceleration PSD enveloped the coupled acceleration PSD at the interface, even if the acceleration at m_5 of the un-coupled structure was not fully enveloped.

The results for the first case are shown in Figs. 6.5 and 6.6. In this case, the Apparent Mass FLV formulation produced a notch of almost 73 dB for the first mode and almost 29 dB for the second mode. Because the mass ratio is close to unity and because the UUT natural frequencies that were notched are close to some of the coupled natural frequencies, the un-notched over-test in this case is mainly due to the impedance differences.

The results for the second case are shown in Figs. 6.7 and 6.8. In this case, the Apparent Mass FLV formulation produced a notch of over 29 dB for the first mode and a small notch of almost 2.4 dB for the second mode. Because of the small mass ratio involved, the impedance mismatch is practically negligible and one might have expected that no notching be produced. The over-test in this case is mainly due to the enveloping process. This case was chosen to show that the Apparent Mass method of FLV can notch a UUT with a small mass and reduce over-testing arising from the enveloping process.

In both cases, the notches produced conservative interface acceleration in that the notched acceleration inputs were equal to or greater than the coupled in-service responses. Thus, Apparent Mass FLV did not produce any under-test situation even with the large notches that were created. The large depth of the notches without resulting in under-testing suggests that in practice, a large over-estimate of the force limits can be used and still end up with significant notching.

Table 6.1 Model parameters

Case	Component	Mass	Stiffness	Damping
1	Source	$m_1 = 0$ kg, m_2 to $m_5 = 1$ kg	k_1 to $k_4 = 3.276E6$ N/m	c_1 to $c_4 = 30$ N s/m
	Load	$m_6 = 0$ kg, m_7 and $m_8 = 1$ kg	k_6 and $k_7 = 1.248E6$ N/m	c_6 and $c_7 = 30$ N s/m
2	Source	$m_1 = 0$ kg, m_2 to $m_5 = 1$ kg	k_1 to $k_4 = 3.276E6$ N/m	c_1 to $c_4 = 30$ N s/m
	Load	$m_6 = 0$ kg, m_7 and $m_8 = 0.02$ kg	k_6 and $k_7 = 24960$ N/m	c_6 and $c_7 = 0.6$ N s/m

Table 6.2 Test input acceleration PSD applied at UUT interface

Base input at m_1						
Frequency	20 Hz	50 Hz	150 Hz	200 Hz	800 Hz	1000 Hz
Base PSD	0.0005 G^2/Hz	0.001 G^2/Hz	0.001 G^2/Hz	0.004 G^2/Hz	0.004 G^2/Hz	0.0001 G^2/Hz
Case 1 Test PSD at m_6						
Frequency	20 Hz	50 Hz	120 Hz	200 Hz	400 Hz	1000 Hz
Test input PSD Case 1	0.02 G^2/Hz	40 G^2/Hz	40 G^2/Hz	1 G^2/Hz	1 G^2/Hz	0.00001 G^2/Hz
Case 2 Test PSD at m_6						
Frequency	20 Hz	80 Hz	120 Hz	200 Hz	400 Hz	1000 Hz
Test input PSD Case 2	0.02 G^2/Hz	60 G^2/Hz	60 G^2/Hz	5 G^2/Hz	5 G^2/Hz	0.00001 G^2/Hz

6.11 Conclusion

The over-testing problem in vibration testing was shown to arise from two sources: the enveloping process and the interface impedance differences between the in-service and test configurations. The Apparent Mass method of performing FLV was derived from first principles and the reduction of the equations of motion to a linear system. The resulting equations were then supplemented with the same assumptions required for a representative test, along with the assumption that the maximum force response occurs at one of the natural frequencies of the coupled structure. The Apparent Mass method can reduce over-testing arising from both the interface impedance difference and the enveloping process. As such, the method is applicable to a UUT of any mass. In addition, since the method uses vibration theory to predict the maximum interface force over a given bandwidth defined by a set of coupled natural frequencies, the force limits produced will never lead to an under-test situation as long as the assumptions used to develop the equations are valid. Equations (6.19) and (6.20) are the key equations of the Apparent Mass method. These equations have been derived with the following assumptions:

- The vibration test is representative
  - Only interface forces act on the unit
  - In service, all points on the interface move with same magnitude and phase
  - In service, the interface moves in a single direction
  - The dynamics of the test structure are the same during testing as in service
  - The test specification envelops the acceleration of the coupled interface acceleration
- The coupled structure behaves linearly
  - Small strains
  - Small displacements
  - Linear material behavior
- In-service interface force peaks occur at the natural frequency of the coupled structure
  - Excitation is broadband and somewhat constant near the frequency we wish to notch
  - Coupled structure is lightly damped

By numerical examples, the method was shown produce significant notches without resulting in under-testing. This was true, even when the mass of the UUT was very small compared to the mass of its in-service supporting structure. One of the main benefits of the method is its ease of use, requiring a minimal amount of information. The method can be used as presented or in conjunction with other methods, especially the Semi-Empirical method, to validate a particular choice of force limit.

References

1. NASA Technical Handbook. Force Limited Vibration Testing. NASA-HDBK-7004C, November 30 2012
2. Scharton, T.D.: In-flight measurements of dynamic force and comparison with methods used to derive force limits for ground vibration tests. In: Proceedings of the European Conference on Spacecraft Structures, Material and Mechanical Testing, ESA SP-428, Paris, 583–588 February 1999
3. Worth, D.B., Kaufman, D.S.: Validation of force-limited vibration testing. J. Inst. Environ. Sci. Technol. **41**(3), 17–23 (1998)
4. Smallwood, D.: The role of force measurements in advanced vibration test methods. SAND-94-2681-C, Presented at the 13th IMAC, Nashville, TN, February 13–16 1994
5. Hackl, K., Goodarzi, M.: An Introduction to Linear Continuum Mechanics. Ruhr University Faculty of Civil and Environmental Engineering, Institute of Mechanics, Bochum, Germany (2010). http://www.vgu.edu.vn/fileadmin/pictures/studies/master/compeng/study_subjects/modules/cm/lecture_notes_-_linear.pdf
6. Wijker, J.: Random Vibrations in Spacecraft Structures Design, Theory and Applications. Springer, New York (2009)
7. McConnell, K.G.: Vibration Testing Theory and Practice. Wiley, New York (1995)

Chapter 7
Harmonic Force Excitation Analysis of a Small-Body Asteroid/Satellite System

Joshua Johnson, William H. Semke, Shankar Nag Ramaseri Chandra, and Ronald Fevig

Abstract A harmonic force excitation analysis is used to determine orbit stability of a small satellite around an asteroid with a complex gravitational field. Harmonic excitation phenomena occurs with both natural and man-made satellites. Jupiter influences asteroid distribution in the main asteroid belt through mean motion resonance where some regions are devoid of asteroids while other regions have an abundance. Simulations of a man-made satellite in orbit around the asteroid Itokawa, which was visited by the Hayabusa Space Mission, have also displayed harmonic excitation phenomena, including regions of high dynamic interactions. Specifically, the influence of the rate of spin of the asteroid and the stability of the orbit was investigated. The radial acceleration of the satellite is used to determine the frequency of gravitational perturbation from the asteroid on the satellite. It has been shown when the satellite is placed in an orbit away from its resonant frequency, the orbit remains stable. An early model was created to study this phenomenon and showed promise to predict regions of stability. From this initial study, further work using a more complex model and an updated harmonic force excitation analysis of the system was shown to be a more accurate predictor of orbital stability.

Keywords Harmonic analysis • Satellite control • Orbital stability • Itokawa • Space mission design

7.1 Introduction

The information in this paper stems from an ongoing collaborative research effort at the University of North Dakota. The Department of Mechanical Engineering and the Department of Space Studies has been working together in order to investigate some unique behaviors regarding the orbit of small-body asteroids. Orbiting Small Near-Earth Objects (NEOs) presents many challenges, but also many interesting opportunities of study. This paper concentrates on harmonic excitation behaviors observed while studying the complexities involved in the satellite-NEO system.

Previous work done by Church and Fevig looked at the feasibility of creating a highly detailed gravity map of small body asteroids (500 m and smaller). This allows for the determination of the internal structure of the asteroid by backing out the mass densities from the gravity model surrounding the asteroid. The process studied calls for two satellites in orbit around a body next to each other. Observations are recorded, from one satellite to the other, of the effects of the gravitational perturbations on the satellites' trajectories. This method has proven to be successful during the GRACE, and GRAIL missions around the Earth and moon, respectively. It was found, however, that two satellites put into orbit next to each other around a simulated model of asteroid 25143 Itokawa quickly diverge in their trajectories [1].

The asteroid Itokawa is not spherical, unlike most planets, which presents more complexity when attempting to maintain a stable orbit due to the changing magnitude of gravity as the satellite progresses through its trajectory. Early in this study it was observed that Itokawa has a highly erratic effect on a satellite's trajectory with minimal changes in initial conditions. Observations were made using AGI's Systems Tool Kit (STK) software package. Initially, two satellites were placed in an orbit next to each other, separated by 5° in true anomaly, the angle between their initial radius vectors. One quickly crashed into the asteroid while the other was ejected from the system. This indicated a significant difference in their behavior despite having similar initial conditions. Therefore, it was noted that the gravitational field surrounding Itokawa differs dramatically depending on the location of the satellite within its orbit.

J. Johnson • W.H. Semke (✉)
Department of Mechanical Engineering, College of Engineering and Mines, University of North Dakota, Grand Forks, ND 58202, USA
e-mail: william.semke@engr.und.edu

S.N.R. Chandra • R. Fevig
Department of Space Studies, John D. Odegard School of Aerospace Science, University of North Dakota, Grand Forks, ND 58202, USA

A. Brandt, R. Singhal (eds.), *Shock & Vibration, Aircraft/Aerospace, Energy Harvesting, Acoustics & Optics, Volume 9*,
Conference Proceedings of the Society for Experimental Mechanics Series, DOI 10.1007/978-3-319-30087-0_7

Later it was observed that retrograde orbits, orbits where the satellite's trajectory is opposite the direction of rotation of the asteroid, are much more stable than prograde orbits, orbits where the satellite's trajectory is in the same direction of the rotation of the asteroid. This observation led to the conclusion that harmonic excitation analysis can be used when studying the satellite-asteroid system. The gravitational perturbation has a much lower frequency of excitation on the satellite than the retrograde orbit, so the excitation becomes amplified as the excitation frequency approaches the natural frequency of the system [2].

7.2 Background

Asteroids or comets are considered Near-Earth Objects when their perihelion distance is less than 1.3 Astronomical Units (AU) [3]. The perihelion distance is the distance between the Sun and the orbiting body at its closest point to the Sun.

There is increasing interest in NEOs as time progresses. First, there is the constant threat of an asteroid striking the Earth. At the time of this writing, there are over 1500 Potentially Hazardous Asteroids (PHAs) that are being tracked and monitored by NASA [4]. The term potentially hazardous does not mean that an impact with Earth is imminent, only that the orbit and size of the object are cause for concern. With continuous study of these PHAs, scientists can better predict their orbit and asses the likelihood of a future close approach to Earth. Also, the resources on Earth are becoming more limited as the population continues to grow. There is a growing potential in mining NEOs for certain elements that can be either shipped back to Earth, or used in long-term space missions.

The satellite control problem in the complex NEO environment is being studied by researchers all around the world. NASA, ESA, and the Japanese Aerospace Exploration Agency (JAXA) have sent unmanned spacecraft that have operated in close-proximity to these objects [5, 6]. What makes the NEO environment so unique is that the gravity field surrounding an NEO is both tenuous and complex. Gravitational force is a function of the spacecraft's and NEO's mass that decays proportionally to the inverse square of the distance between these elements. This is defined by Newton's Law of Gravity. With an irregularly shaped asteroid being orbited by a satellite, the distance between the two bodies is constantly changing, and the magnitude of the gravity exerted on the satellite along with it.

In order to capture the irregularities of the gravity model in three dimensions, a spherical harmonic model of the gravitational potential is used, as shown in Eq. 7.1.

$$U = \frac{\mu}{r}\sum_{n=0}^{\infty}\sum_{m=0}^{n}\left\{\left\{\left(\frac{R_{CB}}{r}\right)^{n} P_{nm} sin\theta\, cos\,(m\lambda)\right\}C_{nm} + \left\{\left(\frac{R_{CB}}{r}\right)^{n} P_{nm} sin\theta\, sin\,(m\lambda)\right\}S_{nm}\right\} \tag{7.1}$$

where U is gravitational potential. The coordinates r, θ, and λ are the radial distance, latitude, and longitude of the spacecraft in a coordinate system fixed to the object's center of mass. R_{CB} is the mean radius for the body and μ is the object's gravitational parameter. The functions P_{nm} are the normalized Legendre polynomials, and C_{nm} and S_{nm} are the gravity coefficients of degree n and order m [7].

Asteroid 25143 Itokawa has been studied extensively and a degree and order four spherical harmonic model was created based on the data collected by JAXA's Hayabusa satellite, which visited Itokawa in 2005 [8]. Studying the natural environment about Itokawa, with no control measures used, it was found that a stable prograde orbit can be achieved using a high-inclination orbit with an orbit radius between 1.0 and 1.5 km. Any closer to the surface of Itokawa and the orbit becomes disrupted by the spherical harmonic gravity model. Outside of this range the orbit becomes disrupted by other effects, such as solar radiation pressure (SRP).

For Itokawa's gravity model, the C_{20}, C_{22}, C_{42}, and C_{44} coefficients are the most significant [9]. These values correspond with the effects of the asteroid's oblateness and its ellipticity. A body's oblateness and its effect on orbits has been characterized extensively. For example, this perturbation is observed in Earth-bound orbits. Due to the Earth's angular velocity as it spins about its polar axis, there is a bulge around equator. The effect this bulge has on Earth-bound orbits, known as the J_2 effect, causes a precession in the orbital plane. In other words, the orbital plane wobbles as it spins. More specifically, the orbit's right ascension of the ascending node (RAAN) (Ω) rotates westward for prograde orbits around the Earth, and the argument of periapsis (ω) rotates in the direction of the spacecraft's motion. Semi-major axis (a), eccentricity (e), and inclination (i) suffer no long-term perturbations from oblateness [10].

A body's ellipticity has more dramatic effects on the orbit and can cause the spacecraft to transition from a safe orbit into an impacting or ejecting orbit within a few periods. The ellipticity of the body causes changes in the orbit semi-major axis, eccentricity, and inclination while effecting both the orbit's energy $\varepsilon = -\mu/(2a)$ and angular momentum $h = [\mu a(1-e)]^{1/2}$ [11]. In previous studies, it has been observed that prograde orbits experience much larger changes in energy and angular momentum for each orbit, where retrograde orbits experience little, if any, changes per orbit [12].

This paper puts all of these ideas together into a harmonic excitation analysis of an asteroid/satellite system. Harmonic excitation occurs naturally in our Solar System. For example, in the main belt asteroid system, there is a mean-motion resonance (MMR) interaction with Jupiter. More specifically, there are clumps of asteroids in certain regions, such as the 3:2 MMR region, and other regions that are practically devoid of asteroids, such as the 3:1 MMR region [13]. With regards to this project, the satellite orbits the asteroid repeatedly. Therefore, the satellite experiences the same gravitational perturbations as it passes over the surface of the satellite over and over again, at a given frequency. It was found by using a model of the system including all of the complexities previously mentioned, there are certain orbital conditions that are more or less stable based on the frequency at which the satellite is subject to these orbital perturbations. This allows for optimal mission planning as the satellite's orbital trajectory can be manipulated in order to make use of the stable orbital excitation conditions.

7.3 System Model

As mentioned, this project focuses on asteroid Itokawa as the asteroid-satellite system's central body. Itokawa was visited in 2005 and a spherical harmonic gravity model has been constructed from the data collected. This allows for a computer simulation of a real-world environment. The shape model used in our simulation is a visual approximation as a triaxial ellipsoid with axial dimensions 0.2741 km × 0.1561 km × 0.1376 km. These were calculated from values of the overall dimensions of Itokawa, given as 0.5481 km × 0.3122 km × 0.2751 km [9]. The rotational period of Itokawa is 12.132 h according the data provided by JPL Horizons [14].

To conduct the study with a high degree of user manipulation and data capture, STK is used to run the simulation and real-time data analysis done in MATLAB. The satellite used in the simulation scenario has a mass of 55 kg. The initial Keplerian elements of the satellite's orbit around Itokawa used in this study are given in Table 7.1.

7.4 Analysis

This project looks exclusively at the effects of the spherical harmonic gravity model. Other effects, such as Solar Radiation Pressure (SRP) and third-body gravitation, are not considered. The equation of motion for a linear spring-mass system is defined as:

$$F = ma_x + kx \tag{7.2}$$

where m is the mass (kg), a_x is the acceleration in the x-direction (m/s^2), k is the spring constant (N/m), and x is the displacement in the x-direction (m).

Newton's Law of Gravity is given as:

$$F_g = G\frac{m_1 m_2}{r^2} \tag{7.3}$$

where F_g is the force of gravity (N), G is the universal gravitational constant ($\approx 6.674 \times 10^{-11}$ N (m/kg)^2), m_1 and m_2 are the mass of the two bodies interacting (kg), and r is the distance between the two bodies' centers of mass (m). The terms G and m_1 (the mass of the orbited body) are combined to make the value μ (km^3/s^2) which is called the body's gravitational parameter.

Table 7.1 Initial conditions for satellite orbit used in the study

Keplerian element	Value
Semi-major axis	0.60 km
Eccentricity	0
Inclination	0° (prograde) or 180° (retrograde)
Argument of perigee	0°
Right-ascension of the ascending node	0°
True anomaly	0°

This can be made to fit into Eq. 7.2 as the spring constant and becomes what is known in orbital mechanics as the simplified two-body problem:

$$F = ma_x + \frac{m\mu}{x^3}x \tag{7.4}$$

Now, some system parameters need to be defined in order to continue. From mechanical vibrations, the undamped natural frequency of a linear spring-mass system is defined as:

$$\omega_n = \sqrt{\frac{k}{m}} \tag{7.5}$$

This is the frequency at which an undamped system will naturally oscillate. This is an important parameter in vibrations because if a system is driven to oscillate at this resonant frequency, small excitations grow into large amplitudes of oscillation.

Replacing the spring constant (k) in Eq. 7.5 with the "non-linear spring" constant, it becomes:

$$\omega_n = \sqrt{\frac{\frac{m\mu}{x^3}}{m}} \tag{7.6}$$

which simplifies to:

$$\omega_n = \sqrt{\frac{\mu}{x^3}} \tag{7.7}$$

where μ is the gravitational parameter (km^3/s^2) of Itokawa and x is the radial distance from the satellite to Itokawa's center (km). This leaves us with the natural frequency of the system in radians per second (rad/s). Scheeres provides that μ for Itokawa is 2.36×10^{-9} km^3/s^2 [9]. This gives a natural frequency of 1.05×10^{-4} rad/s when evaluated at our chosen initial orbital radius of 0.60 km.

Using STK to run the simulations, we can manually adjust the rotational rate of Itokawa in order to study the harmonic excitation. With the spherical harmonic model, the relative frequency between the satellite and the asteroid can be expressed as:

$$\omega_{sat/asteroid} = (\omega_{sat} \pm \omega_{asteroid}) \tag{7.8}$$

where the frequencies add in the retrograde case, and subtract in the prograde case. This results in the excitation frequency:

$$\omega = m\omega_{sat/asteroid} \tag{7.9}$$

where m is based on the degree and order of the spherical gravity model coefficients significantly impacting the excitation. In the case of Itokowa, the C_{20}, C_{22}, C_{42}, and C_{44} coefficients are the most significant. In looking at the shape of the corresponding Legendre polynomials these shapes primarily have two lobes, such that $m = 2$, in Eq. 7.9, for this asteroid.

The amplification ratio at which the oscillations grow when an undamped system is excited at it natural frequency is given as:

$$\frac{X}{\delta_{st}} = \frac{1}{1 - \left(\frac{\omega}{\omega_n}\right)^2} \tag{7.10}$$

where X/δ_{st} is the ratio of the dynamic to the static amplitude of motion, and ω/ω_n is the ratio of the excitation frequency to the natural frequency of the system [15]. Figure 7.1 is a plot of ω/ω_n versus X/δ_{st} for an undamped linear system. From this we expect to see greater instabilities in the spacecraft's orbit as the excitation frequency experienced by the spacecraft from the rotating asteroid approaches the natural frequency of the system.

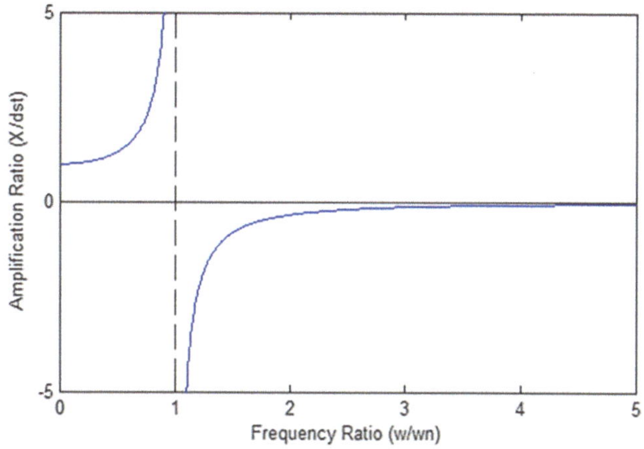

Fig. 7.1 Amplification ratio of an undamped linear system

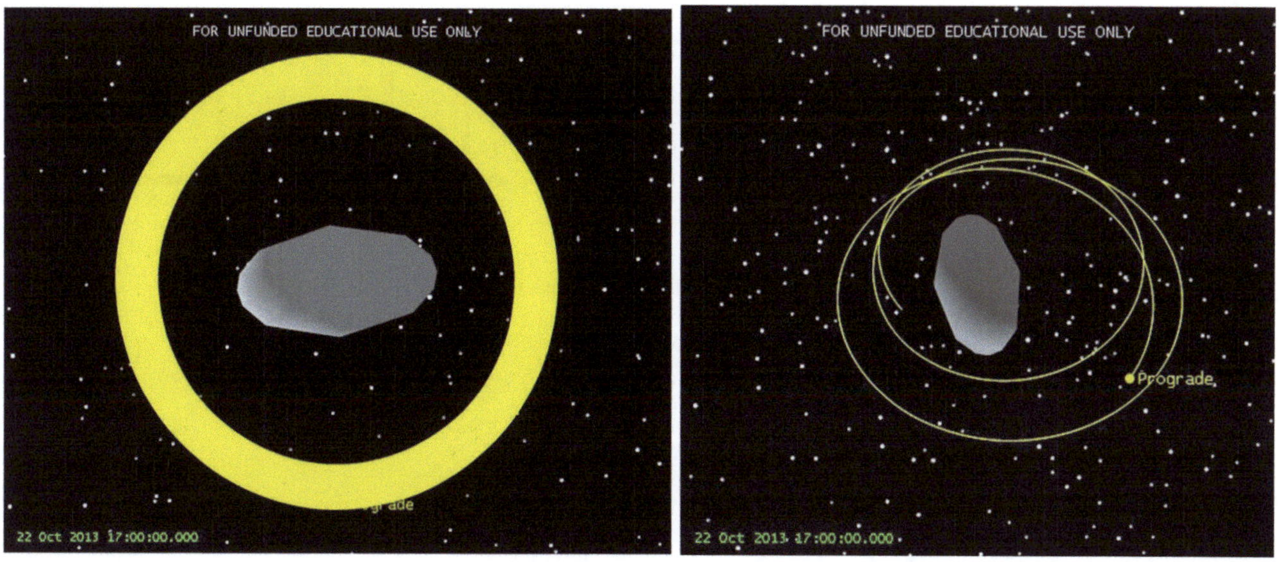

Fig. 7.2 (*Left*) A stable retrograde orbit that maintains a mostly circular orbit, but experiences precession around the central body for upwards of 6 months. (*Right*) An unstable prograde orbit that is perturbed enough to cause the orbit to diverge from its intended circular orbit and crash into the asteroid

Multiple scenarios were simulated in STK varying the rate of rotation of the asteroid and the direction of the orbit. This allows for the study of the interaction between the excitation frequency (ω), which changes according to Eqs. 7.8 and 7.9, and the natural frequency (ω_n), which remains constant for a given spacecraft and altitude. These allowed for the assessment of the validity of the simplified model in a complex environment.

Using the models described and the STK simulations, five spin rates multiples of the natural Itokawa spin rate were investigated; ⅓, ½, 1, 2, & 3. Figure 7.2 shows representative examples of the orbital path for a stable and an unstable orbit over time. Note that even the stable orbit changes over time due to a slight eccentricity in the orbital path. Figures 7.3, 7.4, 7.5, 7.6, 7.7, 7.8, 7.9, 7.10, 7.11 and 7.12 show the time history plots of the radial acceleration indicating the excitation levels and FFTs of the data to identify the significant driving frequencies.

In the FFT plots there are multiple excitation frequencies observed, but in general two frequencies stand out as having a greater influence than the others. Table 7.2 shows the frequencies for select scenarios with the most significant frequencies highlighted. Some prograde scenarios had no significant frequencies due to the presence of orbital instability from the start of the orbit. In other words, the satellite did not complete enough revolutions before crashing in order to observe any excitation frequencies. In these scenarios the radial acceleration grows exponentially larger as the satellite approaches the asteroid prior to impact or the satellite escaped the stable orbit and the radial acceleration goes to zero.

Now it's important to make note of something about each of the above data sets from the scenarios. The first significant frequency in all of these scenarios is quite similar. This can be explained by a brief discussion of orbital mechanics.

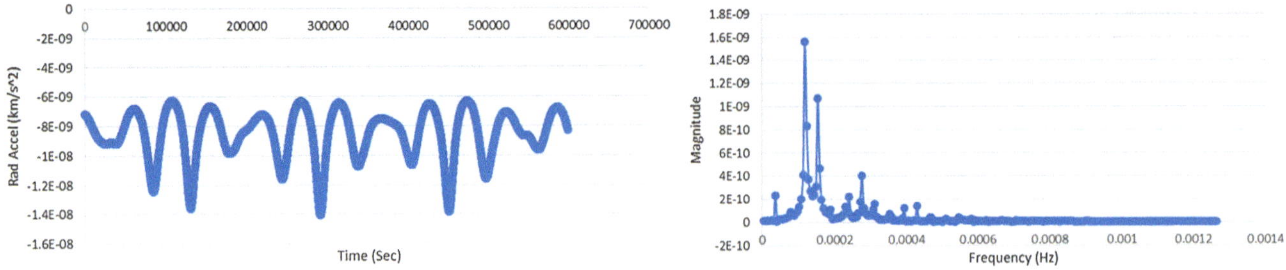

Fig. 7.3 ⅓× natural spin rate in prograde orbit. (*Left*) Radial acceleration vs. time. (*Right*) FFT of time history data

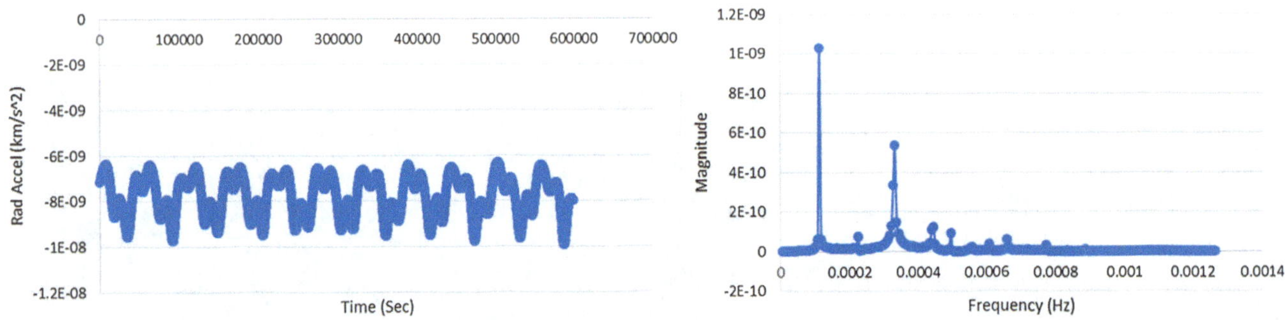

Fig. 7.4 ⅓× natural spin rate in retrograde orbit. (*Left*) Radial acceleration vs. time. (*Right*) FFT of time history data

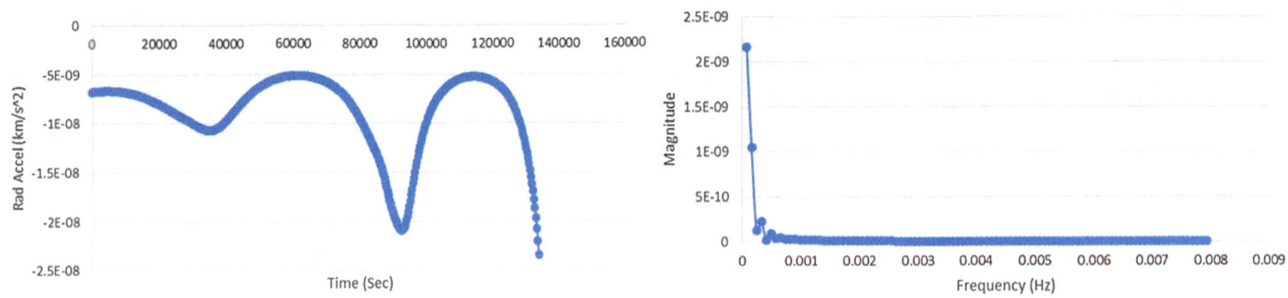

Fig. 7.5 ½× natural spin rate in prograde orbit. (*Left*) Radial acceleration vs. time. (*Right*) FFT of time history data

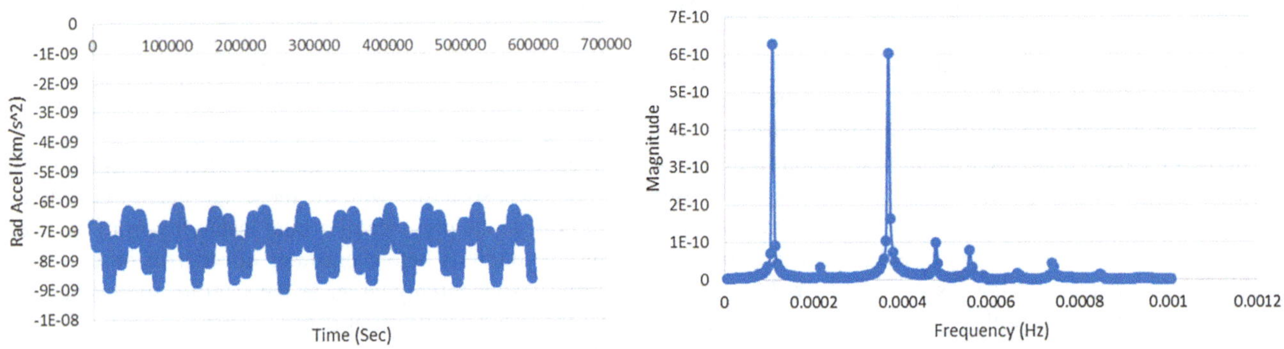

Fig. 7.6 ½× natural spin rate in retrograde orbit. (*Left*) Radial acceleration vs. time. (*Right*) FFT of time history data

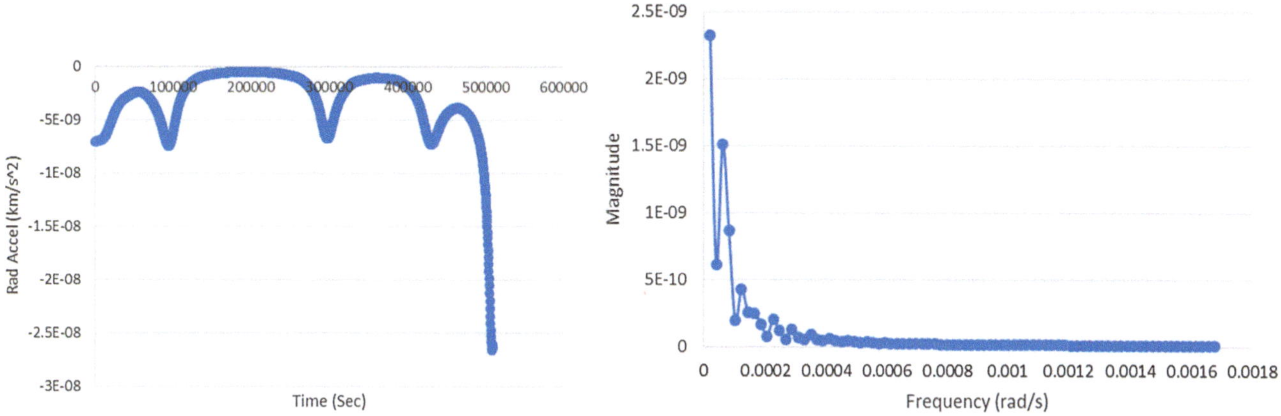

Fig. 7.7 1× natural spin rate in prograde orbit. (*Left*) Radial acceleration vs. time. (*Right*) FFT of time history data

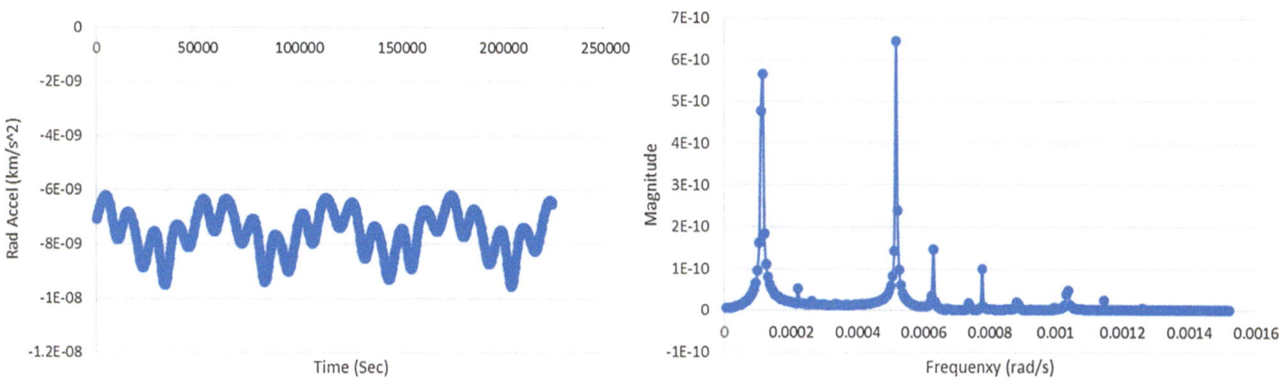

Fig. 7.8 1× natural spin rate in retrograde orbit. (*Left*) Radial acceleration vs. time. (*Right*) FFT of time history data

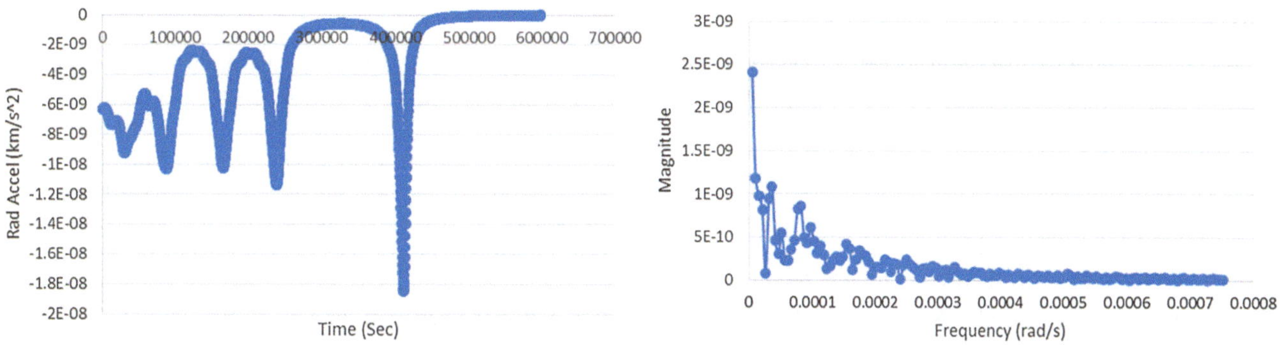

Fig. 7.9 2× natural spin rate in prograde orbit. (*Left*) Radial acceleration vs. time. (*Right*) FFT of time history data

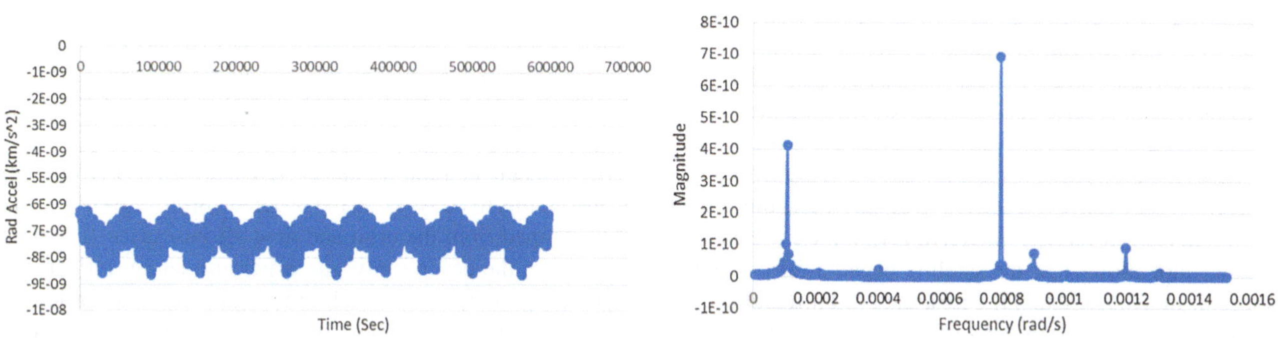

Fig. 7.10 2× natural spin rate in retrograde orbit. (*Left*) Radial acceleration vs. time. (*Right*) FFT of time history data

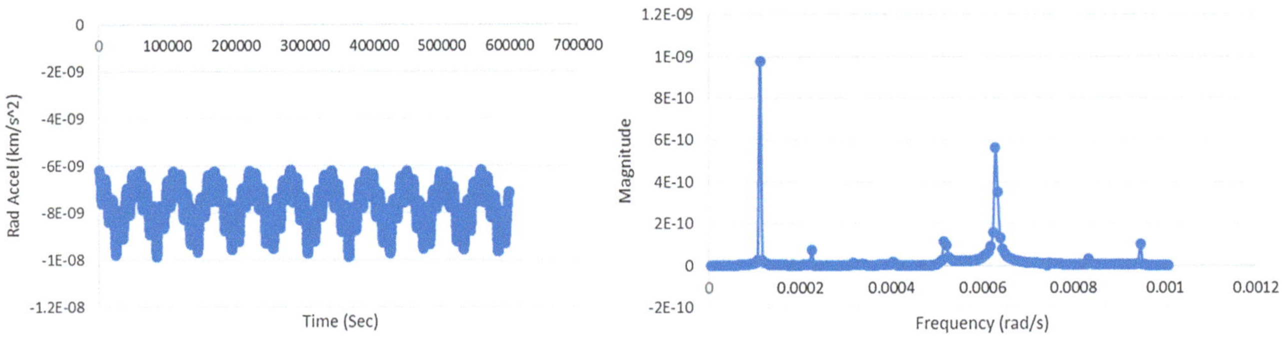

Fig. 7.11 3× natural spin rate in prograde orbit. (*Left*) Radial acceleration vs. time. (*Right*) FFT of time history data

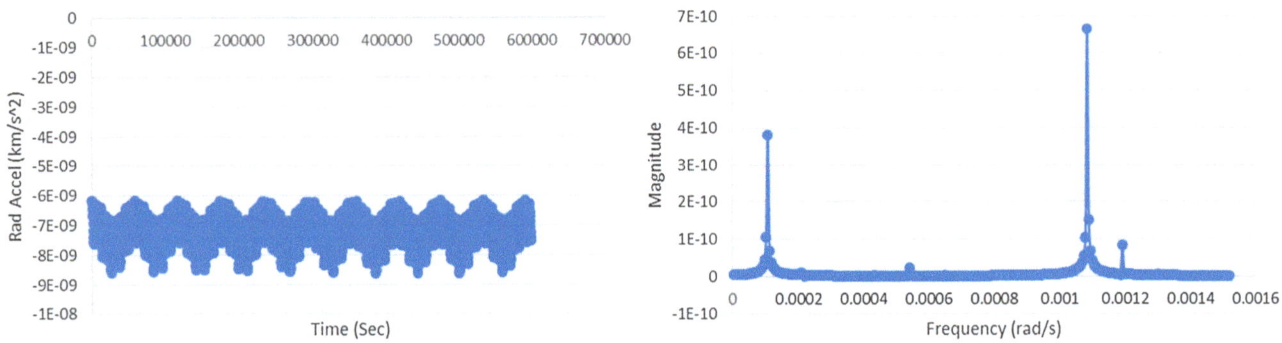

Fig. 7.12 3× natural spin rate in retrograde orbit. (*Left*) Radial acceleration vs. time. (*Right*) FFT of time history data

Table 7.2 Observed prograde frequencies for various Itokawa rotation rates

	Asteroid spin rate multiplier	Observed prograde frequencies (rad/s)		Observed retrograde frequencies (rad/s)	
	1/3	1.18E-04	1.53E-04	1.1249E-04	3.3236E-04
	1/2	–	–	1.0738E-04	3.6816E-04
Natural rate	1	–	–	1.1249E-04	5.1644E-04
	2	–	–	1.0738E-04	7.9767E-04
	3	1.12E-04	6.29E-04	1.0738E-04	1.0840E-03

When the scenario is created in STK, an initial velocity is calculated within the program for a point-mass, two-body model. These scenarios are not a point-mass, two-body scenario and have a complex gravity model. So the initial velocity that is given from STK is the initial velocity needed to maintain a circular orbit around a sphere, not the initial velocity required to maintain a circular orbit around an oblate and eccentric body like Itokawa. This discrepancy in initial velocities results in a slight eccentricity in the satellite's orbit from the beginning of the scenario. This slight eccentricity is what is causing the first significant excitation frequency, and explains why it is similar in all scenarios despite the fact that the rotation rate of the central body is changing significantly.

With this in mind, we can look at the FFT in order to filter out this frequency caused by the eccentric orbit. This will give excitation frequencies caused solely by the significant complex gravity model coefficients, and not by the eccentricity of the orbit. Using the filtered frequencies from the STK scenarios, the results are in Table 7.3

In all the retrograde orbit cases shown the resulting orbit quality is stable and the frequency ratios are all nearly three or higher. These orbit are predicted to be stable and the simulations proved this to be valid. In the prograde orbit cases below six show signs of instability. Interestingly, in the 2× case the satellite escaped from the orbit and flew off into space. The time history plot shows that the satellite began a fairly stable orbit, but then the eccentricity got larger and larger until it was ejected. In the ½× and 1× cases, the orbital radial acceleration levels grew until the satellite flew nearer the asteroid and eventually crashed into the surface. The predicted prograde frequency ratios show negative values when the satellite is

Table 7.3 Orbit scenario observations with varying asteroid spin rates and changing orbital direction (prograde and retrograde)

	Asteroid spin rate multiplier	Prograde simulated (ω/ωn)	Prograde predicted (ω/ωn)	Observed orbit quality	Retrograde simulated (ω/ωn)	Retrograde predicted (ω/ωn)	Observed orbit quality
	1/3	1.45	−1.08	*Stable*	3.18	2.91	*Stable*
	1/2	–	−0.63	**Unstable**	3.52	3.37	*Stable*
Natural rate	1	–	0.75	**Unstable**	4.94	4.75	*Stable*
	2	–	3.5	***Semi-stable***	7.63	7.50	*Stable*
	3	5.99	6.25	*Stable*	10.37	10.25	*Stable*

moving faster that the asteroid, however the magnitude of the ratio is still the defining parameter. It was observed that low values of the frequency ratios result in the unstable behavior. This corresponds to the regions where the amplification ratio is greater than one, as shown in Fig. 7.1. The simulated frequency ratios for the retrograde cases match very well and the behavior is well predicted. In the prograde cases the values show more difference, but still are fairly close. However, the correlation with the amplification ratio is not as well behaved. This behavior is not well understood by the authors, but may have to do with the relatively low frequency ratios and the complex shape of the asteroid. Other Legendre polynomial terms may be playing a larger role in these orbits. However, the general trend still is captured and the relative speed of the satellite and rotational speed of the asteroid can provide good correlation to areas where greater orbit stability can be expected.

7.5 Conclusion

Harmonic excitation analysis can be used as a tool when planning out a space mission as long as some information is known about the target body. This project used a linear approach in analyzing a nonlinear system and produced promising results. The FFT allowed us to filter out different excitation frequencies in order to study their effects on the system. When considering the complex gravity model by itself, a harmonic excitation analysis allowed for a fairly accurate prediction of the satellite's behavior in a complex system. If the mathematical model of the dynamic system is precisely known, accurate results may be obtained with STK simulations. However, without accurate information about unexplored asteroids, a retrograde orbit with a high excitation frequency produces a more stable obit.

Acknowledgements This research was supported in part by the North Dakota Space Grant Consortium, the North Dakota NASA EPSCoR grant, the UND Seed/Planning Grant for Collaborative Research, the National Science Foundation (NSF Grant #EPS-081442), and AGI with STK Educational Licenses.

References

1. Church, C., Fevig, R.: A feasibility study on the characterization of the internal structure of small NEOs with small spacecraft. In: Lunar and Planetary Science Conference (LPSC XLIV), Houston, March 2013
2. Johnson, J., Semke, W., Zimmer, M., Fevig, R.: Orbit stability determination of satellites using harmonic force excitation analysis. In: Proceedings of the International Modal Analysis Conference (IMAC) XXXIII: A Conference and Exposition on Structural Dynamics, 2015
3. Yeomans, D.: NEO groups, from http://neo.jpl.nasa.gov/neo/groups.html, 2014
4. Yeomans, D. Orbit diagrams, from http://neo.jpl.nasa.gov/orbits/, 2014
5. Netting, R.: NEAR-shoemaker, from http://science.nasa.gov/missions/near/, 2014
6. Kawaguchi, J., Fujiwara, A., Uesugi, T.: Hayabusa (MUSES-C) – rendezvous and proximity operation, IAC-05-A3.5.A.01. In: International Astronautical Congress, International Astronautical Federation, Fouduoka, 2005
7. Montenbruck, O., Gill, E.: Satellite Orbits: Models, Methods, and Applications, pp. 53–116. Springer, Heidelberg (2000)
8. Church, C.: A feasibility study on the implementation of satellite-to-satellite tracking around a small near-earth object. M.S. Thesis, Department of Space Studies, University of North Dakota (2014)
9. Scheeres, D.J., Gaskell, R., Abe, S., Barnouin-Jha, O., Hashimoto, T., Kawaguchi, J., Kubota, T., Saito, J., Yoshikawa, M., Hirata, N., Mukai, T., Ishiguro, M., Kominato, T., Shirakawa, K., Uo, M.: The actual dynamical environment about Itokawa, AIAA 2006-6661. In: AIAA/AAS Astrodynamics Specialist Conference and Exhibit, American Institute of Aeronautics and Astronautics, Inc., Keystone, 2006
10. Scheeres, D.J., Ostro, S.J., Hudson, R.S., Werner, R.A.: Orbits close to asteroid 4769 Castalia. Icarus **121**, 67–87 (1996). Elsevier Limited
11. Scheeres, D.J., Marzari, F., Tomasella, L., Vanzani, V.: Rosetta mission: satellite orbits around a cometary nucleus. Planet. Space Sci. **36**(6/7), 649–671 (1998). Elsevier Limited

12. Scheeres, D.J., Williams, B.G., Miller, J.K.: Evaluation of the dynamic environment of an asteroid: applications to 433 eros. J. Guid. Control Dyn. **23**(3), 466–475 (2000)
13. Nesvorny, D., Ferraz-Mello, S., Holman, M., Morbidelli, A.: Regular and chaotic dynamics in the mean-motion resonances: implications for the structure and evolution of the asteroid belt. Asteroids III **1**, 379–394 (2002)
14. JPL HORIZON System, telnet://horizons.jpl.nasa.gov:6775
15. Rao, S.S.: Mechanical Vibrations, 5th edn, pp. 259–362. Upper Saddle River, Prentice Hall (2011)

Chapter 8
A Study on the Dynamic Interaction of Shock Response Fixtures and Test Payload

Jesus M. Reyes and Peter Avitabile

Abstract Shaker system physical characteristics (e.g. shake tables, modal shakers, and shock fixtures) vary from one laboratory to another. Parameters such as size and payload capacity, make every system's dynamic characteristics unique. The commonly used control accelerometer cannot adequately adjust for the actual dynamic interaction/dynamic coupling effects between the test article and excitation fixture. Therefore, the payload/test article deformation will vary from test fixture to test fixture, making every test different—even with the same Shock Response Spectrum (SRS).

In order to properly address the dynamic coupling and dynamic interaction that results, and to subject the test article to a vibration level that provides the proper excitation, customization of the shock response input is necessary.

This work focuses on the identification as to how the article under test is affected by the attachment to the excitation fixture. Models are presented to show how the test article response is distorted by the dynamic interaction with the test fixture.

Keywords Shock response spectrum • Mechanical shock environment • Shock • Mechanical excitation • Qualification testing • Dynamic coupling • Vibration exciter

8.1 Introduction

Spacecraft structures, electronic components, packaging designs, automotive assemblies and military weaponry are a few examples of systems which are subjected to, in early design stages, dynamic testing because they are created/manufactured to accomplish great-risk tasks in high-energy environments. For instance, impact shock, transportation shocks, and, particularly, pyroshocks from multi-stage separation vehicles, naval and ballistics shock due to the extreme amounts of energy that they release [1]. Shock response data is obtained from these service environments and used for vibration testing; the frequency response of these environments is characterized by the Shock Response Spectrum (SRS) and is used to calculate the maximum dynamic response of structures. However, in practice, because the acceleration profiles are acquired while the structure is in service, the acceleration data does not contain any information regarding the dynamic characteristics of the excitation fixture. Depending on the particular dynamic characteristics of the test fixture (slip plate/shock plate) and if there are resonances in or near the frequency range of interest, the response of the Device Under Test (DUT) may be seriously altered [2].

In the past several decades, efforts have been made to account for the dynamic coupling and dynamic interaction between the DUT and the test fixture. McConnell [3] stated that a general theory of vibration testing is necessary, because very different laboratory results are acquired when field data is used as the excitation input; collected field data is needed for design engineers but the dynamic interaction between the test article and shaker/test fixture must be properly addressed. Several years later, Varoto [4] defined which measured responses and modal parameters are required to be used by the design engineers to account for the dynamic interaction between the test article and shaker/fixture system such that the actual field vibration is properly represented in the test.

Oftentimes, in vibration testing, the correct level of response at the DUT is not obtained by just using the measured acceleration profile recorded during the shock event. In addition, the SRS can be formed from many different signals and may not expose the test article to the proper in-situ environment. Also, the DUT dynamic characteristics can be altered due to any dynamic interaction between the DUT and the shaker/test fixture; the changes, if any, in the dynamic characteristics

J.M. Reyes (✉) • P. Avitabile
Structural Dynamics and Acoustic Systems Laboratory, University of Massachusetts Lowell, One University Avenue, Lowell, MA 01854, USA
e-mail: jesus_reyesblanco@student.uml.edu

A. Brandt, R. Singhal (eds.), *Shock & Vibration, Aircraft/Aerospace, Energy Harvesting, Acoustics & Optics, Volume 9*, Conference Proceedings of the Society for Experimental Mechanics Series, DOI 10.1007/978-3-319-30087-0_8

must be understood and adjusted otherwise the test may not provide valid results. This dynamic interaction must first be understood and then approaches and methodologies to correct for this problem must be developed. This paper addresses the problem of dynamic interaction showing the differences in response and the follow-on work will develop the approaches to address this issue. The work in this paper presents several analytical models which show the differences in the DUT responses depending on the shaker/fixture characteristics and how much dynamic coupling exists between the DUT and the shaker/fixture system.

8.2 Methodology

In order to identify how the Test Article (TA) response is affected by the attachment to the shock fixture, four main cases will be analyzed. Each main case presents two Finite Element (FE) models. The first FE model simulates the response due to a force pulse of a TA attached to a shock plate. The second model reproduces the behavior of the TA attached to a slip table while being subjected to a base acceleration. Throughout this work, these FE models will be referred as the Force Input Response (FIR) model and the Base Excitation Response (BER) model, respectively. In addition, a variation of each main case will be studied. These variations will show how the TA response may not be an accurate representation of the test environment if the acceleration profile used as an input does not account for the dynamic interaction between the test fixture and the structure under test.

The overall process to point out how the response of a TA is altered by the dynamic behavior of the shock fixture starts with the application of a force pulse at the base of the FIR model. Then, the response of the TA is calculated. Particularly, the acceleration response due to the input force at the base is critical for this study because this acceleration will be used as the input acceleration in the BER model. Thus, the response of the TA is also computed. Finally, the responses of the TA from both models will be compared and discussed.

8.3 Description of Analytical Models

In order to study the interaction between a test article and the shock fixture, two main models are used. These FE models were developed to analyze the dynamic behavior of the test article on two different test fixtures. In this study, the models are representations of a shock plate or a slip table with the attachment of a 3-DOF test article. The shock plate model will be analyzed from a force input response perspective; the slip plate model will be solved from a base excitation perspective. Figure 8.1 displays the schematics of the Force Input Response (FIR) model shown on the left and the Base Excitation Response (BER) model on the right.

Initially, four variants of the FIR model will be used. This first FE model is created by attaching the test article (m_2 through m_4) to a shock plate (m_1) using a shear spring of stiffness value k_2. Then, the following cases will be studied:

- *Case A*—The mass of the shock plate will be a hundred times more massive than any of the remaining lumped masses ($m_1 = 100\ m_2$).

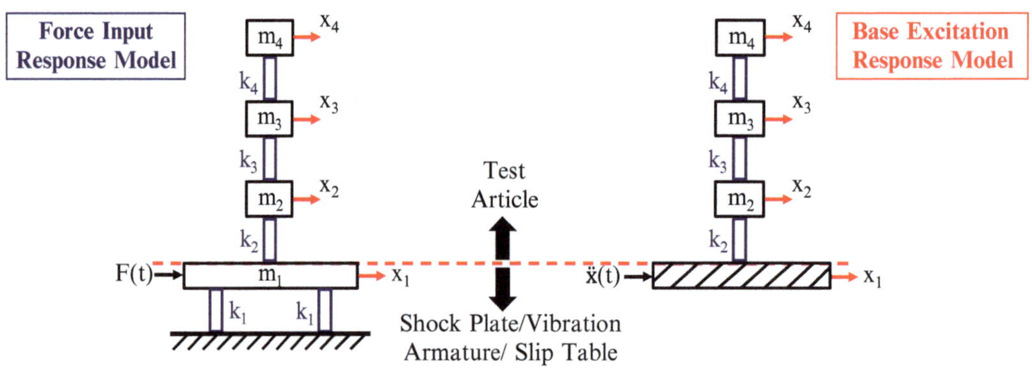

Fig. 8.1 Force input response model (*left*) and base excitation response model (*right*)

Table 8.1 Physical parameters of the FE models

Mass [blob]	Case A	Case B	Case C	Case D	Stiffness [lb/in]	Case A	Case B	Case C	Case D
m_4	0.037	0.037	0.037	0.037	k_4	2560	2560	2560	2560
m_3	0.037	0.037	0.037	0.037	k_3	2560	2560	2560	2560
m_2	0.037	0.037	0.037	0.037	k_2	2560	2560	2560	2560
m_1	3.730	0.373	0.037	0.037	k_1	2560	2560	256000	25600

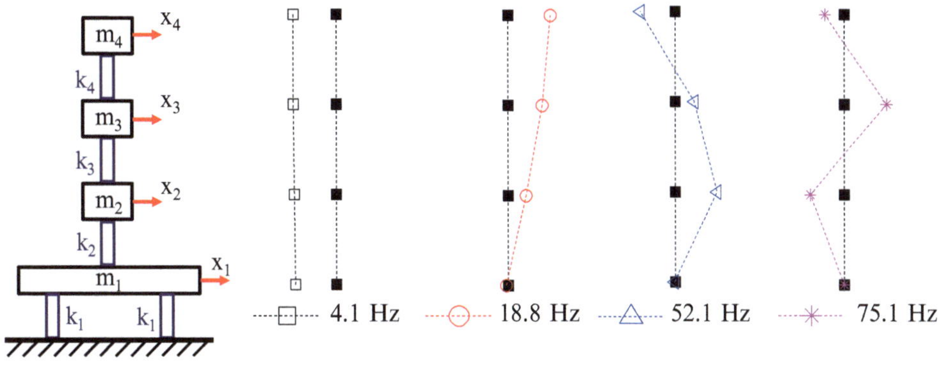

Fig. 8.2 Frequencies and mode shapes of the FIR model case A (100× more massive)

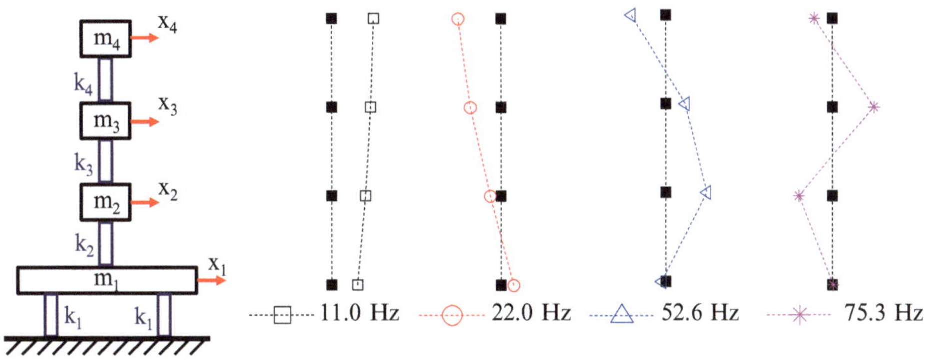

Fig. 8.3 Frequencies and mode shapes of the FIR model case B (10× more massive)

- *Case B*—The value of m_1 from Case A is decreased by a factor of ten; becoming ten times more massive than the rest of the lumped masses ($m_1 = 10\ m_2$).
- *Case C*—Masses from m_1 to m_4 remain with constant values, although there is a significant change in stiffness of the spring supporting the shock plate (k_1); the spring at the base is a hundred times stiffer than Case A and B ($k_1 = 100\ k_2$).
- *Case D*—The value of k_1 used in Case C is decreased by a factor of ten ($k_1 = 10\ k_2$).

Finally, the computed acceleration response from the FIR model at the shock plate location will be the input base excitation for the BER model. The TA will be subjected to the base acceleration profile and its response will be compared to responses obtained from the FIR model. The physical parameters of the shock plate and test article for every case are displayed in Table 8.1.

These mass and stiffness changes made to the FIR model and the comparison to the BER model will demonstrate that reliable vibration tests can be achieved if the shock fixture is sufficiently massive or stiff in comparison to the test article. This data will be shown later in the Results section. The motion of the shock plate and slip table are represented by x_1 for each case shown in Fig. 8.1, respectively. Figures 8.2, 8.3, 8.4, and 8.5 display the FIR model's modal parameters for every case.

The responses that will be obtained from the FIR and BER models will help to understand that there is an interaction between the shock plate/shock fixture and TA and is a function of the physical parameters of the structures. For instance, if the resonant behavior of the shock plate is not in the frequency range of the TA, then the TA response will not be affected; however, if this is not true, then the dynamic response of the TA will be distorted.

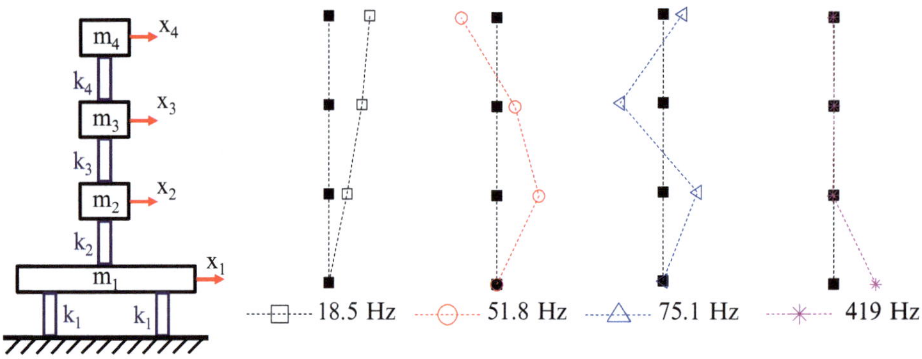

Fig. 8.4 Frequencies and mode shapes of the FIR model case C (100× stiffer)

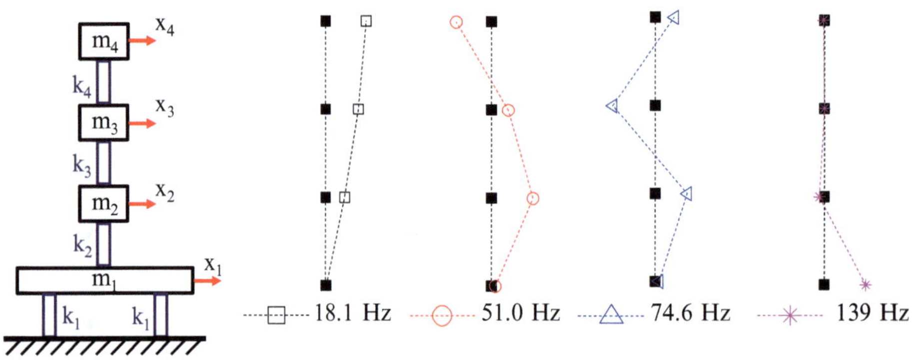

Fig. 8.5 Frequencies and mode shapes of the FIR model case D (10× stiffer)

8.4 Discussion and Results

In order to identify how the dynamic behavior of the test article is affected by being attached to the shock fixture, the four FIR models described above are used to compute the response at particular DOFs due to an input force applied at the location of the shock plate. Then, these responses will be compared to the responses obtained from the BER model and discussed. All the following computations were performed using MATLAB [5].

8.4.1 Force Input Response and Based Excitation Models

8.4.1.1 Case A: Mass Modification #1 (100× More Massive)

Applying a 2-lbf-half-sine pulse as input force at m_1 of the FIR model (shock plate) is the first step in order to compute the response of the system. After that, the responses of the FIR model due to the force pulse at m_1 and m_4 for all cases are computed using Newmark's method of direct integration. For all computations, 5 % of critical damping was assumed for every mode. Now, the BER model will have as input base acceleration, the acceleration response of the FIR model computed at m_1. Then, the response at m_4 of the BER model is computed and compared to the corresponding response of the FIR model. The responses of the FIR model and BER model are displayed on the left (blue curves) and in the middle (red curves) of Fig. 8.6, respectively. A comparison of the responses computed at m_4 for both models are displayed on the right of Fig. 8.6. These curves match almost flawlessly, which indicates that the test setups produce the same dynamic response in the test article. The reason is that the resonant behavior of the shock plate does not significantly contribute to the responses of the TA.

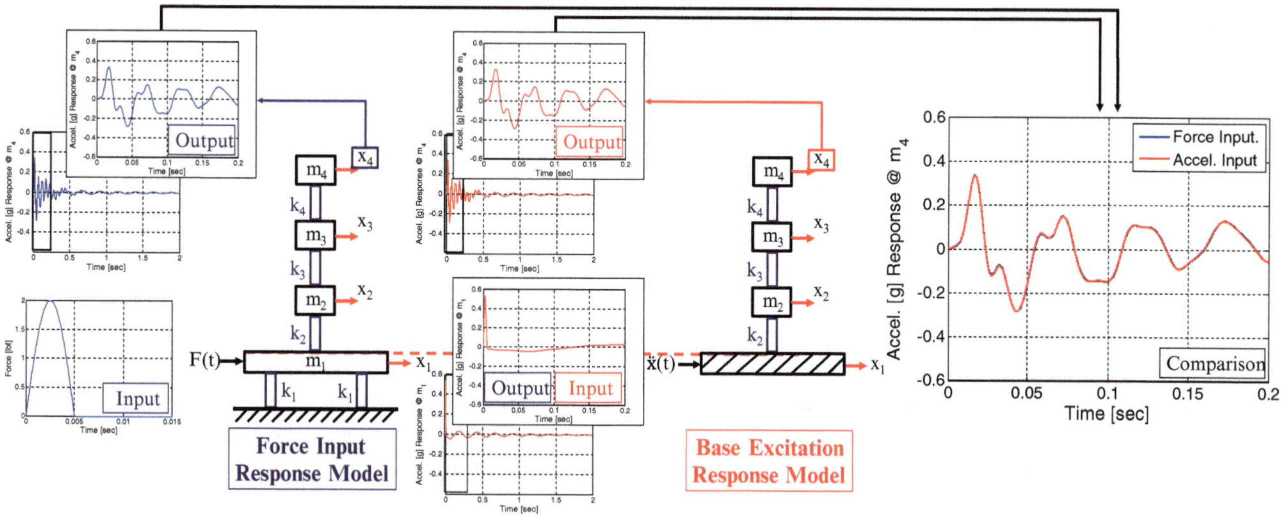

Fig. 8.6 Computation of output at m_1 and m_4 of FIR model (*left*) for case A due to the half-sine pulse with test article mounted. (*Dashed line*) Computation of output at m_4 of BER model (*middle*) using as input base acceleration the output at m_1 from FIR model. (*Dashed line*) Output response at m_4 from BER model is compared to output response at m_4 from the FIR model with test article (*right*)

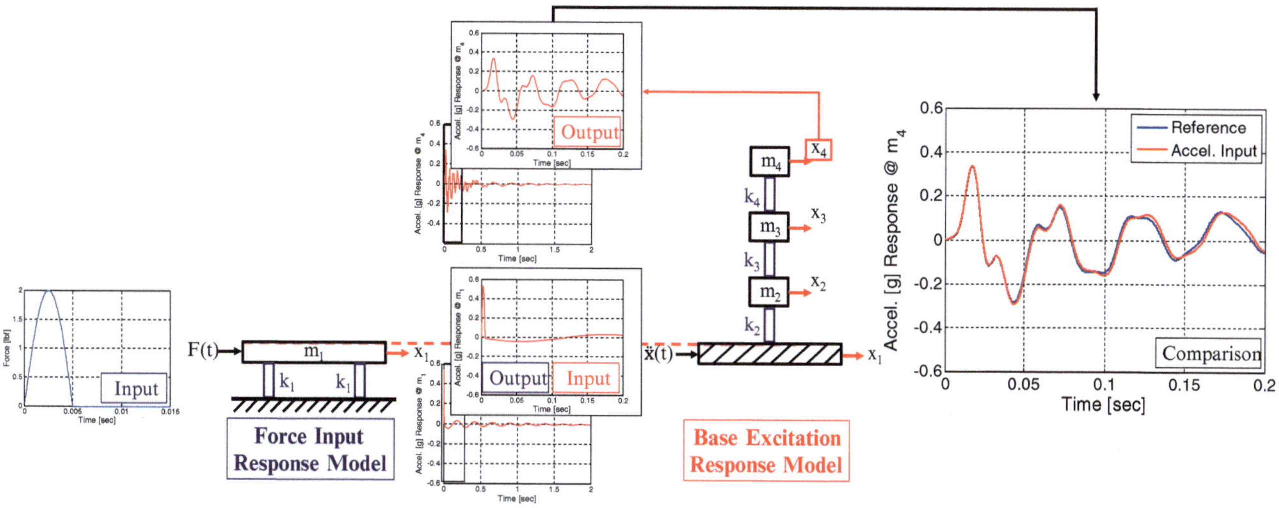

Fig. 8.7 Computation of output at m_1 and m_4 of FIR model (*left*) for case A due to the half-sine pulse without test article mounted. (*Dashed line*) Computation of output at m_4 of BER model (*middle*) using as input base acceleration the output at m_1 from FIR model. (*Dashed line*) Output response at m_4 from BER model IS compared to output response at m_4 from the FIR model with test article (*right*)

In order to show the effects of the test article on the shock plate, the test article will be removed and responses will be computed in a similar fashion as previously described. The bare shock plate will be excited with the half pulse sine and its response will be obtained. The acceleration at the bare shock plate will be the input to the BER model and the response at m_4 of the test article will be computed and compared to the response of the FIR model. The response for both models are shown in Fig. 8.7.

The comparison plot in Fig. 8.7 indicates that the response at m_4 of the FIR model (blue curve), which has information of the full system, and the computed response at m_4 using the BER model which has as input the acceleration response of the bare shock plate caused by the half-sine pulse (red curve), are very similar. The reason is that, m_1 is a hundred times more massive than any other mass in the system. With this stated, the shock plate is so massive that the lack of the test article does not contribute critically to the response of the test article.

8.4.1.2 Case B: Mass Modification 2 (10× More Massive)

For Case B, the responses were computed and compared identically the same as in the previously described Case A. The main and most important difference is that m_1 is only ten times more massive than any lumped mass belonging to the test article. Case A shows the situation where the mass is very large but may not be possible to achieve in a laboratory environment. For this reason, Case B uses a smaller mass.

The FIR model is excited at m_1 by the half-sine pulse and responses at m_1 and m_4 are computed. Thus, the acceleration response at m_1 is used as input base acceleration in the BER model. A new response is calculated at m_4 of the BER model (red curve) which is compared to the corresponding response from the FIR model (blue curve). Figure 8.8 shows the comparison between the models.

Barely noticeable differences can be observed as was expected and showed similarly in Case A, when the TA is mounted on the shock plate. Now, the test article will be removed and the bare shock plate will be excited by the half-sine pulse. The response of the bare shock plate will be used as the input base excitation in the BER model. The response at m_4 of the BER model is compared to the corresponding response of the FIR model. This comparison is showed in Fig. 8.9. The large differences in responses is because the shock plate resonant behavior contributes to the response of the TA causing an increase of magnitude in the vibration of the TA.

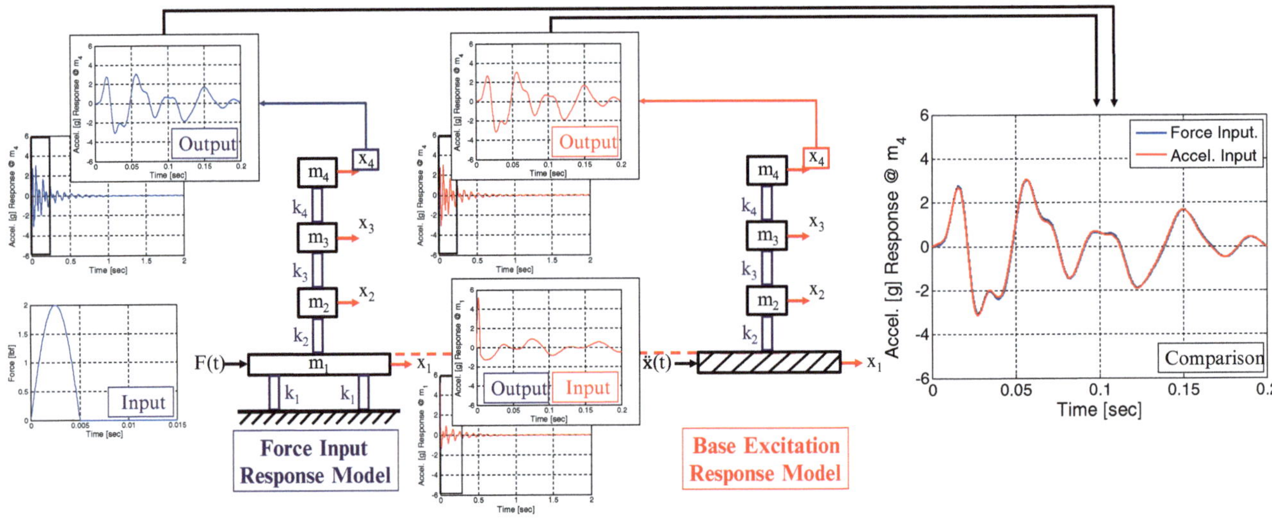

Fig. 8.8 Computation of output at m_1 and m_4 of FIR model (*left*) for case B due to the half-sine pulse with test article mounted. (*Dashed line*) Computation of output at m_4 of BER model (*middle*) using as input base acceleration the output at m_1 from FIR model. (*Dashed line*) Output response at m_4 from BER model is compared to output response at m_4 from the FIR model with test article (*right*)

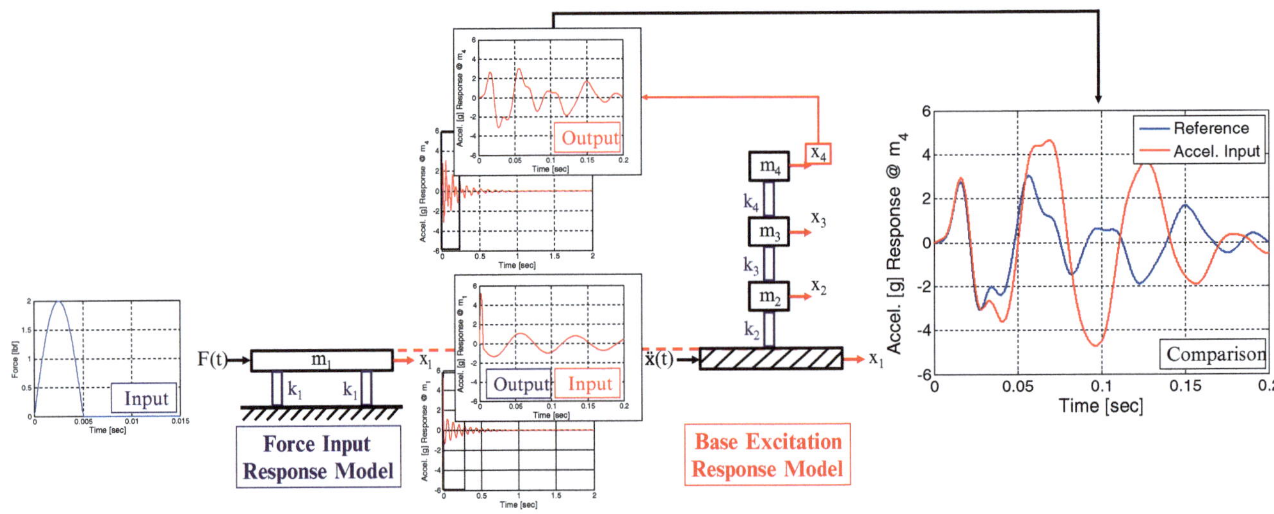

Fig. 8.9 Computation of output at m_1 and m_4 of FIR model (*left*) for case B due to the half-sine pulse without test article mounted. (*Dashed line*) Computation of output at m_4 of BER model (*middle*) using as input base acceleration the output at m_1 from FIR model. (*Dashed line*) Output response at m_4 from BER model is compared to output response at m_4 from the FIR model with test article (*right*)

8.4.1.3 Case C: Stiffness Modification 1 (100× Stiffer)

For Cases A and B, the effects of changing the shock plate mass were studied. This time, for Case C, the mass of the shock plate will remain the same but its stiffness will be modified. As previously shown in Table 8.1, the stiffness value of spring k_1 will be a hundred times stiffer than any other spring in the system. A half-sine pulse is applied at the shock plate of the FIR model and responses at m_1 and m_4 are computed. The acceleration output at m_1 of the FIR model will be used as the input in the BER model to excite the TA; its response at m_4 (red curve) is computed and compared to the response at m_4 of the FIR model (blue curve). Figure 8.10 shows the responses for both models.

The comparison in Fig. 8.10 indicates that responses from both models are very similar. This is the same behavior that was observed in Cases A and B when the TA was mounted on the shock plate. Now, the TA will be removed from the shock plate and the response due to the half-sine pulse will be used to compute the acceleration at m_1 of the FIR model. Figure 8.11 displays the responses for both models and the comparison of the responses at m_4 of the test article. The absence of the TA on the shock plate for this case had no effect in the response of the TA based on the comparison in Fig. 8.11. Responses of the FIR and BER models matched very well because of the highly stiff shock plate.

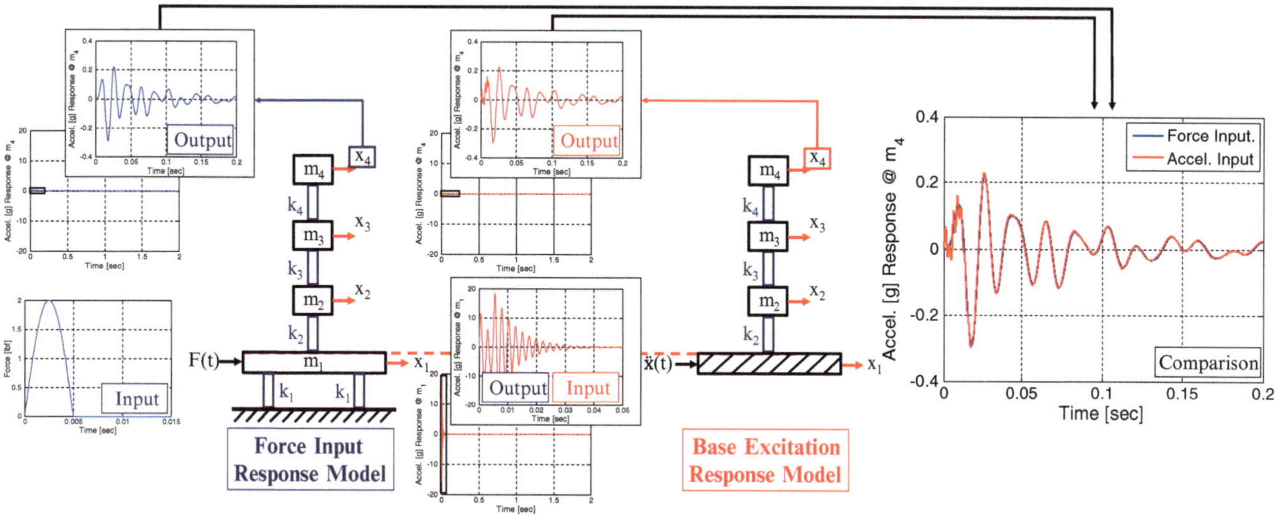

Fig. 8.10 Computation of output at m_1 and m_4 of FIR model (*left*) for case C due to the half-sine pulse with test article mounted. (*Dashed line*) Computation of output at m_4 of BER model (*middle*) using as input base acceleration the output at m_1 from FIR model. (*Dashed line*) Output response at m_4 from BER model is compared to output response at m_4 from the FIR model with test article (*right*)

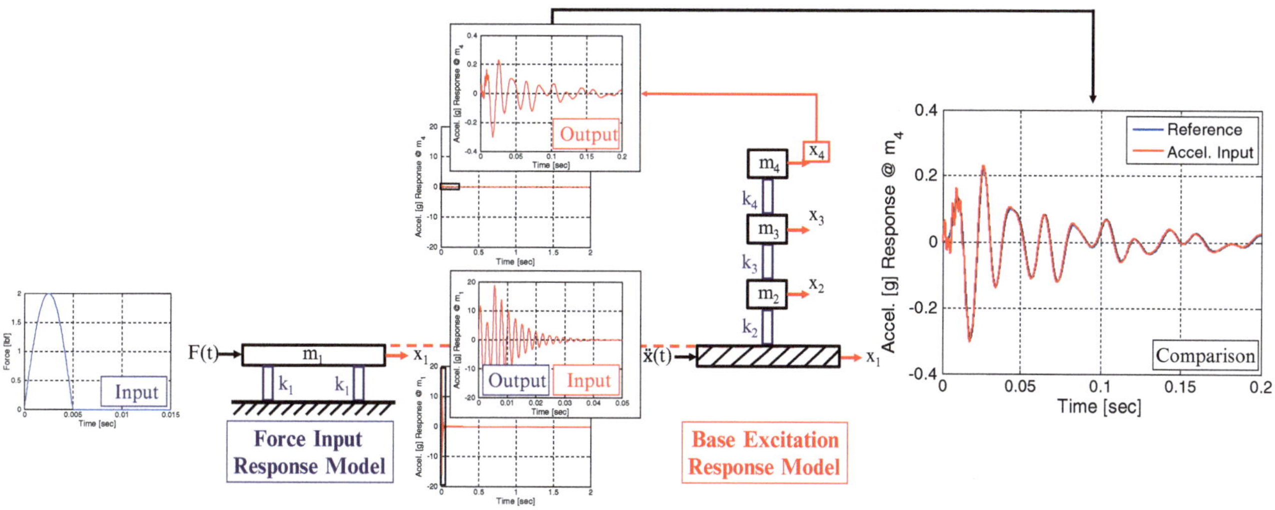

Fig. 8.11 Computation of output at m_1 and m_4 of FIR model (*left*) for case C due to the half-sine pulse without test article mounted. (*Dashed line*) Computation of output at m_4 of BER model (*middle*) using as input base acceleration the output at m_1 from FIR model. (*Dashed line*) Output response at m_4 from BER model is compared to output response at m_4 from the FIR model with test article (*right*)

8.4.1.4 Case D: Stiffness Modification 2 (10× Stiffer)

The same steps shown for Case C are performed in Case D—but, the stiffness value of spring k_1 will be only ten times stiffer than the rest of the springs in the system. Responses due to the half-sine pulse are computed using the FIR model; the output at m_1 of the FIR model is used as the input for the BER model. The response of the TA at m_4 due to the base acceleration is computed and compared to the response at m_4 of the FIR model. Figure 8.12 shows the comparison between responses.

Minimal differences can be observed in the comparison plot in Fig. 8.12 meaning that the effect of the contribution of the shock plate in the overall response of the TA is almost insignificant because the TA was mounted on the plate. Now, similarly to Cases A, B and C the response of the bare shock plate will be computed and used as the input acceleration for the BER model. The response at m_4 of the BER model (red curve) will be compared to the response of m_4 of the FIR model (blue curve). Figure 8.13 shows an overlay of these responses.

The lack of the TA on the shock plate, as shown in Fig. 8.13, does affect the response of the TA due to the resonant behavior of the shock plate. The shock plate used in this case is not as stiff as the one from Case C where the shock plate contribution to the response of the TA is almost negligible.

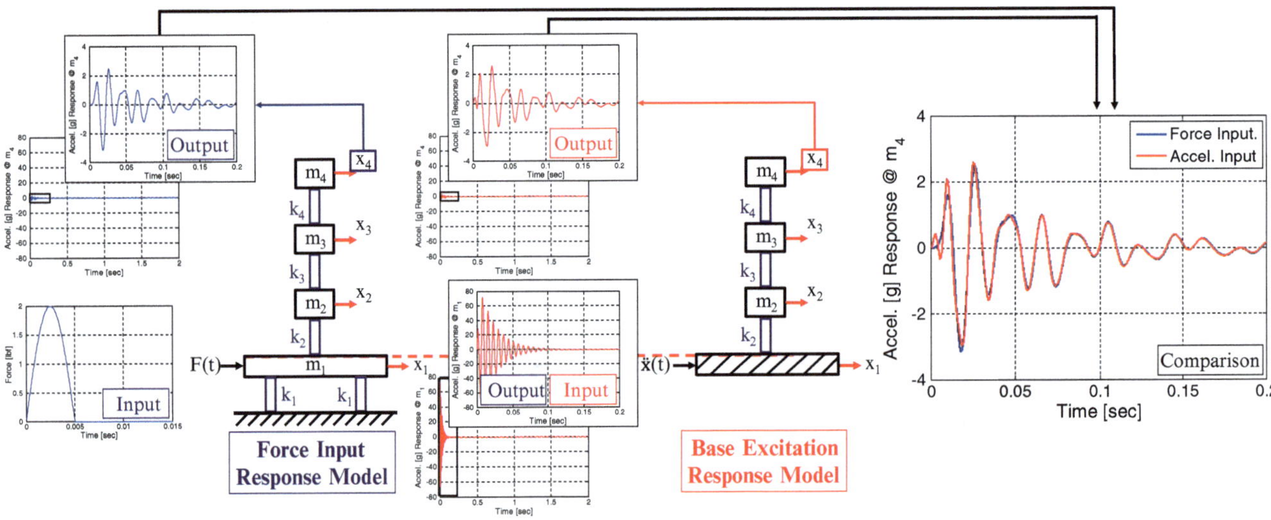

Fig. 8.12 Computation of output at m_1 and m_4 of FIR model (*left*) for case D due to the half-sine pulse with test article mounted. (*Dashed line*) Computation of output at m_4 of BER model (*middle*) using as input base acceleration the output at m_1 from FIR model. (*Dashed line*) Output response at m_4 from BER model is compared to output response at m_4 from the FIR model with test article (*right*)

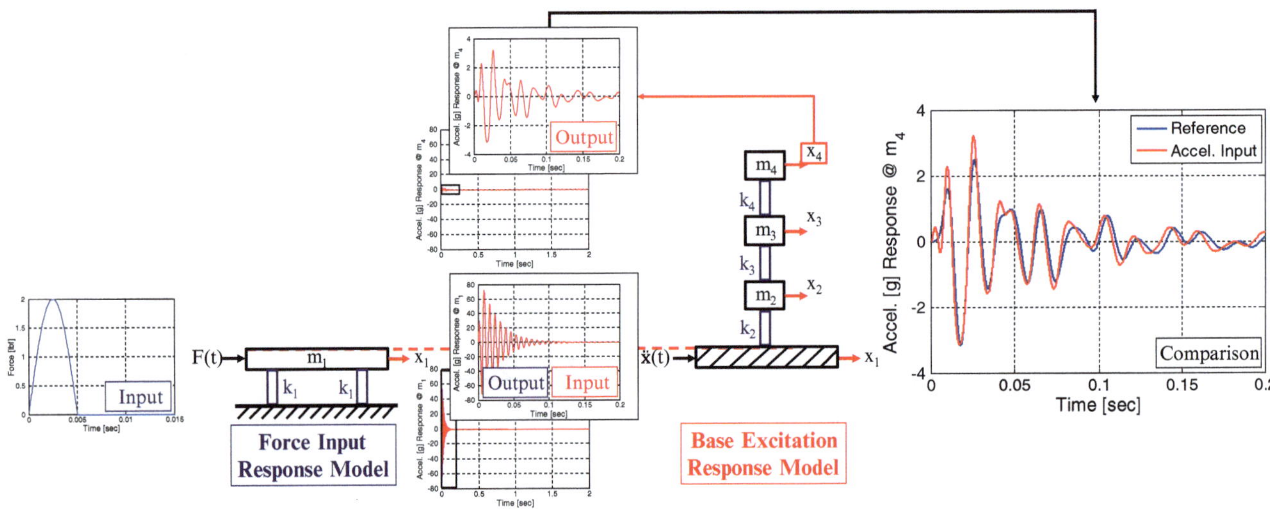

Fig. 8.13 Computation of output at m_1 and m_4 of FIR model (*left*) for case D due to the half-sine pulse without test article mounted. (*Dashed line*) Computation of output at m_4 of BER model (*middle*) using as input base acceleration the output at m_1 from FIR model. (*Dashed line*) Output response at m_4 from BER model is compared to output response at m_4 from the FIR model with test article (*right*)

8.4.2 *Final Remarks*

The previous results presented four cases where vibration testing leads the test article to a proper level of excitation and showed situations where additional attention is required due to the large discrepancies between responses. Within the studied cases, two conditions/circumstances can be observed which lead to the proper level of vibration in the structure under test. The first condition appears in the scenarios where the input base acceleration profile is acquired while the test article is mounted on the shock plate. For this condition, the captured acceleration at the shock plate due to the shock pulse, retained the information of the dynamic behavior of the test article while being coupled to the shock plate. This condition could be fulfilled if the structure under test can be subjected to any kind of excitation without consequences of damage or high risk of failure. The second condition leads to the proper excitation on the test article when the shock plate is much more massive than the test article. Also, simulations were run where stiffness changes were made, instead of increasing the mass, and showed the same behavior. The right level of vibration can be achieved if the shock plate is stiff enough in comparison to the test article. Finally, some situations revealed that, if the shock plate is not massive or stiff enough and the input base acceleration profile was obtained in the absence of the test article, the test would be meaningless. The response will not match.

For further explanation on how the dynamic characteristics of the shock fixture interacts with the dynamic behavior of the test article, Figs. 8.14 and 8.15 provide more information from a frequency domain perspective. In Fig. 8.14, the red curve belongs to the test article's FRF in a built-in condition. This condition is used because is the behavior that the test engineer wants to achieve while performing vibration testing. The black curves represents the resonance frequencies of two 1-DOF

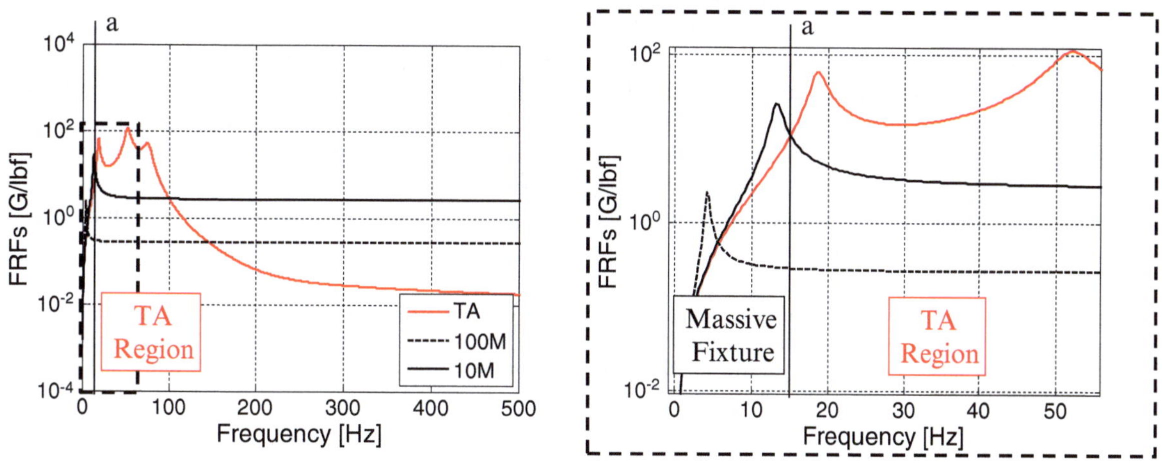

Fig. 8.14 Built-in test article and mass related variants

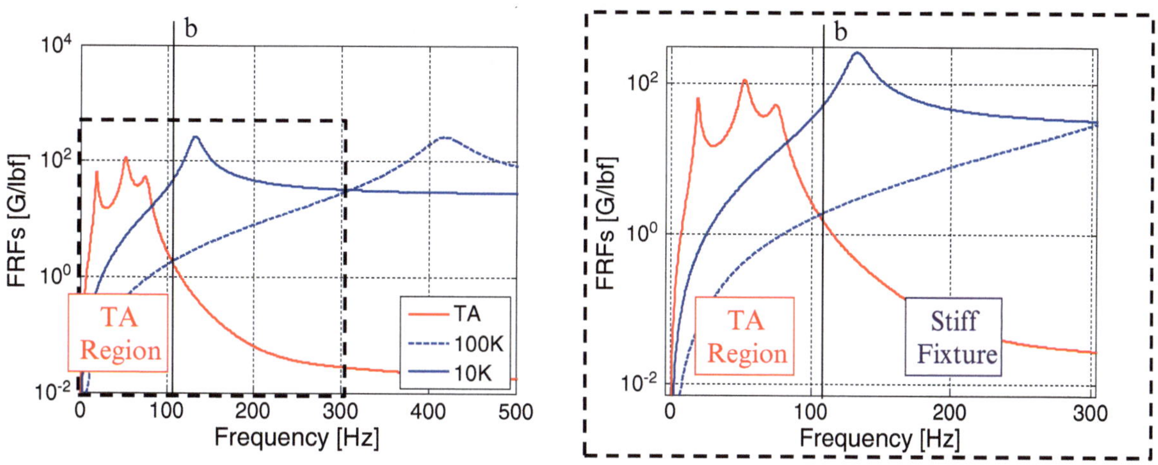

Fig. 8.15 Built-in test article and stiffness related variants

shock plate more massive than the test article. The resonant frequency of the more massive shock plate (black-dotted line) is not close (far removed) to the first resonant frequency of the TA (lower end of the test frequency range)—meaning that the shock plate resonant behavior may not significantly contribute to the dynamic response of the test article during the test. The responses for this scenario were presented in Case A.

Nonetheless, this is not true when the shock plate is not massive enough (Fig. 8.14, black solid line) and its resonance frequency is close to test article's resonant frequency. The dynamic behavior of the shock plate contributes to the response of the test article's first mode. Case B displayed the response of the overall system when the shock plate's resonant frequency almost aligns to the first resonant frequency of the test article.

In a similar fashion, two highly stiff shock plates were modeled and their FRFs are shown in Fig. 8.15 (blue curves). Both of these fixtures' resonant frequencies are very high in comparison to the test article's resonant frequencies. This means that the contribution of each shock plate to the total response of the overall system will not be as significant as if the shock plate resonant frequency were closer to the test article frequency range. In other words, the dynamic coupling between the test article and the shock plate may not influence the response of the system drastically.

8.5 Conclusion

Finite element models were developed to show the difference in response while using different shock fixtures during vibration testing—4-DOF representations simulated the behavior of a test article mounted on a shock plate (Force Input Response model) and a slip table (Base Excitation Response model). The discrepancies in the level of vibration of a generic test article were discussed for different shock fixtures. Two main differences between the cases were highlighted: running the test with and without the article mounted on shock fixtures with different physical characteristics.

The results show differences in response due to the dynamic interaction of the test article and shock fixture. Scenarios where the test article response is being affected by dynamic characteristics of stiff and massive shock fixtures were analyzed and discussed. The data indicates that information about the dynamic coupling between the test article and shock fixture is necessary when the shock fixture is not massive or stiff enough in comparison to the article under test. Current work is being developed to provide a method to modify the input base acceleration in the absence of the test article on the shock fixture. This work will be the subject of future papers.

Acknowledgements Any opinions, findings, and conclusions or recommendations expressed in this material are those of the authors and do not necessarily reflect the views of the particular funding agency. The authors are grateful for the support obtained. Also, the support given by La Secretaria de Educacion Publica, El Gobierno Mexicano, El Consejo Nacional de Ciencia y Tecnologia CONACYT and La Universidad Autonoma de Sinaloa to the first author is appreciated.

References

1. Alexander, J.E.: Shock response spectrum-a primer. Sound Vib. **43**, 6–15 (2009)
2. Avitabile, P.: Why can't you ignore those vibration fixtures. Sound Vib. **3**, 20–26 (1999)
3. McConnell, K.G.: From field vibration to laboratory simulation. Exp. Mech. **34**, 181–193 (1994)
4. Varoto, P.S.: The rules for the exchange and analysis of dynamic information in structural vibration. Ph.D. thesis, Iowa State University of Science and Technology, Ames, IA, (1996)
5. MATLAB – Matrix Analysis Software, The MathWorks, Inc., Natick, MA

Chapter 9
Modal Analyses and Experimental Verifications of Joined-Wing Configurations

Berkan Alanbay, Melin Şahin, and Güvenç Canbaloğlu

Abstract In recent days, wing kits are extensively used to convert conventional munitions into guided munitions. The wing kit integration enables munitions to gain standoff attack capability, extends their range and together with a laser guidance system, adds moving target tracking capability. Typical wing kits are usually composed of two main wings, however, some of them have a joined-wing configuration where the aft wings can support the main wings in order to increase maneuverability, range performance and structural performance. In this study, since the geometry and sweep angle of front wings are kept fixed due to aerodynamic effects in operating conditions, joined-wing configurations are investigated by changing two key parameters, namely; aft wing sweep angle and location of the joint. Response surface methodology is used in order to determine how these parameters affect the vibration characteristics of the joined-wing configurations. In numerical analyses, the natural frequencies and the corresponding mode shapes for the configurations are obtained by using ANSYS software. Then the results of selected configurations are verified via modal testing so as to achieve accurate finite element models of the joined-wing configurations.

Keywords Joined-Wing • Design of experiment • Response surface • Modal testing

9.1 Introduction

Understanding both the natural frequencies and the corresponding mode shapes and knowing their effects on the structure helps engineers to design better structures [1]. So, in this study in order to investigate dynamic characteristics of joined-wing configurations; first, sampling points are determined through design of experiments (DOE) methods. Secondly, modal analyses are performed for all of the sampling points and natural frequency information is stored. Thirdly, response surfaces are constructed by using kriging model. Then, the accuracy of the created responses is checked through an error criterion, if the requirement of the criterion is not satisfied, a refinement point is inserted into design space and the previous steps are repeated. This procedure continues until the error criterion is fulfilled. Finally, in order to verify response surface models, verification points are inserted and analyzed. The procedure is summarized in Fig. 9.1.

Thereafter, the finite element (FE) models of two selected configurations are validated through classical modal analyses techniques comprising both impact hammer and shaker tests. The selected points are defined as Design Point 1 (Joint location: 350 mm, Aft wing sweep angle: 0) and Design Point 2 (Joint location: 750 mm, Aft wing sweep angle: 20) throughout this paper. As an outcome of these tests, it is aimed to get the validated theoretical models.

B. Alanbay (✉) • G. Canbaloğlu
Middle East Technical University, 06800 Ankara, Turkey

MGEO Business Sector, ASELSAN Inc., 06750 Ankara, Turkey
e-mail: balanbay@aselsan.com.tr; gcanbal@aselsan.com.tr

M. Şahin
Middle East Technical University, 06800 Ankara, Turkey
e-mail: msahin@metu.edu.tr

A. Brandt, R. Singhal (eds.), *Shock & Vibration, Aircraft/Aerospace, Energy Harvesting, Acoustics & Optics, Volume 9*,
Conference Proceedings of the Society for Experimental Mechanics Series, DOI 10.1007/978-3-319-30087-0_9

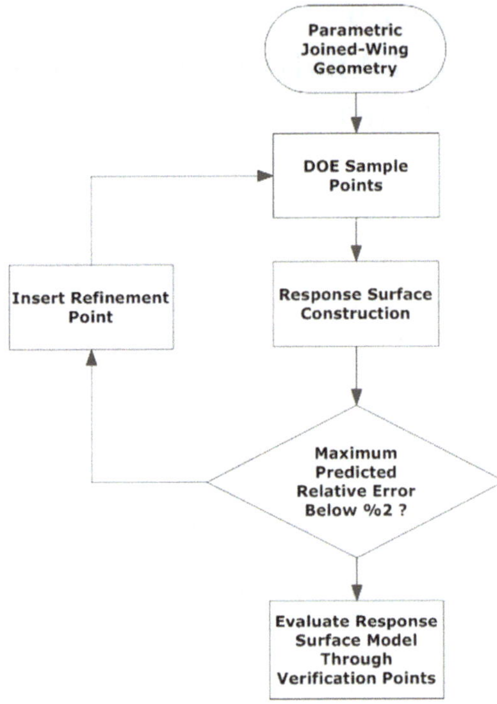

Fig. 9.1 Summary of the response surface generation

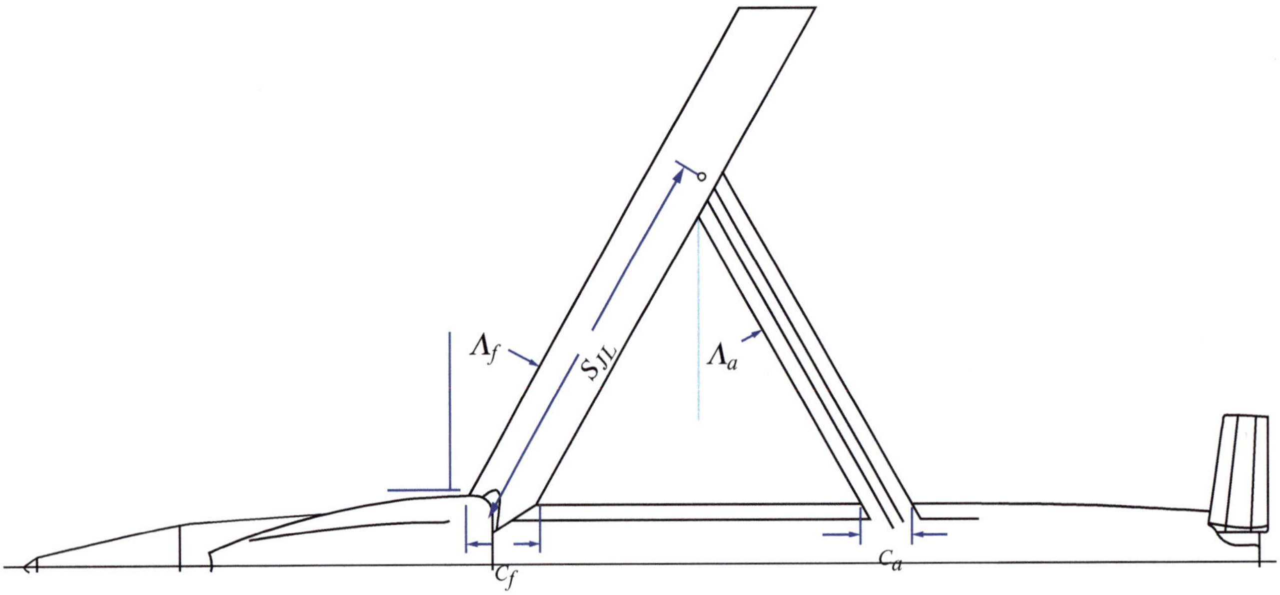

Fig. 9.2 Design variables of the joined-wing

9.2 The Joined Wing Geometry and Associated Variables

Each geometric configuration is defined by two key independent design variables; namely the aft wing sweep angle (Λ_a) and the joint location (S_{JL}). Figure 9.2 illustrates the typical joined-wing configuration used in this study and Table 9.1 summarizes the related geometric variables used in order to settle the range of configurations. In these configurations, the key parameters; the aft wing sweep angle (Λ_a) is defined as the angle the aft wing creates with fuselage and the joint location (S_{JL}) is defined as the parallel distance between axis of the intersection point where the front wing coincides with the aft wing and the intersection point where the front wing is connected to fuselage.

Table 9.1 Definition of design variables of the joined-wing

Variable	Name	Size
S_{JL}	Joint location	Varies
Λ_f	Front wing sweep	30°
Λ_a	Aft wing sweep	Varies
c_f	Chord at front wing	103.9 mm
c_a	Chord at aft wing	69.3 mm

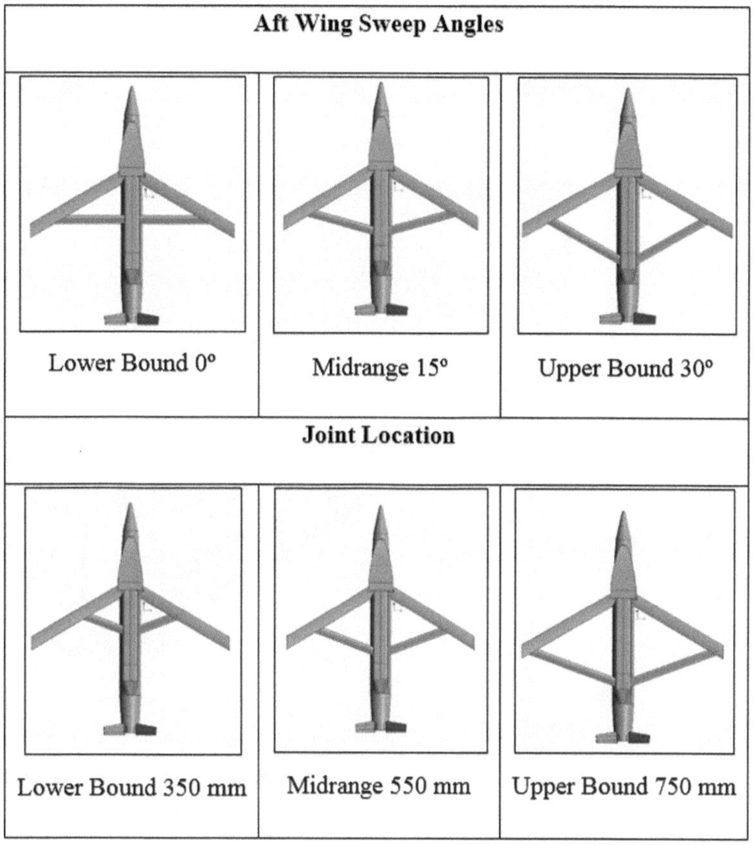

Fig. 9.3 Pictorial illustrations of the upper and lower bounds

In order to reduce the analysis domain within a conceivable scope, the "aft wing sweep angle" and the "joint location" are allowed to change from 0° to 30° and from 350 to 750 mm, respectively. Pictorial illustrations of upper and lower bounds for the parameters are shown in Fig. 9.3.

9.3 Modal Analyses of Joined-Wings

9.3.1 Design of Experiments

DOE can be defined as a scientific way to locate the sample design points in the space design [2]. The common characteristics of the DOE methods are to pinpoint these sampling points such that the relation between random input variables and the output variables is explored in most efficient way. In order to generate initial set of sampling points, a Latin Hypercube Sampling (LHS) [3] design is used to generate nine sample design points. Then modal analyses are performed for each of those sampling points. The sampling points and their modal analyses results are given in Table 9.2.

Table 9.2 DOE- LHS points

	Input design variables		Output design variables			
Design point #	Joint location (mm)	Aft wing sweep (degree)	1st Mode (Hz)	2nd Mode (Hz)	3rd Mode (Hz)	4th Mode (Hz)
1	505.56	1.67	22.84	22.97	86.94	87.24
2	638.89	3.89	21.23	21.35	94.79	95.26
3	372.22	0.56	21.42	21.52	83.63	83.91
4	594.44	5.00	21.96	22.09	94.39	94.99
5	416.67	8.33	22.15	22.26	81.56	81.78
6	727.78	2.78	19.43	19.54	84.50	84.72
7	550.00	9.44	22.41	22.53	89.53	90.33
8	683.33	7.22	20.27	20.37	90.39	90.76
9	461.11	6.11	22.69	22.81	82.97	83.25

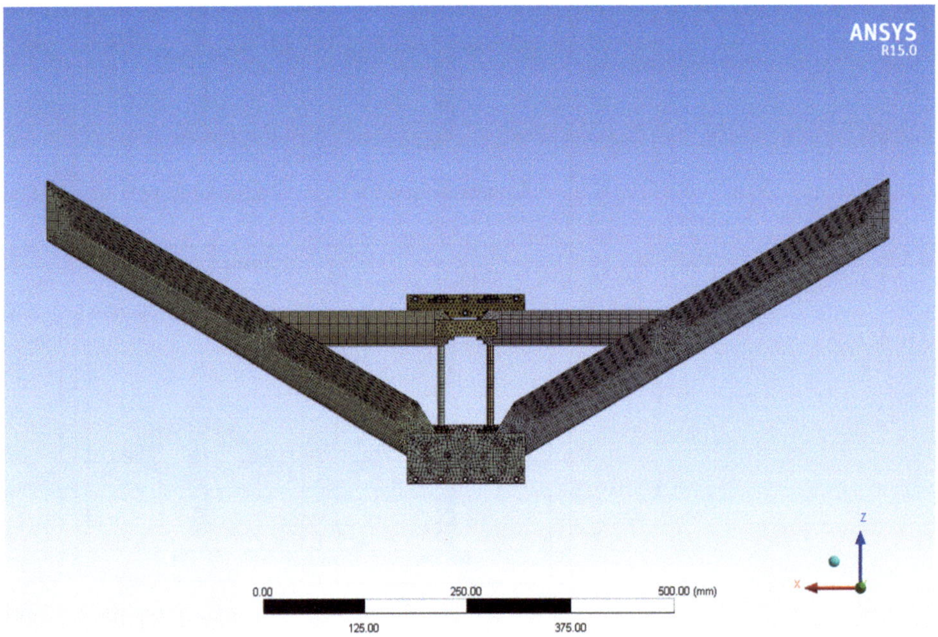

Fig. 9.4 Mesh of the sample joined-wing geometry

9.3.2 *FE Modeling of Joined-Wings*

In order to investigate dynamic characteristics of the joined-wing configurations, FE method is used. ANSYS® 15.0.7, commercial FE code [4], is used for modelling the selected joined-wing configurations. 3D parametric computer aided design (CAD) model is constructed in Creo 2.0, an advanced CAD program [5]. Then, ANSYS® and Creo are connected each other directly in such a way that the geometric parameters are changed in CAD model, FE model is automatically updated.

In the wings, aluminum 6061-T6 alloy and in the guidance shafts, steel is used as a material. The properties of these materials are defined in the analysis from the material library of ANSYS®. Twenty node hexagonal (Solid 186) and ten node tetrahedron (Solid 187) higher order 3D solid elements are utilized. Average number of elements and nodes used in the FE modelling of joined-wings are 275231 and 462653, respectively. All body to body contacts such as connection between front wing and aft wing is defined as bonded and in this contact type, two bodies are assumed to be connected to each other rigidly. In addition to the body to body contacts, in order to simulate the connection interface between the munition and the joined-wing assembly, fixed boundary condition is designated to the mounting interface of the assembly. Typical mesh for the joined-wings is shown in Fig. 9.4.

9.3.3 Modal Analyses of the Selected Joined-Wing Configurations

After constructing FE model, modal analyses are performed for the first four modes of the joined-wings. The first four natural frequencies and the corresponding mode shapes are then found for both the Design Point 1 and the Design Point 2. For the Design Point 1 and 2, the natural frequencies are listed in Tables 9.3 and 9.4 respectively and the corresponding mode shapes are shown in Figs. 9.5 and 9.6, respectively.

Table 9.3 Natural frequencies of the joined-wing for Design Point 1 and their descriptions

Mode	Frequency (Hz)	Description of the free vibration modes
1	20.89	Front wing first anti-symmetric out-of-plane bending, Aft wing first anti-symmetric out-of-plane bending
2	20.98	Front wing first symmetric out-of-plane bending, Aft wing first symmetric out-of-plane bending
3	84.84	Front wing second anti-symmetric out-of-plane bending, Aft wing mixed
4	85.19	Front wing second symmetric out-of-plane bending, Aft wing mixed

Table 9.4 Natural frequencies of the joined-wing for Design Point 2 and their descriptions

Mode	Frequency (Hz)	Description of the free vibration modes
1	17.99	Front wing first symmetric out-of-plane bending, Aft wing first symmetric out-of-plane bending
2	18.00	Front wing first anti-symmetric out-of-plane bending, Aft wing first anti-symmetric out-of-plane bending
3	78.86	Front wing second anti-symmetric out-of-plane bending, Aft wing mixed
4	78.89	Front wing second symmetric out-of-plane bending, Aft wing mixed

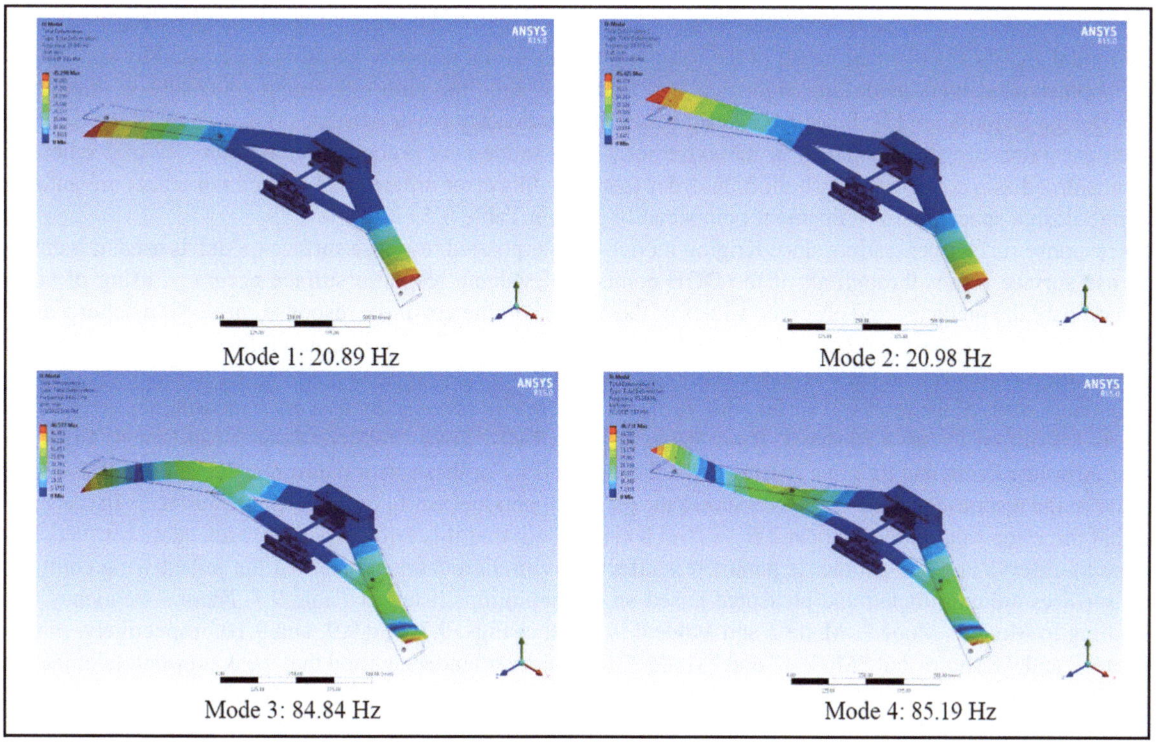

Fig. 9.5 Corresponding mode shapes of the of joined-wing for Design Point 1

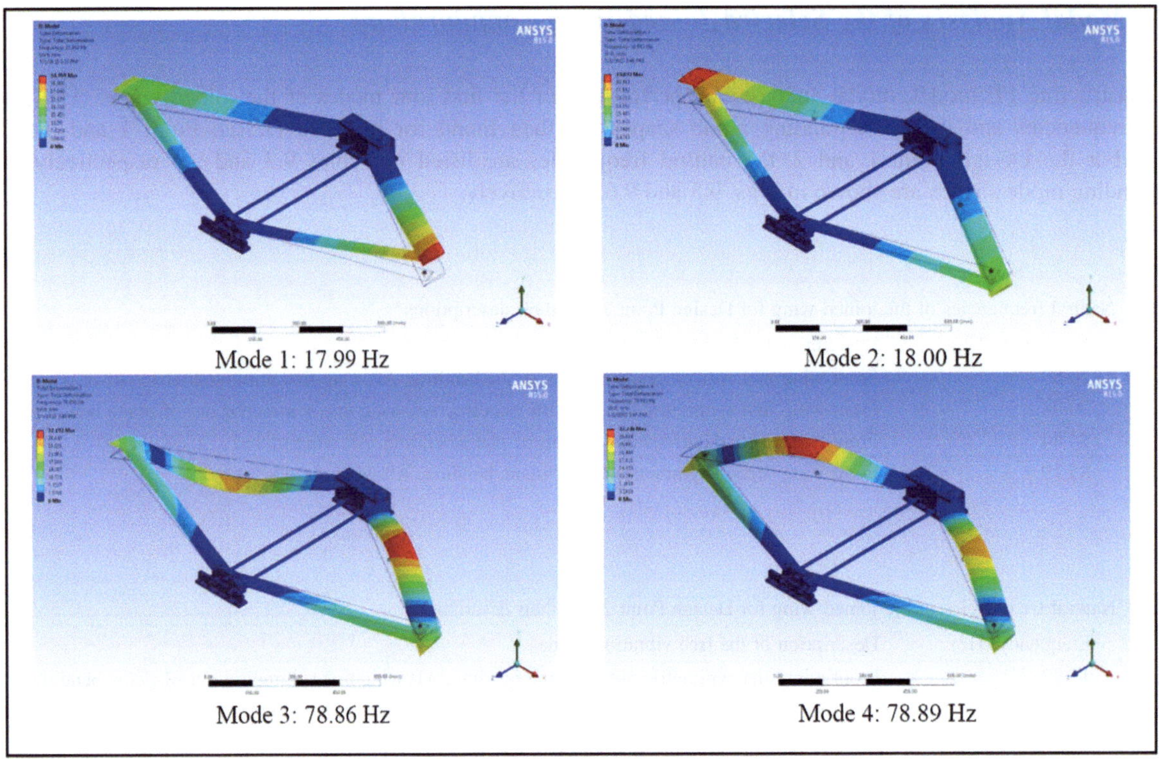

Fig. 9.6 Corresponding mode shapes of the of joined-wing for Design Point 2

9.3.4 *Response Surface Models*

After the modal analyses performed for all of the sampling points, initial response surfaces are created by applying Kriging model, a sophisticated meta-modelling algorithm [6]. Following this, the auto-refinement procedure is implemented by ANSYS® Design Exploration [7]. Since LHS design does not necessarily cover extremes (i.e. corners of the design space), the refinement points are chosen mainly at the extremes. In the refinement procedure, maximum %2 of predicted related error is considered as a convergence criterion. In order to satisfy this error criterion 23 refinement points are automatically inserted into design space. These refinement points can be seen in Table 9.5.

In this response surface generation, since Kriging model, an interpolated response surface model, is used, it is ensured that the response surface passes through all of the DOE points. To evaluate response surface accuracy, using of verification points to compare the predicted and observed values of the output parameters of the response surface is a better way. For this aim, eight verification points are placed where the distance from existing DOE points and refinement points are maximum. These verification points can be seen in Table 9.6.

The goodness of fit of the response surface for verification points is assessed by five error measures in ANSYS® Design Explorer as Maximum Relative Residual, Root Mean Square (RMS) Error, Relative Root Mean Square Error, Relative Maximum Absolute Error, and Relative Average Absolute Error [7]. For these error measures, the value of 0 % indicates the best quality of the response surface. Table 9.7 shows the results of goodness of fit of the response surfaces. In this case, it can be seen that the error values are less than 2.5 %. And it can be said that this error is perfectly adequate for this case.

In order to observe how the geometric parameters affect the vibration characteristics of the joined-wing configurations, response surfaces are constructed and presented based on the definitions listed in Table 9.7. Natural frequency responses corresponding to Mode 1, Mode 2, Mode 3 and Mode 4 is given in Figs. 9.7, 9.8, 9.9, and 9.10, respectively.

Figures 9.7 and 9.8 shows that "Mode 1" and "Mode 2" have similar tendencies and they peak up when joint location is in between 450 mm and 550 mm. If the influence of the aft wing sweep angle is considered, it can be seen that it has almost no effect on the natural frequencies.

As it can be seen from Figs. 9.9 and 9.10, Mode 3 and Mode 4 have similar tendencies; however, their characteristics are completely different from Mode 1 and Mode 2. From these figures, it is observed that joint location has highest influence on Mode 3 and Mode 4. The frequency values are at its absolute minimum and maximum value when the joint location is in

Table 9.5 Refinement design points

	Input design variables		Output design variables			
Design point #	Joint location (mm)	Aft wing sweep (degree)	1st Mode (Hz)	2nd Mode (Hz)	3rd Mode (Hz)	4th Mode (Hz)
1	750.00	0.00	18.95	19.06	81.22	81.41
2	750.00	10.00	18.74	18.81	80.12	80.29
3	350.00	10.00	20.71	20.80	83.69	83.88
4	432.71	0.00	22.51	22.63	83.03	83.14
5	595.34	0.00	22.00	22.13	95.40	95.79
6	462.07	10.00	22.61	22.73	82.06	82.55
7	622.05	10.00	21.34	21.45	94.09	94.84
8	695.93	0.00	20.10	20.22	89.61	89.87
9	549.45	0.00	22.54	22.68	91.65	91.96
10	350.00	0.00	20.87	20.96	84.60	84.96
11	509.40	10.00	22.68	22.80	85.32	86.01
12	715.99	10.00	19.47	19.55	85.50	85.76
13	350.00	5.00	20.80	20.89	84.36	84.63
14	383.13	10.00	21.50	21.60	81.99	82.18
15	750.00	6.27	18.89	18.98	80.56	80.74
16	534.56	5.08	22.65	22.79	89.06	89.58
17	413.60	3.78	22.19	22.30	82.47	82.51
18	646.49	0.00	21.09	21.21	94.96	95.30
19	480.69	0.00	22.85	22.98	85.17	85.36
20	658.80	10.00	20.66	20.75	92.53	93.07
21	576.94	2.47	22.22	22.37	93.75	94.22
22	391.17	0.00	21.83	21.93	83.04	83.28
23	350.00	7.88	20.75	20.84	84.03	84.21

Table 9.6 Verification points

	Input design variables		Output design variables			
Design point #	Joint location (mm)	Aft wing sweep (degree)	1st Mode (Hz)	2nd Mode (Hz)	3rd Mode (Hz)	4th Mode (Hz)
1	505.56	1.67	22.84	22.97	86.94	87.24
2	638.89	3.89	21.23	21.35	94.79	95.26
3	372.22	0.56	21.42	21.52	83.63	83.91
4	594.44	5.00	21.96	22.09	94.39	94.99
5	416.67	8.33	22.15	22.26	81.56	81.78
6	727.78	2.78	19.43	19.54	84.50	84.72
7	550.00	9.44	22.41	22.53	89.53	90.33
8	683.33	7.22	20.27	20.37	90.39	90.76

Table 9.7 Goodness of fit for verification points

	1st Mode	2nd Mode	3rd Mode	4th Mode
Maximum relative residual (Best value = 0 %)	0.13	0.14	0.11	0.12
Root mean square error (Best value = 0)	0.01	0.02	0.04	0.04
Relative root mean square error (Best value = 0 %)	0.06	0.07	0.05	0.04
Relative maximum absolute error (Best value = 0 %)	2.31	2.39	1.93	2.03
Relative average absolute error (Best value = 0 %)	0.82	0.97	0.68	0.43

between 350–450 mm and 550–650 mm, respectively. On the other hand, the aft wing sweep angle appeared to have a minimal influence on the natural frequencies when compared to the influence of joint location.

The local sensitivity graph for the mode shape response surfaces of the joined-wings is given in Fig. 9.11. As observed from Fig. 9.11, the aft wing sweep angle does not have any influence on the Mode 1 and Mode 2. However, in the Mode 3 and

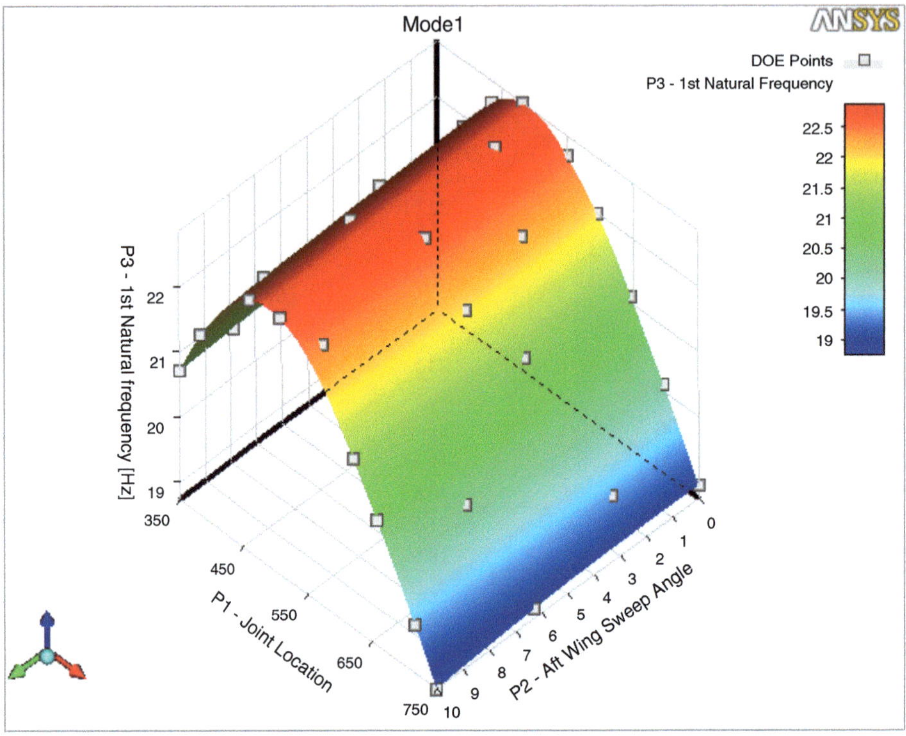

Fig. 9.7 Natural frequency responses corresponding to Mode 1 as a function of joint location and aft wing sweep angle

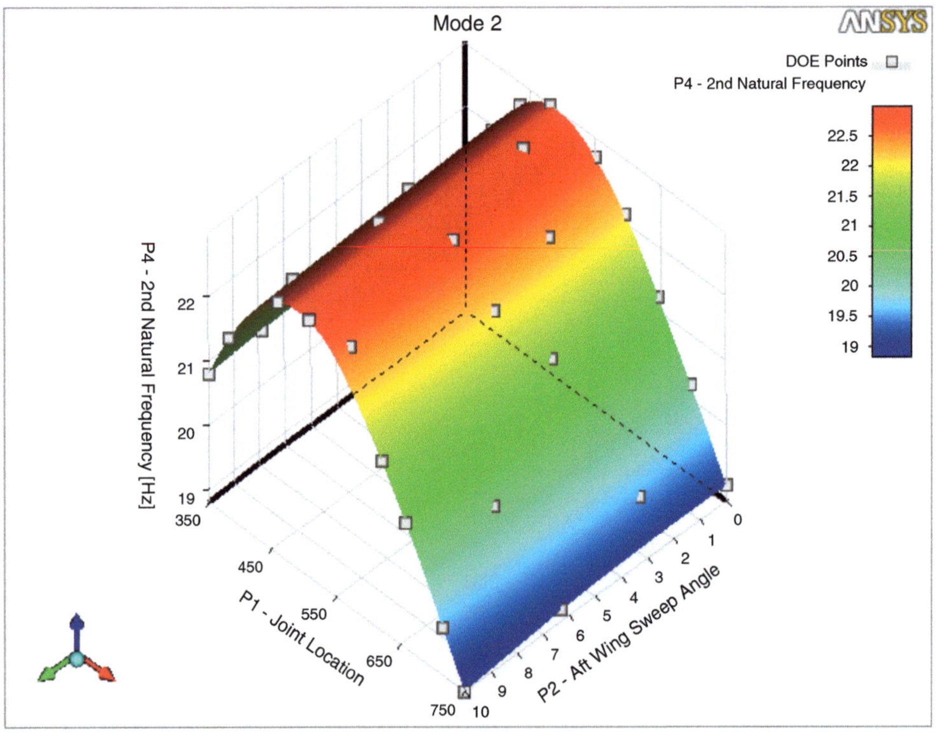

Fig. 9.8 Natural frequency responses corresponding to Mode 2 as a function of joint location and aft wing sweep angle

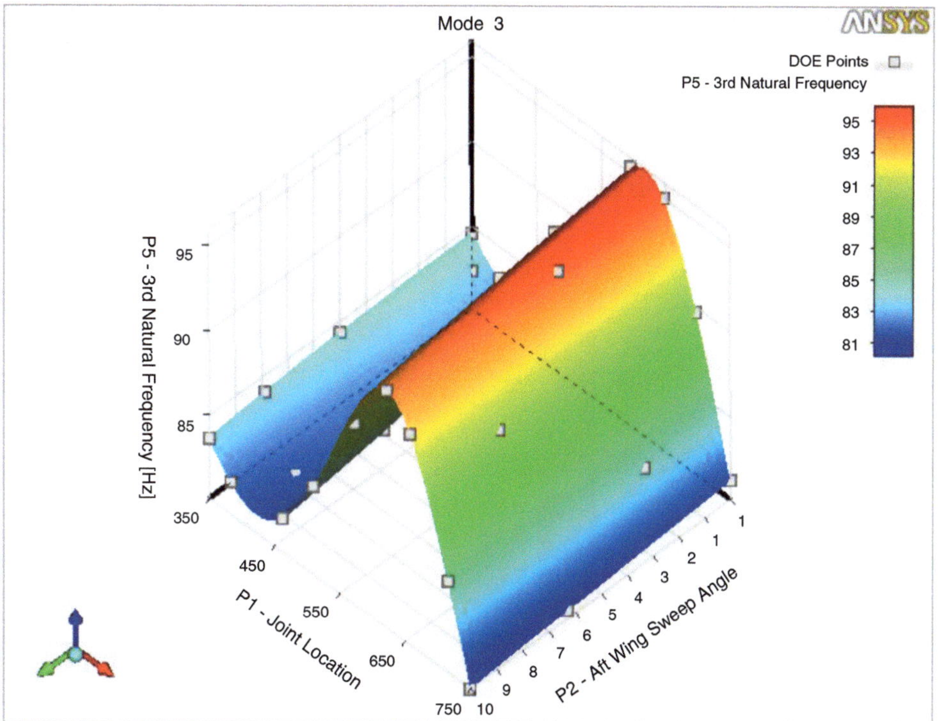

Fig. 9.9 Natural frequency responses corresponding to Mode 3 as a function of joint location and aft wing sweep angle

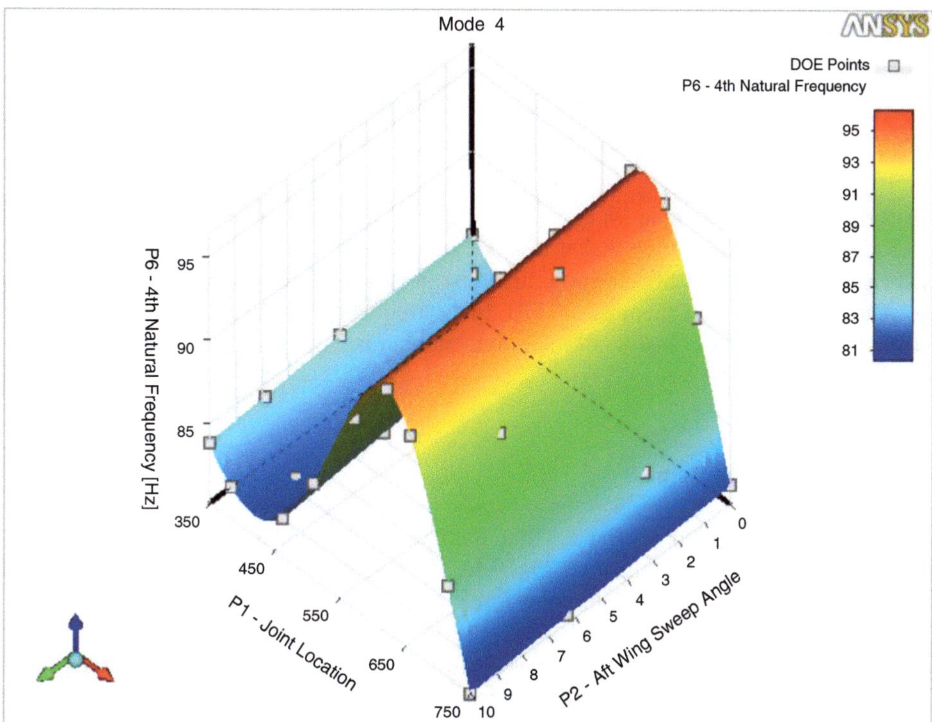

Fig. 9.10 Natural frequency responses corresponding to Mode 4 as a function of joint location and aft wing sweep angle

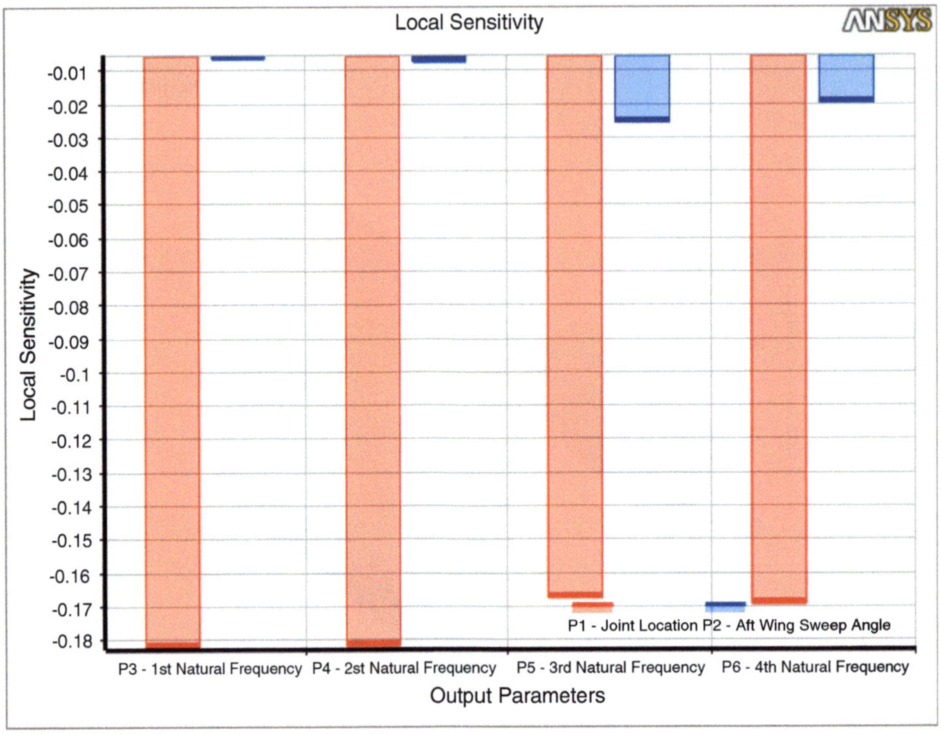

Fig. 9.11 Local sensitivity graph for the mode shape response surfaces of the joined-wings

Mode 4, the sensitivity intensity of the aft wing sweep angle increases. Although, the aft wing sweep angle is getting more effective when compared to its effectiveness in the first two mode, it can be easily seen that the joint location has the highest influenced intensity on all the modes.

9.3.5 Experimental Modal Analyses of the Selected Joined-Wing Configurations

In this part, classical modal analyses techniques are applied on two different test configurations. For both configurations, first, impact hammer tests are performed then shaker tests are conducted. Modal Test Consultant of PULSE Labshop 16.0 software [8] is used for all tests. In PULSE Labshop, first; test geometries are created, then measurement and excitation points are defined. After analysis setup (number of FFT lines, frequency span, and measurement lengths) is arranged, vibration measurements are completed. Finally, all measurement data is exported to the ME'scope software [9] for modal analysis.

9.3.5.1 Shaker Test for the Design Point 1

Design Point 1 configuration of the joined wing attached to the fixture with modal shaker is shown in Fig. 9.12. B&K 4524 triaxial accelerometer is roved to 101 points shown in Fig. 9.13.The joined-wing is excited by using a modal shaker which is attached to node 73. As an excitation type random noise force input is applied by the shaker.

After the shaker test is completed in the Pulse platform, the measured data is exported to ME'scope software for modal analysis. In the ME'scope, modal parameter estimation is performed to predict the modal parameters of the structure from set of frequency response functions (FRFs). In the calculation and plotting of receptance FRFs, a MATLAB script is used. Frequency response curves for 101 different measurements are displayed in Fig. 9.14. Mode shapes obtained experimentally and their definitions are given in Fig. 9.15 and Table 9.8, respectively. The comparison of the natural frequencies obtained from FEA and the impact hammer test are presented in Table 9.9. The results corresponding to the first four modes are in good agreement with small deviations.

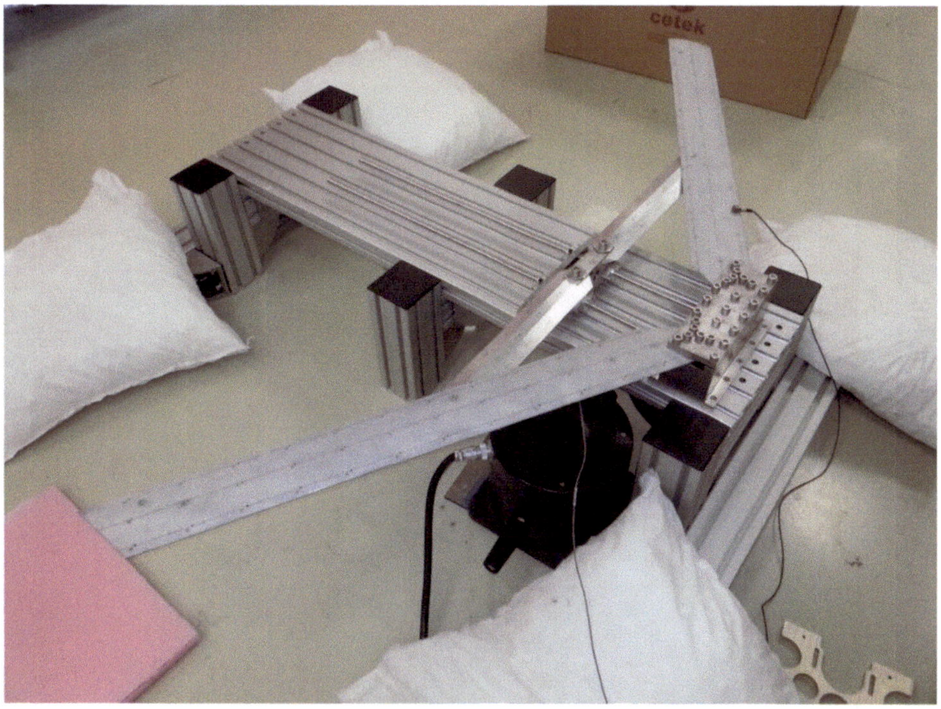

Fig. 9.12 Test setup for shaker test of the Design Point 1

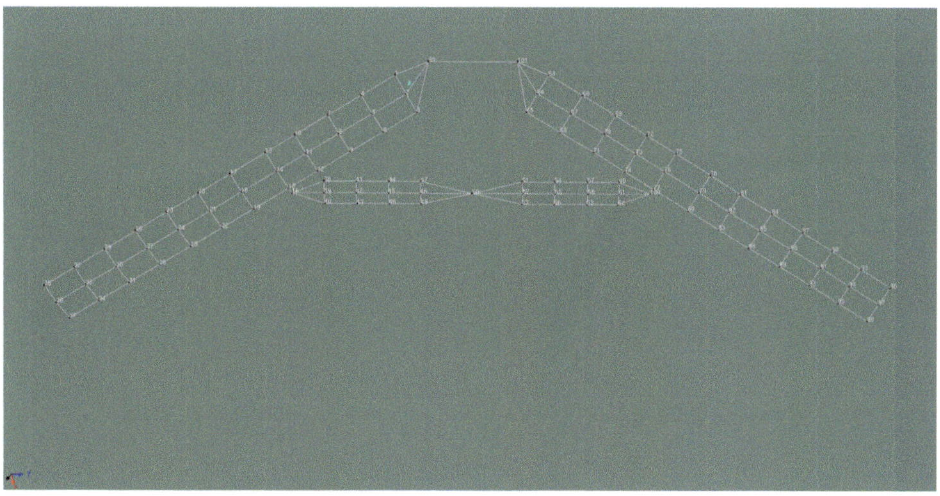

Fig. 9.13 Measurement points of the Design Point 1

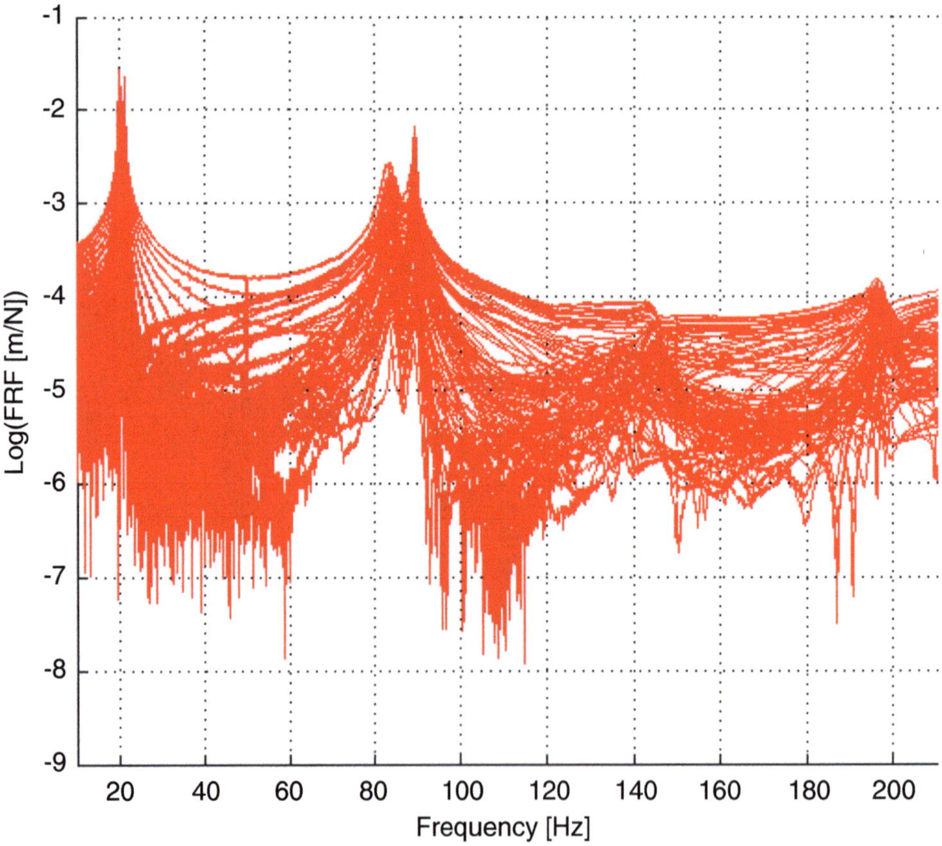

Fig. 9.14 FRF plots of the shaker test for the Design Point 1

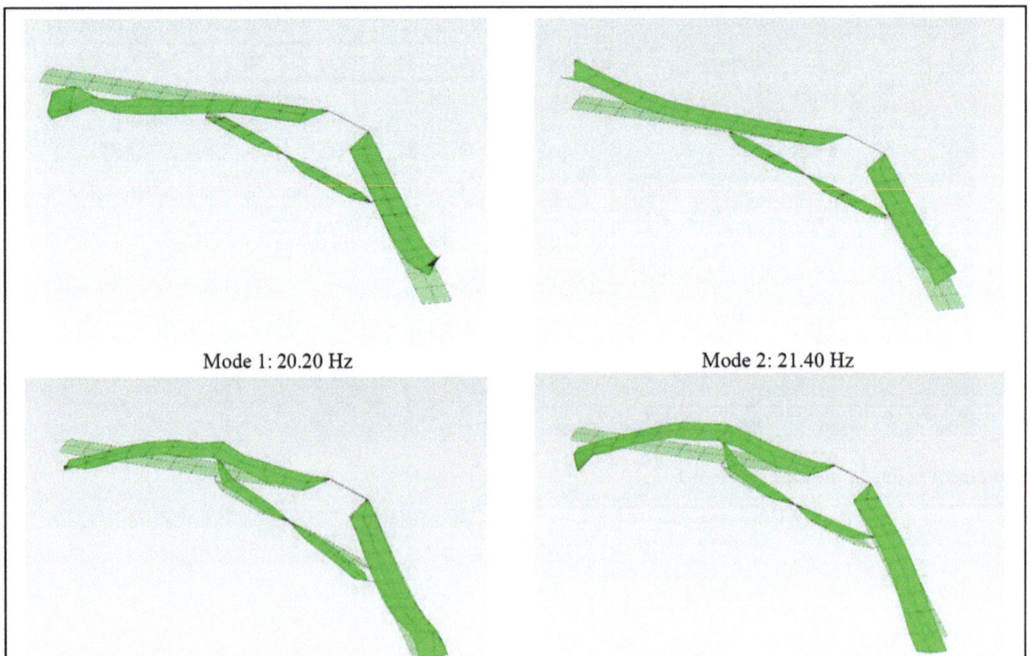

Fig. 9.15 The first four mode shapes of the Design Point 1 from shaker test

Table 9.8 First four vibration modes of the Design Point 1 from shaker test

Mode	Frequency (Hz)	Description of the free vibration modes
1	20.20	Front wing first anti-symmetric out-of-plane bending, Aft wing first anti-symmetric out-of-plane bending
2	21.40	Front wing first symmetric out-of-plane bending, Aft wing first symmetric out-of-plane bending
3	83.40	Front wing second anti-symmetric out-of-plane bending, Aft wing mixed
4	89.40	Front wing second symmetric out-of-plane bending, Aft wing mixed

Table 9.9 Comparison of FEA and shaker test results for the Design Point 1

	Frequencies (Hz)		
Mode	Finite element analysis	Shaker test	Difference (%)
1	20.89	20.20	−3.31
2	20.98	21.40	+2.01
3	84.84	83.40	−1.70
4	85.19	89.40	+4.94

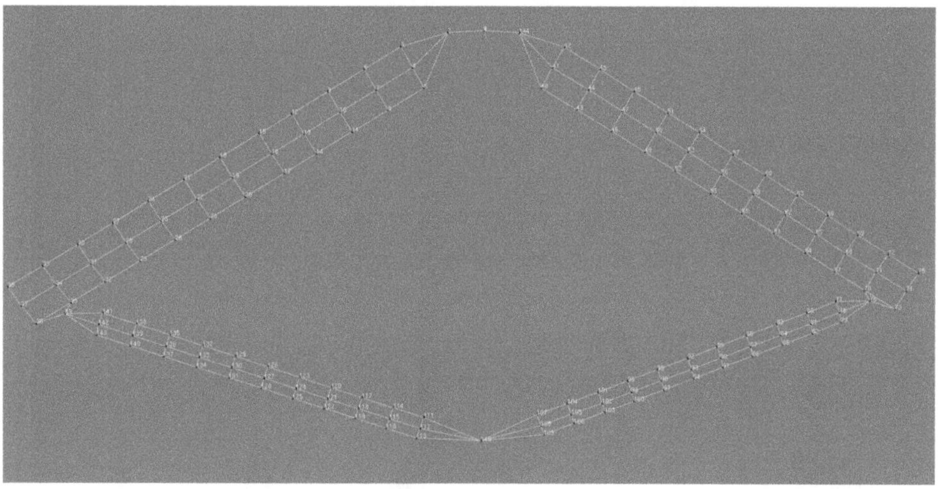

Fig. 9.16 Measurement points of the Design Point 2

9.3.5.2 Shaker Test for the Design Point 2

Test procedure for the Design Point 2 is similar to the previous test. Figure 9.16 shows the measurement points for Design Point 2. Figure 9.17 shows the joined-wing assembly attached to the test fixture and the modal shaker. In this case, the model shaker is attached to the node 47 and the accelerometer is roved to 144 points. Frequency response curves for 144 different measurements are shown in Fig. 9.18. Mode shapes obtained experimentally and their definitions are given in Fig. 9.19 and Table 9.10, respectively. Finally, the comparison of the natural frequencies obtained from FEA and the impact hammer test are presented in Table 9.11. The comparison results show that the FEA and test outcomes are in close agreement.

Fig. 9.17 Test setup for shaker test of the Design Point 2

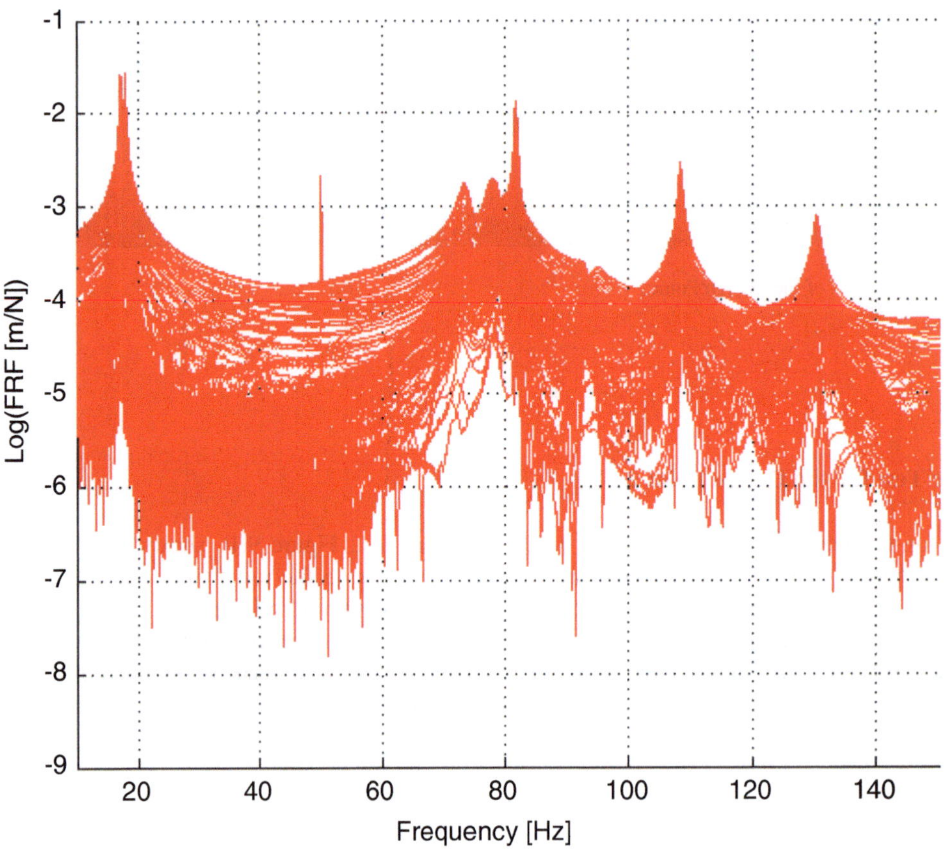

Fig. 9.18 FRF plots of the shaker test for the Design Point 2

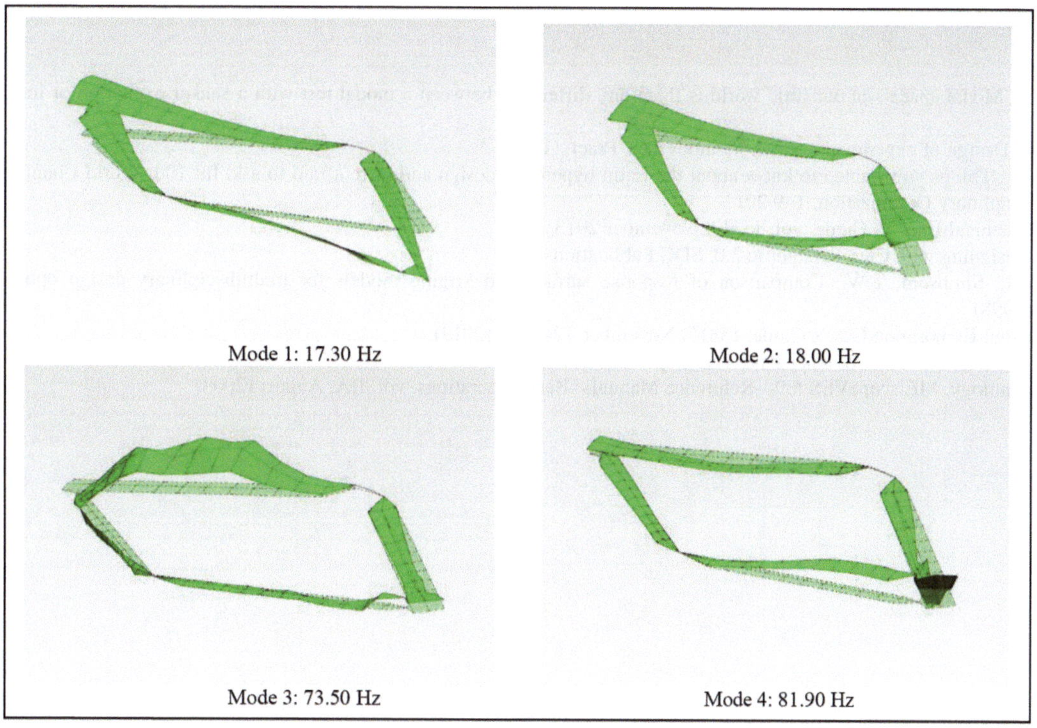

Fig. 9.19 The first four mode shapes of the Design Point 2 from shaker test

Table 9.10 First four vibration modes of the Design Point 2 from shaker test

Mode	Frequency (Hz)	Description of the free vibration modes
1	17.30	Front wing first anti-symmetric out-of-plane bending, Aft wing first anti-symmetric out-of-plane bending
2	18.00	Front wing first symmetric out-of-plane bending, Aft wing first symmetric out-of-plane bending
3	73.50	Front wing second anti-symmetric out-of-plane bending, Aft wing mixed
4	81.90	Front wing second symmetric out-of-plane bending, Aft wing mixed

Table 9.11 Comparison of FEA and shaker test results for the Design Point 2

	Frequencies (Hz)		
Mode	Finite element analysis	Shaker test	Difference (%)
1	18.00	17.30	−3.89
2	17.99	18.00	+0.06
3	78.86	73.50	−6.80
4	78.89	81.90	+3.82

9.4 Discussion and Conclusion

In this study, dynamic characteristic of the joined-wing configurations which are defined by two geometric parameters, namely; the joint location and the aft wing sweep angle are investigated through series of FE modelling and analyses. In order to understand how these geometric parameters affect the vibration characteristics of the joined-wing configurations, response surface methodology is used. It is seen that the joint location has the highest influenced intensity. Then, the results of selected configurations, Design Point 1 and Design Point 2, are verified via classical modal testing. The comparison studies show that the FE analyses results and test results have a good agreement. Finally, a high fidelity dynamical FE models are constructed for the design points which will be used for the further dynamic analysis.

References

1. Avitabile, P.: Modal space—ɪn our little world is there any difference between a modal test with a shaker excitation or impact excitation?. (1998)
2. Fisher, R.A.: Design of experiments. Anal. Methods Des. Pract. (1935)
3. Viana, F.A.C.: Things you wanted to know about the Latin hypercube design and were afraid to ask. In: 10th World Congress on Structural and Multidisciplinary Optimization, 1–9 2013
4. ANSYS Mechanical User ' s Guide, vol. 15317, November 2013
5. Rider, M.: Designing with Creo Parametric 2.0. SDC Publications
6. Mauery, T.M., Simpsom, T.W.: Comparison of response surface and kriging models for multidisciplinary design optimization. Mech. Eng. 1–11 (1998)
7. ANSYS. Design Exploration User's Guide. 15317, November 724–746 (2013)
8. Forbes, G., Caresta, M.: Brüel & Kjær Pulse Labshop Primer. (2009)
9. Vibrant Technology. ME'scopeVES 5.0—Reference Manual—Basic Operations. vol. IIA, August (2008)

Chapter 10
Modal Testing of James Webb Space Telescope (JWST) Optical Telescope Element (OTE)

Douglas J. Osterholt, David Cloutier, Timothy Marinone, and Reem Hejal

Abstract A modal survey was conducted on the James Webb Space Telescope optical telescope element (OTE) to obtain dynamic characterization of the system to be used in finite element model validation and loads assessments. Modal testing was performed by ATA Engineering, Inc., (ATA) and Northrop Grumman Systems Corporation (NGSC) for two configurations (stowed and deployed), which are representative of the launch and on-orbit orientations, respectively. The modal test of the OTE is essential in order to correlate and update the finite element model (FEM) that will be used for predicting launch loads and on-orbit performance, including wave-front error and line-of-sight calculation for optimal performance. Fixed base testing was conducted for both configurations. Pretest analysis activities were performed by NGSC, and testing was performed by ATA. Unanticipated flexibility in the stowed configuration fixture necessitated additional testing and analysis to validate the model boundary condition. The deployed configuration testing was successfully performed with 188 modes extracted up to 100 Hz. This paper provides details on the overall test effort and insight into the performance of a modal test on complex spacecraft flight hardware with high modal density.

Keywords Modal survey • Mode enhancement • Fixed base • Spacecraft

Abbreviations

ATA	ATA Engineering, Inc.
CMIF	Complex mode-indicator function
DOF	Degree of freedom
FEM	Finite element model
FRF	Frequency response functions
Hz	Hertz
IMAT	Interface between MATLAB, analysis, and test
JWST	James Webb space telescope
MAC	Modal assurance criteria
MMIF	Multivariate mode-indicator function
NGSC	Northrop Grumman Systems Corporation
OSS	OTE support structure
OTE	Optical telescope element
PSMIF	Power spectrum mode-indicator function
RMS	Root mean square
TAM	Test analysis model
TDM	Test display model
TEDS	Transducer electronic data sheet

D.J. Osterholt (✉) • D. Cloutier • T. Marinone
ATA Engineering Inc., 13290 Evening Creek Drive South, Suite 250, San Diego, CA 92128, USA
e-mail: doug.osterholt@ata-e.com

R. Hejal
Northrop Grumman Systems Corporation, Aerospace Systems, One Space Park, Redondo Beach, CA 90278, USA

A. Brandt, R. Singhal (eds.), *Shock & Vibration, Aircraft/Aerospace, Energy Harvesting, Acoustics & Optics, Volume 9*, Conference Proceedings of the Society for Experimental Mechanics Series, DOI 10.1007/978-3-319-30087-0_10

10.1 Introduction

Spacecraft models are becoming larger and more complex every year. Performing a modal test on the full assembly can be challenging and require significant resources. To reduce risk and cost, detailed modal tests of subassemblies are being performed to improve high-fidelity models that can be used with confidence in the final assembly. The challenge, in this case, was to accurately measure a modally dense structure up to higher frequencies than similar modal test programs in the past.

NASA is developing a next-generation space telescope, the James Webb Space Telescope (JWST), and Northrop Grumman Systems Corporation (NGSC) is building the optical telescoping element (OTE). As part of the development, a FEM was built for loads assessment and optical performance, or line of sight. A modal survey was performed to obtain the dynamic characteristics of the OTE and will be used to correlate and update the FEM. ATA Engineering (ATA) performed the modal survey of the JWST OTE in two fixed base configurations: stowed and deployed. These two configurations represent the launch and on-orbit orientations of the JWST. This structure has many modes in the frequency band of interest, which provides a unique challenge in test planning, performance, and data analysis.

The JWST OTE, shown in Fig. 10.1 in the stowed configuration and in Fig. 10.2 in the deployed configuration, has dimensions of 3 m × 4 m × 8 m when stowed and dimensions of 10 m × 6 m × 10 m when deployed. The deployed configuration was mounted to the OTE support structure (OSS), which consists of a large frame structure and the struts used to connect to the OTE. The deployed configuration also utilized tuned mass dampers (TMDs) mounted to the three main struts. These small TMDs were mounted to minimize the lateral motion of the struts in the 6 Hz frequency range; 9.5 Hz TMDs were also evaluated but are not discussed in this paper.

The objectives of the modal survey were to measure transfer functions up to 250 Hz and identify all primary structural modes below 50 Hz in the stowed configuration and significant modes below 100 Hz in the deployed configuration.

Fig. 10.1 JWST OTE stowed configuration

Fig. 10.2 JWST OTE deployed configuration

This majority of the modal analysis focused on the deployed configuration; however, both the deployed and stowed configurations are discussed in this paper. For the deployed configuration, the analysis predicted a total of 349 modes below 100 Hz. Many of these modes were local modes that are not of interest. The final test results were 188 structural modes below 100 Hz. The stowed configuration presented its own challenge in that the fixture did not provide an ideal boundary condition; one direction was less stiff than predicted, so static testing was performed to verify and update the model boundary condition.

10.2 Pretest Analysis

Northrop Grumman performed a pretest analysis to identify the required accelerometer locations and directions. The final set of measurement locations for the stowed configuration totaled 341 fixed degrees of freedom (DOF) at 145 node locations. The final set of measurement locations for the deployed configuration totaled 354 fixed DOF at 151 node locations. The test measurement DOF were used to make a test display model (TDM). This model, used for visualizing the mode shapes, consists of nodes and "dummy" elements. The TDM is shown for the stowed configuration in Fig. 10.3 and for the deployed configuration in Fig. 10.4, where each arrow represents a DOF or accelerometer. A large number of channels represent the primary mirror bays. These measurements are important for on-orbit performance evaluation of the telescope.

The pseudo-orthogonality for the deployed configuration showed that the test measurements do a good job of describing the mode shapes up to approximately 50 Hz. Above 50 Hz the modes were identified but with the limited number of accelerometers, unique mode shapes could not be described. Once the pretest analysis was finalized, the test analysis model (TAM) was computed using Guyan reduction. The TAM consists of the mass, stiffness, and constraint matrices. These were used to perform cross-orthogonality and back-expansion during the test. The back-expansion uses the TAM constraint matrix to interpolate missing DOF for purposes of visualization only.

10.3 Test Performance

The first challenge with setup of this test was cleaning all the equipment and cables to accommodate space flight hardware in a 10 k clean room. Every cable had to be scrubbed by hand. Installation required using superglue and two layers of Kapton tape for attachment of the accelerometers and blocks. Barcodes and TEDS technology were used to map the channels to the finite element model node numbers [1]; using the FEM nodes as the test measurement locations removes any steps to map from test to analysis to perform cross-orthogonality and cross-modal assurance criteria (cross-MAC) calculations. The accelerometer installation was performed by ATA personnel in parallel with NGSC personnel performing final hardware

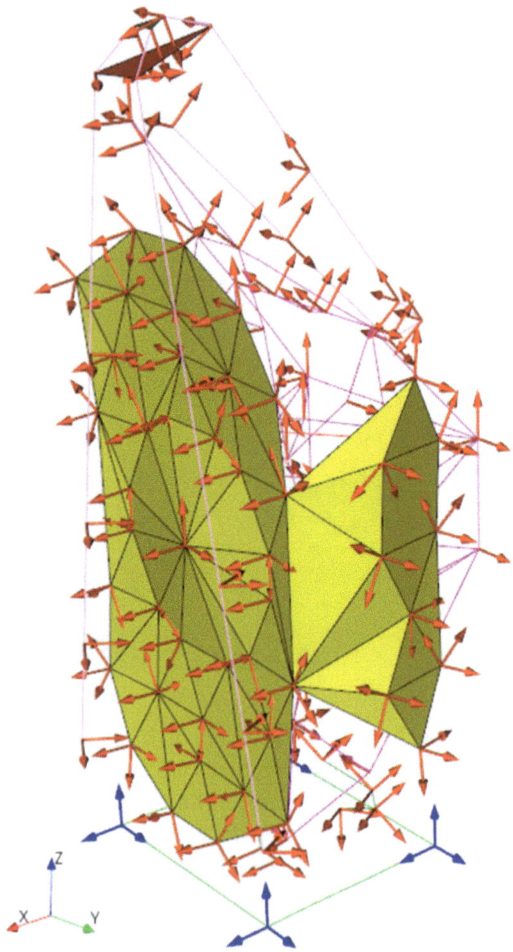

Fig. 10.3 JWST OTE TDM with measurement locations in the stowed configuration

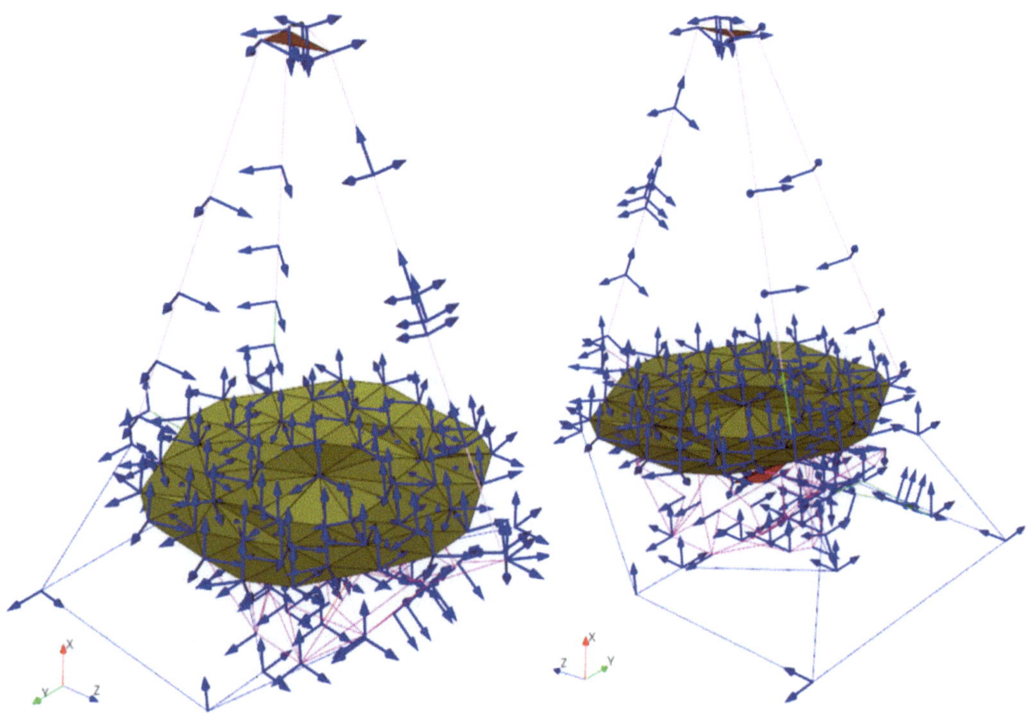

Fig. 10.4 JWST OTE TDM with measurement locations in the deployed configuration

Fig. 10.5 Stowed configuration shaker locations

configuration. A majority of the accelerometers were installed in local coordinate systems to allow accurate orientation. The installation of accelerometers, cables, and shakers was done intermittently over a 6-day period.

Shakers were selected using the analysis predictions to excite all the modes in a single setup. The stowed configuration utilized five Modal-110 shakers and is shown in Fig. 10.5. The deployed configuration utilized seven Modal-110 shakers and is shown in Fig. 10.6, which provides a front and side view of the deployed configuration. The shakers were supported using suspension cages and a variety of support stands and one overhead crane. Since the test article boundary condition was fixed, the shakers were supported in a free-free boundary condition. Most of the stands used were provided by ATA and consist of modified tripods allowing easy adjustment and alignment.

Test performance started with ambient background noise checks and a linearity assessment at each shaker location. Linearity was assessed using multiple force levels. The "best" force level, based upon data quality, for each shaker was then used in a multi-reference burst-random run. For the stowed configuration, the best levels ranged from 1.5 to 7 lbs RMS. Some locations were lower because they were attached to flexible points while others were at hard points. The deployed configuration required lower force levels and ranged from 0.5 to 3 lbs RMS. All the data sets were recorded with a maximum frequency of 250 Hz (sample rate of 640 Hz) and processed with a delta F of 0.029 Hz to provide accurate resolution for observing modes throughout the 0–250 Hz range.

Preliminary data analysis was performed as soon as the frequency response functions (FRFs) were available for processing. This allowed a quick study of the dynamic characteristics of the JWST OTE and also allowed the data quality to be verified through the use of coherence, complex mode-indicator functions (CMIFs), multivariate mode-indicator functions (MMIFs), power spectrum mode-indicator functions (PSMIFs), and FRF quality. Quick processing of the data into mode shapes permitted the MAC, orthogonality, and self-orthogonality to be computed between the test shapes, which helped to verify the independence of the extracted modes. Cross-orthogonality was also used to compare the test shapes to the analysis shapes.

An example drive point FRF and coherence plot for the stowed configuration is provided in Fig. 10.7. The drive point FRFs for all six shakers in the deployed configuration are provided in Fig. 10.8. As can be seen, the deployed configuration has a high modal density below 100 Hz. Linearity assessment was performed using the PSMIF. Because the PSMIF is the summation of all FRFs multiplied by each FRF conjugate, it is an ideal function to allow a global study of nonlinearity.

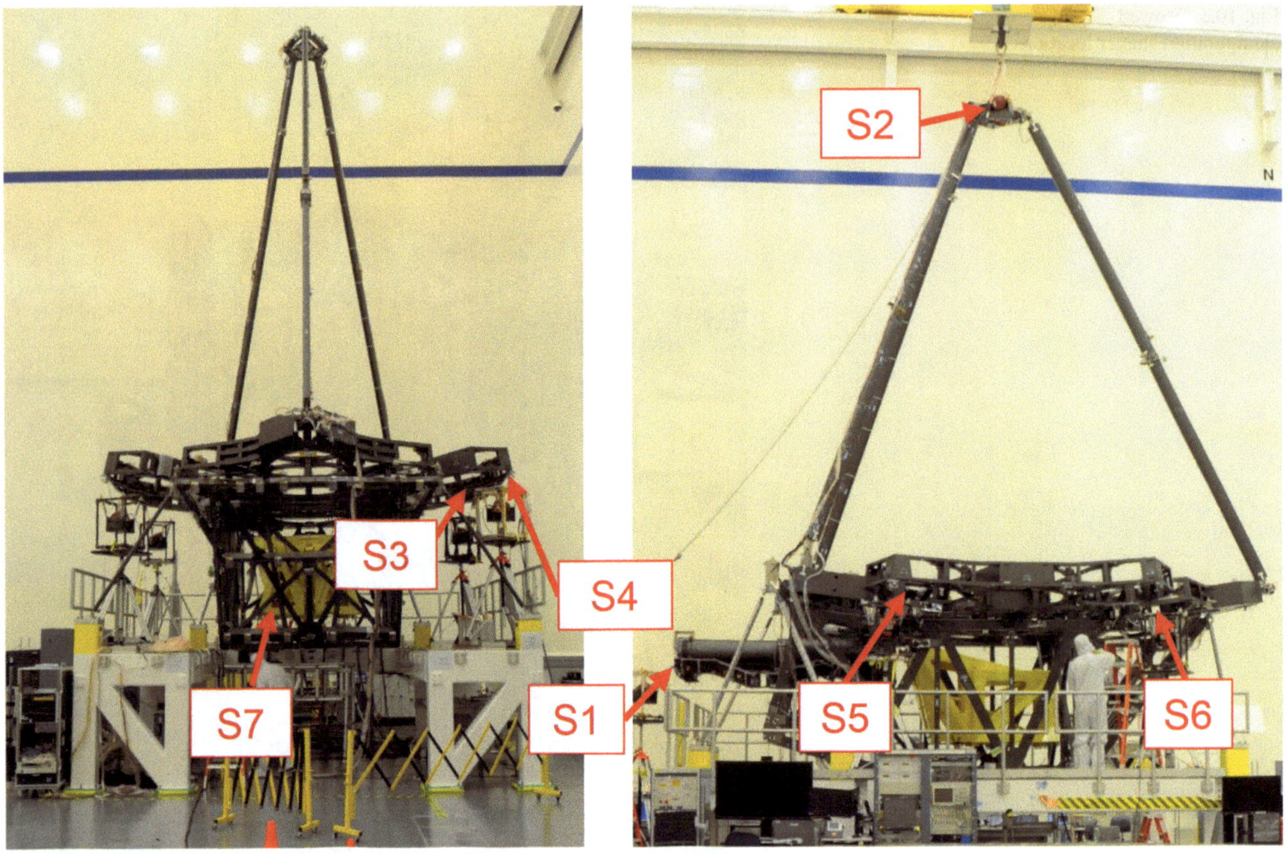

Fig. 10.6 Deployed configuration shaker locations

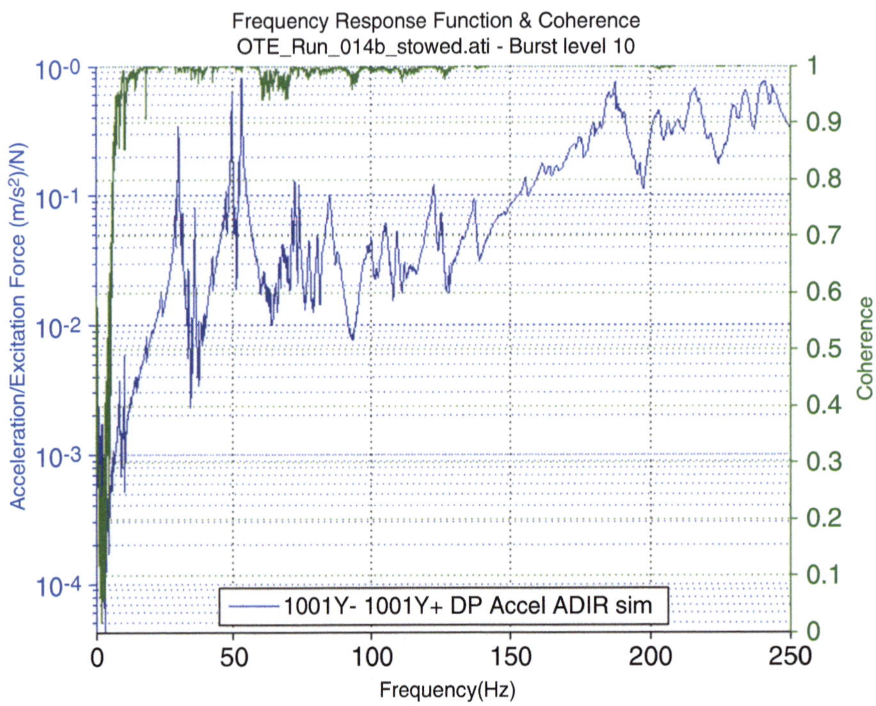

Fig. 10.7 Stowed configuration drive point FRF and coherence for one shaker, 0–250 Hz

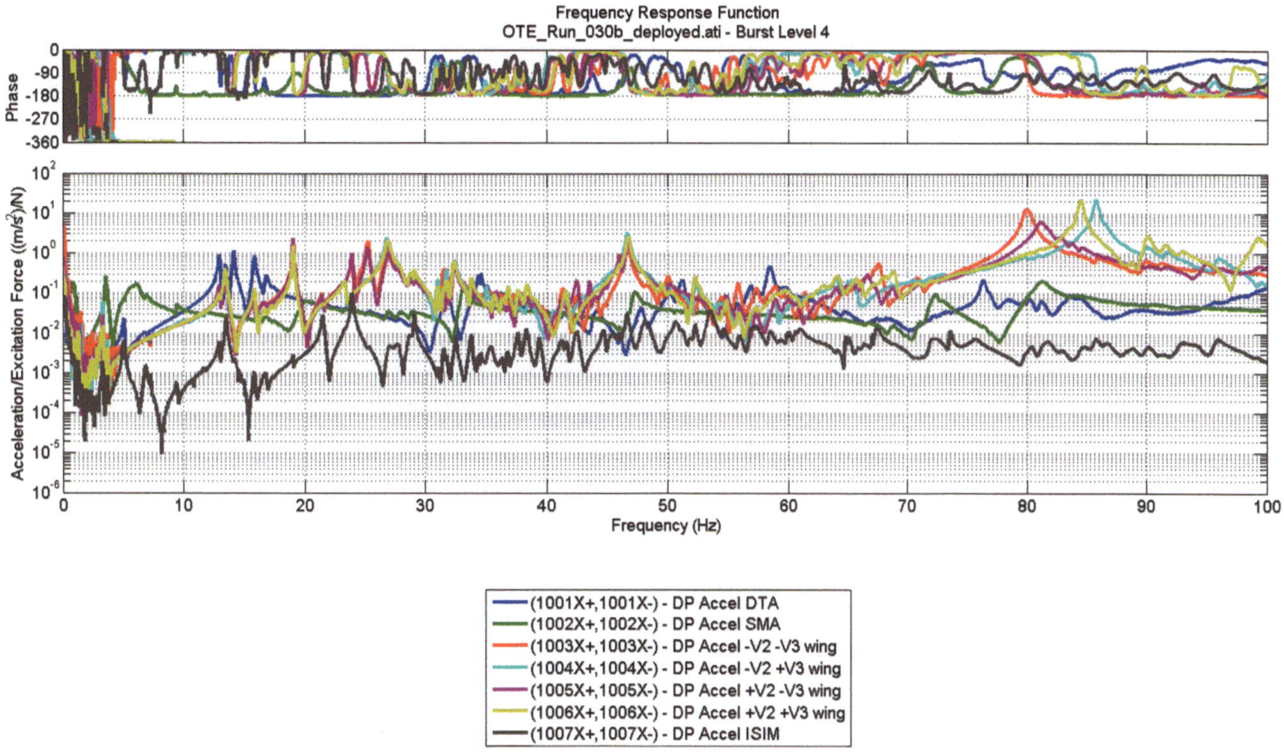

Fig. 10.8 Deployed configuration drive point FRFs, 0–100 Hz

10.4 Stowed Test Results

The stowed configuration test was performed first. During the test, it was noticed that the first bending mode in one direction was nonlinear and was 40 % below the analysis prediction. Fig. 10.9 shows the nonlinear response of the first mode. These PSMIF plots range from extremely low level (0.7 lbs RMS) to the highest level (7 lbs RMS). The nonlinear response and comparison to the analysis indicate that the support structure was not as stiff as preferred. A complete modal survey was completed, and while the test article was reconfigured for the deployed configuration, additional testing was performed on the stowed fixture to identify any differences in the stiffness from the prediction.

Initial modal impact testing of the fixture shows that there were differences in the stiffness of the four mounts. Figure 10.10 shows the three impact directions at each mount location to access and compare the stiffness. Figure 10.11 shows the drive point FRF in one direction at each of the 4 mounts. The acceleration/force (inertance) at low frequency is inversely proportional to stiffness. As can be seen in the overlay, one of the mounts is much stiffer than the other three. This quick modal test provided enough insight to show that the mounts were different. NGSC then performed additional static testing to experimentally measure the mount stiffness, which was then used to update the FEM used in the modal predictions. After this update, the analysis matched the test data within 20 % and the stowed configuration test was deemed complete. The static test was performed in parallel with the deployed configuration so that schedule was not affected.

A total of 76 modes were extracted up to 50 Hz for the stowed configuration. Using the static test data and updating the FEM predictions allowed the team to finalize the stowed configuration test without impacting schedule. The stowed configuration will be used to access launch loads. The higher-frequency content (above 40 Hz) is not critical for launch load assessment.

10.5 Deployed Test Results

The final mode set for the deployed configuration consisted of 188 modes up to 100 Hz. To get the best set of independent modes, several methods were implemented. First, five or six shakers were used simultaneously, providing burst-random input, and having multiple references allowed ATA to identify closely spaced modes using MMIF and CMIF plots. Second,

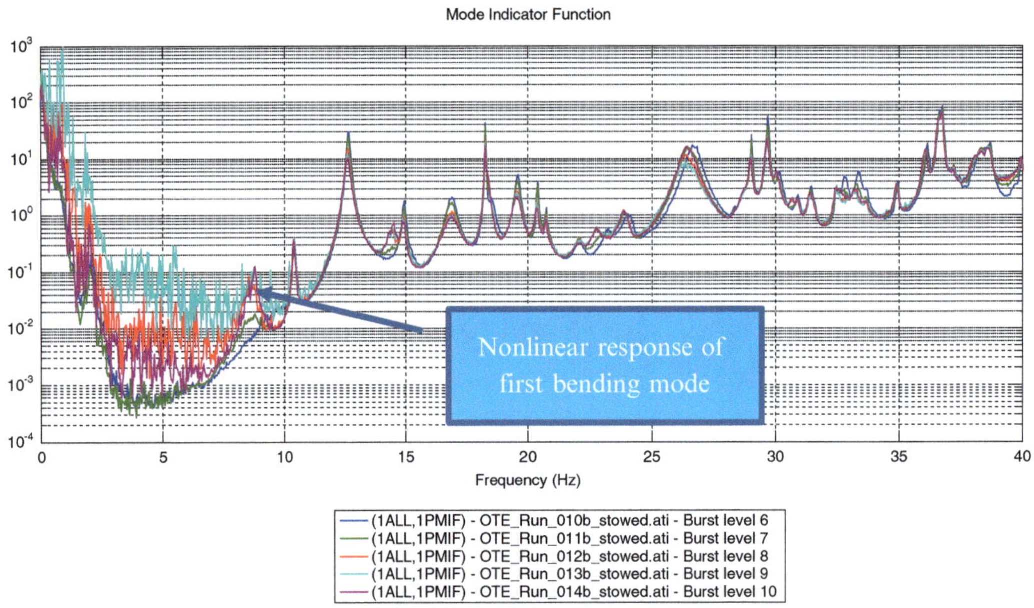

Fig. 10.9 Stowed PSMIF overlay to access linearity

Fig. 10.10 Fixture impact directions

impact tests at each primary mirror simulator were combined into one data set. There were 18 locations in total, all very similar to each other, with closely spaced modes. By combining the FRFs from each location, a multi-reference impact data set was used to fit the frequency bands of the closely spaced modes. These modes were then combined with the burst-random modes and sorted for the best match to the analysis. Repeated modes were discarded. Photos of two locations are shown in Fig. 10.12. The transfer functions between the primary mirrors and the secondary mirror were also an important measurement. These transfer functions will be used to evaluate the mirror performance and line of sight.

The third method was using multiple types of curve-fitting algorithms to obtain low off-diagonal terms in the orthogonality matrix. ATA's IMAT software and alias-free polyreference (AFpoly^{TM}) [2] were used as the primary curve-fitting algorithm. The off-diagonal terms in the test self-orthogonality matrix from modes from AFpoly were then reviewed. There were a couple frequency bands that contained higher off-diagonal terms than desired—for example between 33 and 37 Hz—but a method called mode enhancement [3] was used to provide better mode extraction and therefore reduce the off-diagonal terms.

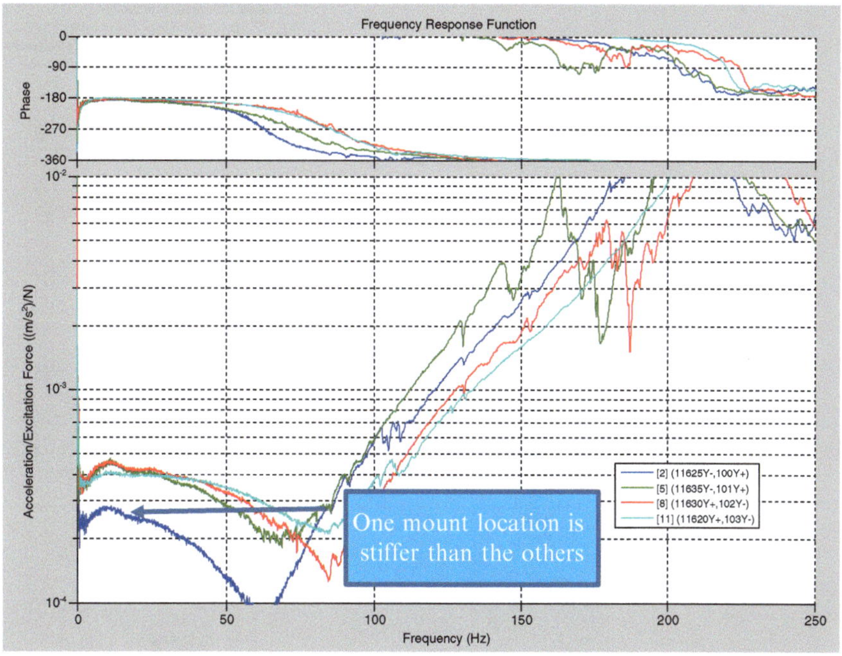

Fig. 10.11 Drive point inertance comparison

Fig. 10.12 Primary mirror mass simulator impacts—example photos of two locations

Mode enhancement techniques using the MMIF and/or CMIF were used to extract modes that were difficult to extract using AFPoly™ and other algorithms. These included bands of modes where there were high levels of modal coupling as defined by high off-diagonal terms in the test self-orthogonality matrix, and frequencies where the synthesized FRFs from modes extracted by AFPoly™ did not match the test data well. In the following example, the 36–38 Hz region contained a high modal density and required mode enhancement. Mode enhancement techniques often work better for sets of test data with a large number of modes over a very small frequency range, such as the modes associated with local deformations of the primary mirror mass simulators. The eigenvectors associated with the minimum value of the MMIF for each mode are defined as the force patterns necessary to drive the structure into a normal mode. Figure 10.13 shows the MMIFs from the random five-shaker run for the stowed configuration. The FRF matrix was post-multiplied by these force patterns to generate a set of mode-enhanced FRFs. The imaginary part of the resulting FRF matrix at resonance is assumed to be the mode shape as it would be for a mode extracted using a normal mode tuning technique. The FRF matrix is then pre-multiplied by this shape to create a single enhanced FRF. The measured drive point FRF and the enhanced FRF are compared in Figs. 10.14 and 10.15, showing that the modes in the 36–38 Hz range are much more prevalent after the enhancement. Damping and natural frequency values were extracted from the enhanced FRF. Figure 10.16 shows the enhanced FRF and the curve-fit data for the first two modes in this frequency range.

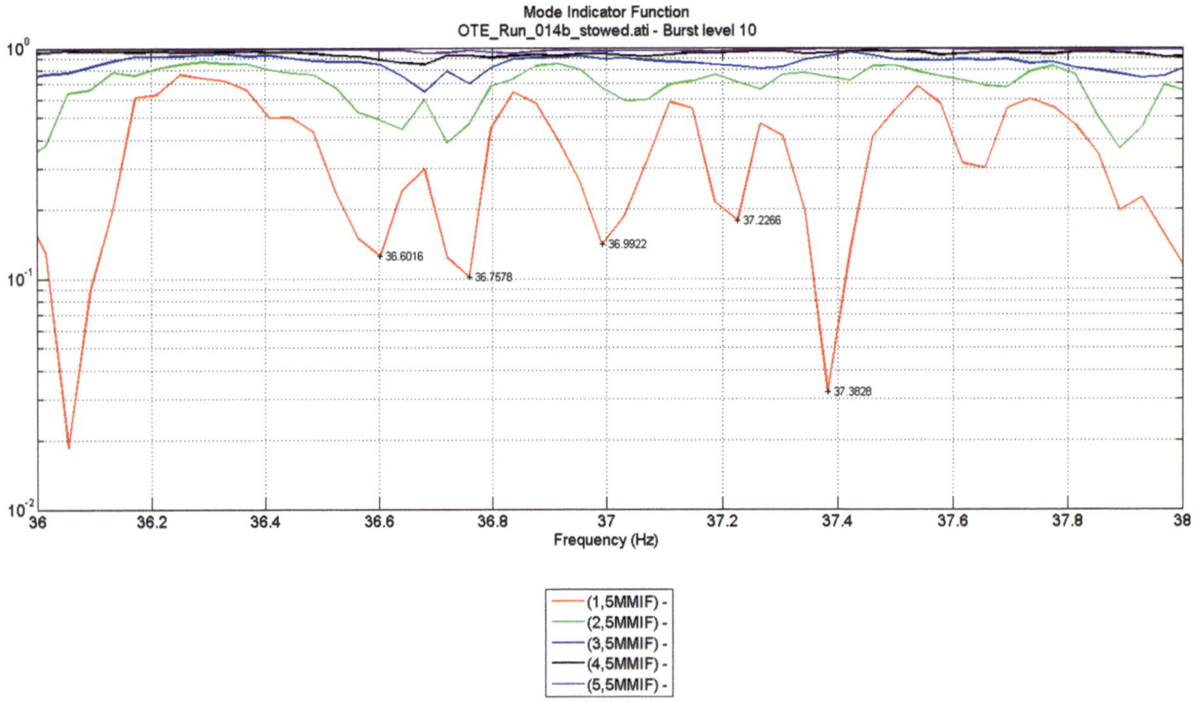

Fig. 10.13 MMIFs used to tag frequencies for mode enhancement

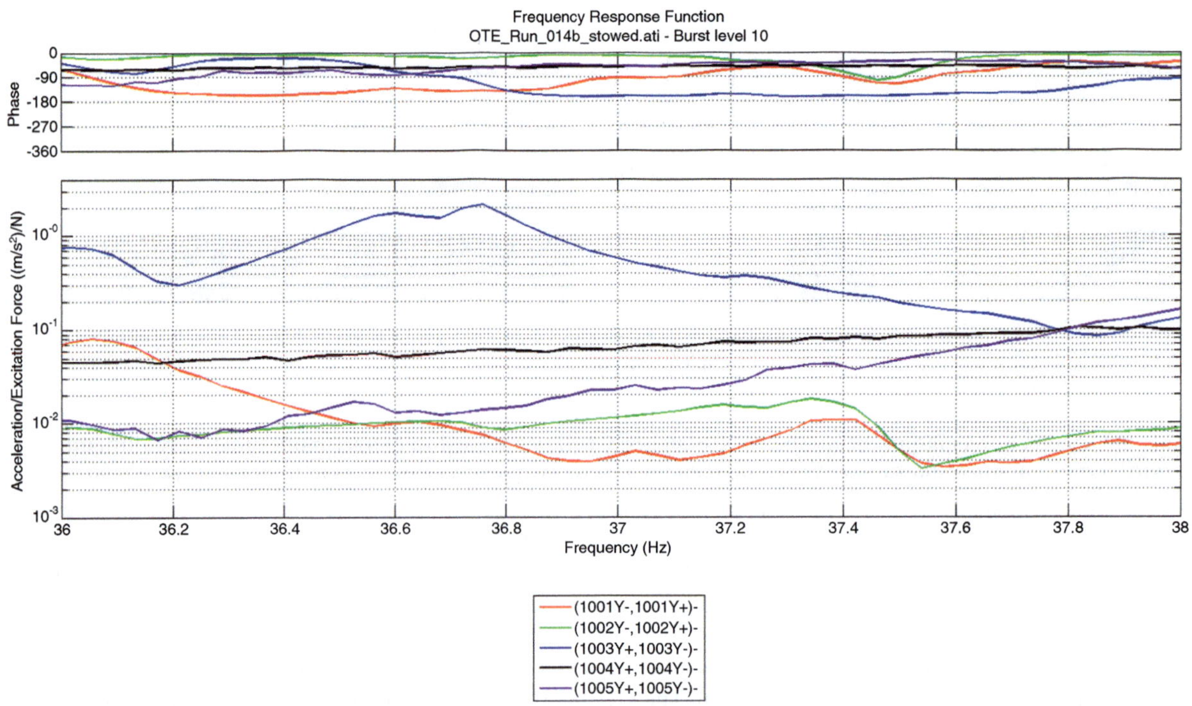

Fig. 10.14 Drive point FRF 36–38 Hz

The self-orthogonality tables are provided in Table 10.1 for the modes from 33 to 37 Hz. The table on the left was created using the measured FRF to perform the curve fits. The table on the right was created using the mode enhancement curve-fitting method. As can be seen in the table, just performing mode enhancement curve fitting reduced the off-diagonal terms by obtaining better mode shape fits from the data. This mode enhancement was repeated for other narrow frequency bands with a high number of modes.

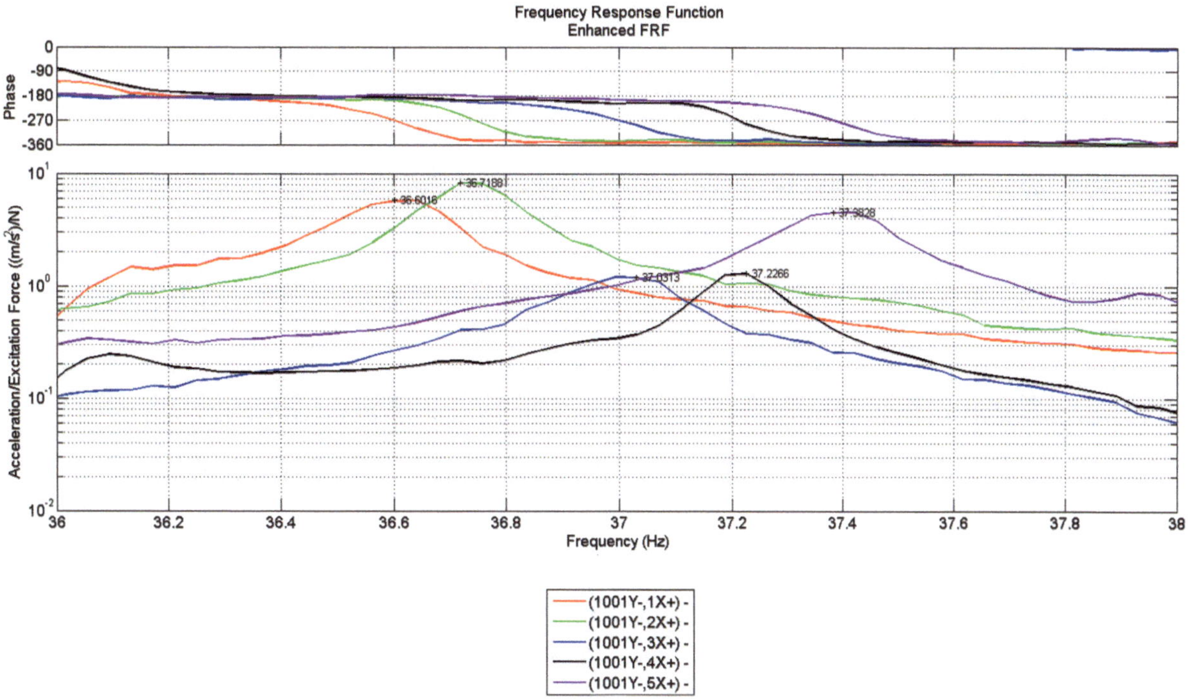

Fig. 10.15 Enhanced FRF 36–38 Hz

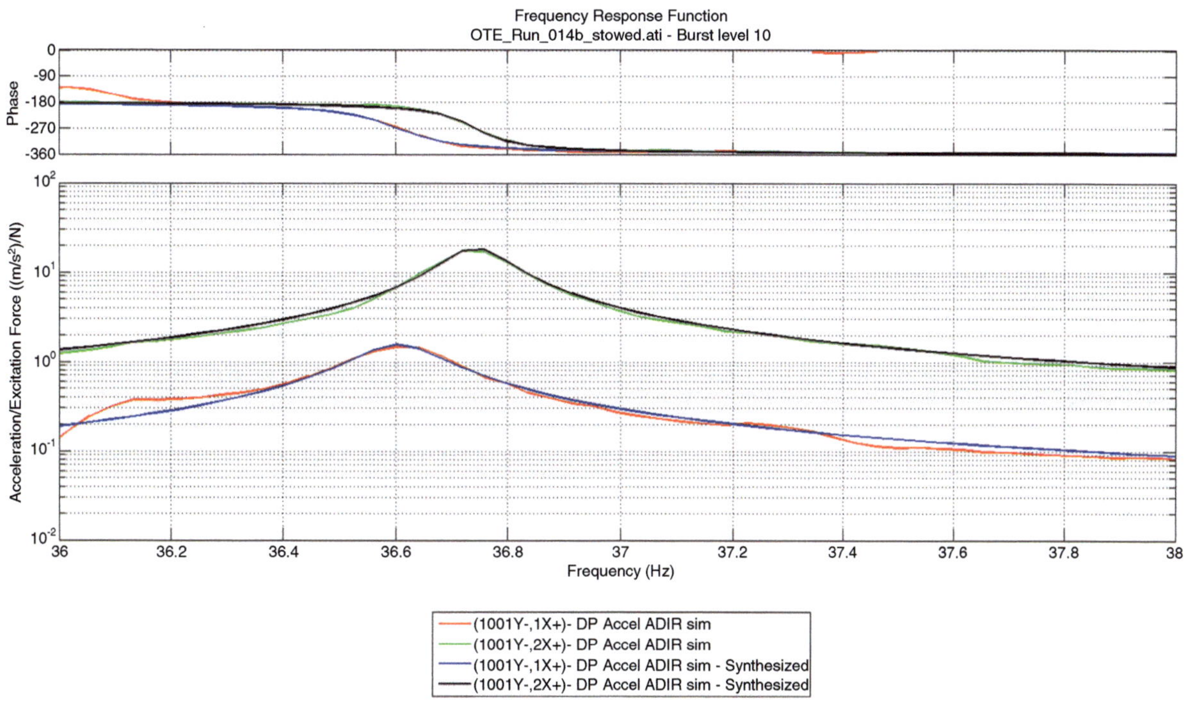

Fig. 10.16 FRF curve fit to enhanced FRF—first two modes shown

The typical fourth step in obtaining the best data for a spacecraft with high modal density is to use impact testing to excite modes that cannot be well excited from the shakers. These could be local component modes that are isolated from the main structure. In this case, the shakers were well placed, so impact testing to identify additional modes was not necessary.

The final self-orthogonality table up to 40 Hz is provided in Table 10.2. As can be seen, low off-diagonals were obtained using the methods and techniques described above. There were two areas where coupling of modes was observed. The first

Table 10.1 Self-orthogonality tables before (left) and after (right) mode enhancement curve fitting

Test Self Orthogonality Table (before)

	Test Shapes	35	36	37	38	39	40	41	42	43	44	45
	Ott	33.8	34.6	35.0	35.5	35.6	35.7	35.8	36.1	36.3	36.7	36.8
35	33.8	1.00										0.06
36	34.6		1.00	0.13	0.07	0.09				0.09	0.08	0.06
37	35.0		0.13	1.00	0.06							
38	35.5		0.07	0.06	1.00	0.44	0.07	0.13				
39	35.6		0.09		0.44	1.00	0.08	0.06	0.07	0.06	0.11	
40	35.7				0.07	0.08	1.00					
41	35.8				0.13	0.06		1.00	0.19		0.10	
42	36.1					0.07		0.19	1.00	0.08	0.09	
43	36.3		0.09			0.06			0.08	1.00	0.16	0.05
44	36.7		0.08			0.11		0.10	0.09	0.16	1.00	0.14
45	36.8	0.06	0.06							0.05	0.14	1.00

Test Self Orthogonality Table (after)

	Test Shapes	35	36	37	38	39	40	41	42	43	44	45
	Ott	33.8	34.6	35.0	35.4	35.6	35.7	35.9	36.1	36.3	36.6	36.8
35	33.8	1.00										0.08
36	34.6		1.00	0.09	0.05	0.08		0.06		0.07	0.08	0.07
37	35.0		0.09	1.00								
38	35.4		0.05		1.00	0.07	0.07	0.07				0.06
39	35.6		0.08		0.07	1.00	0.08	0.06				
40	35.7				0.07	0.08	1.00				0.05	
41	35.9		0.06		0.07	0.06		1.00	0.08	0.05	0.11	
42	36.1							0.08	1.00		0.09	
43	36.3		0.07					0.05		1.00	0.13	
44	36.6		0.08				0.05	0.11	0.09	0.13	1.00	0.08
45	36.8	0.08	0.07		0.06						0.08	1.00

Table 10.2 JWST OTE deployed final test modes: self-orthogonality (59 modes from 0 to 40 Hz)

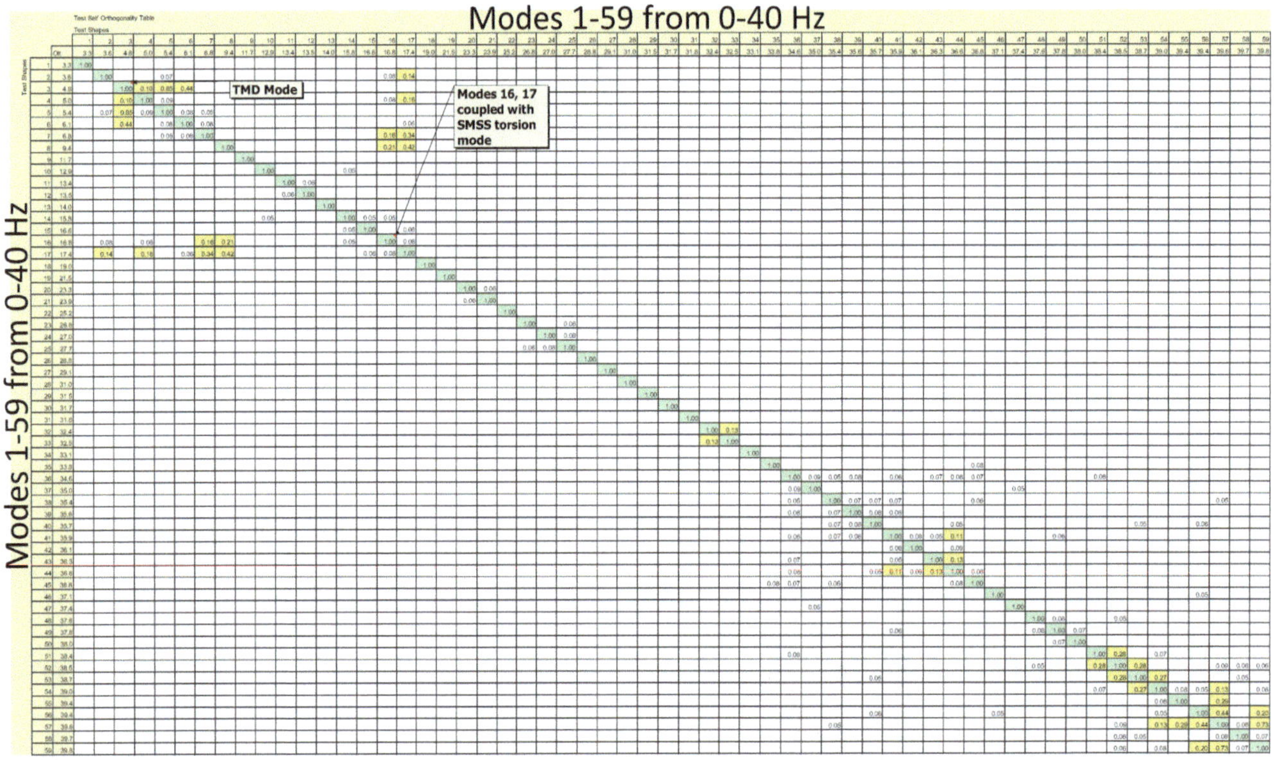

was the TMDs near 5 Hz. The model did not include the TMD modes, so the mass matrix did not include the TMDs. The TMD mode couples with the main strut modes, per their design. The second set of modes that coupled was with a strut torsion mode: there was a single accelerometer mounted off the centerline of the single main strut, and this single DOF picked up the torsion mode of the strut. Since there was only one DOF, the mode shape was not completely defined and coupled with other strut modes.

An example mode shape, mode 11 at 13.39 Hz, is provided in Fig. 10.17 for the deployed configuration. Mode shapes were fairly well defined up to 40 Hz. Beyond this range, shapes were computed up to 100 Hz and will be used in performance calculations. As part of the final check of the mode fit quality, FRFs were synthesized using the final mode shapes. These computed FRFs were then compared to the measured FRFs. The first three orders of the CMIF overlay between the synthesized FRFs and the measured FRFs are provided in Fig. 10.18 up to 100 Hz. The closely matched CMIFs verify quality mode extractions.

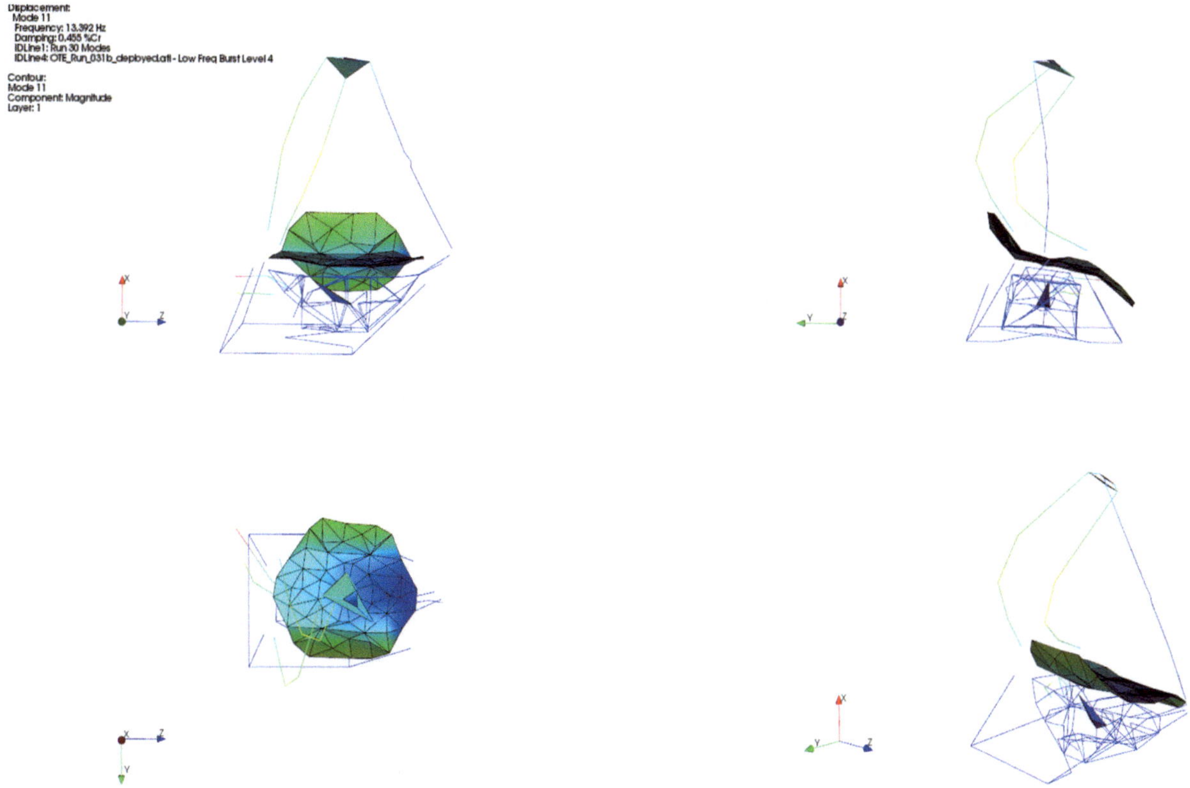

Fig. 10.17 Example mode shape of the deployed OTE

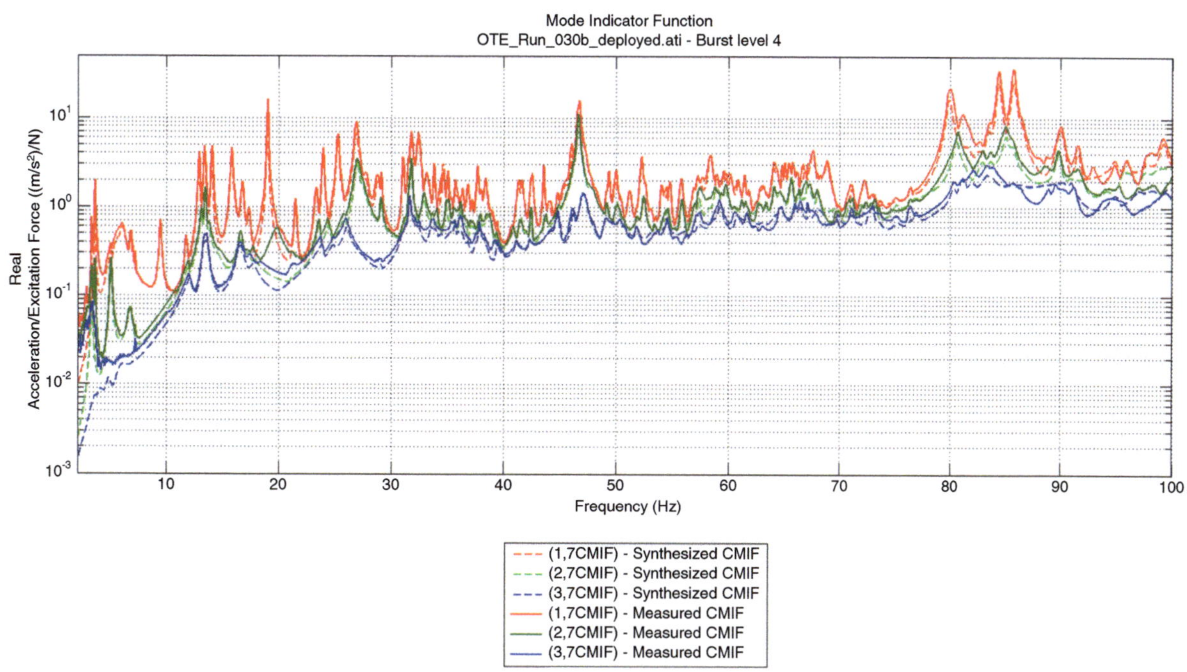

Fig. 10.18 Synthesized CMIF and measured CMIF used to verify mode shape extraction quality

10.6 Summary

The JWST OTE structure is a large and complex spacecraft. With proper test planning and execution and on-the-fly analysis, the modal test was successfully performed, obtaining critical transfer functions out to 250 Hz and modal parameters up to 100 Hz. This modally dense structure has several frequency bands with closely spaced modes, and several techniques were used to identify and verify unique modes. The fixture interface for the stowed configuration posed challenges; however, these were overcome by performing testing and analysis in parallel with those for the deployed configuration.

References

1. Brillhart, R., Dillion, M.: Automated test setup in modal testing. In: 10th International Modal Analysis Conference, Los Angeles, CA, February 2002
2. Brillhart, R., Napolitano, K., Osterholt, D.: Utilization of alias free polyreference for mixed mode structures. In: 26th International Modal Analysis Conference, Orlando, FL, February 2008
3. Napolitano, K.: Using FRF interpolation to help separate closely spaced modes. In: 24th International Modal Analysis Conference, St. Louis, MO, 2006

Chapter 11
Quantification of Dynamic Differences Between Boundary Conditions for Environment Specification Improvement

Julie M. Harvie and Randy Mayes

Abstract Qualification of complex systems typically involves testing the components individually in shock and vibration environments before assembling them into the system. When the components are secured to a fixture on the shaker table, the mechanical impedance of the boundary condition is quite different from that of the next level of assembly. Thus the modes of the component under test are not excited in the same way that they are excited in the system using the typical methods for defining input specifications. Here, the boundary condition impedance is investigated and quantified using substructuring techniques. Also, fixture inputs are derived to overcome the impedance differences and excite a component in the same way it is excited in the next level of assembly.

Keywords Substructuring • Boundary conditions • Craig Bampton

Nomenclature

DOF	Degree of freedom
EOM	Equations of motion
ω	Frequency (rad/s)
ζ	Critical damping ratio
q	Free modal displacements of the component/fixture assembly
p	Modal displacements of the component with fixed boundary
s	Free modal displacements of the fixture
x	Physical displacements
a	Physical force vector
f	Modal force vector
K	Modal stiffness matrix
M	Modal mass matrix
I	Identity matrix
T	Transformation matrix
L_{fix}	Reduction matrix applying fixed boundary constraint to experimental EOM
Φ	Free mode shape matrix for the component/fixture assembly
Ψ	Free mode shape matrix for the fixture
Γ	Eigenvectors resulting from fixed boundary constraint of experimental EOM
b	Subscript for the boundary DOF
fix	Subscript for the fixed boundary modes of the component
$free$	Subscript for the free modes of the component/fixture assembly
$+$	Superscript indicating the Moore-Penrose pseudo-inverse of a matrix

Sandia National Laboratories is a multi-program laboratory managed and operated by Sandia Corporation, a wholly owned subsidiary of Lockheed Martin Corporation, for the U.S. Department of Energy's National Nuclear Security Administration under contract DE-AC04-94AL85000.

J.M. Harvie (✉)
Program & Test Integration Department, Sandia National Laboratories, P.O. Box 5800 – MS 0840, Albuquerque, NM 87185, USA
e-mail: jharvie@sandia.gov

R. Mayes
Experimental Mechanics, NDE, and Model Validation Department, Sandia National Laboratories,
P.O. Box 5800 – MS 0840, Albuquerque, NM 87185, USA
e-mail: rlmayes@sandia.gov

A. Brandt, R. Singhal (eds.), *Shock & Vibration, Aircraft/Aerospace, Energy Harvesting, Acoustics & Optics, Volume 9*,
Conference Proceedings of the Society for Experimental Mechanics Series, DOI 10.1007/978-3-319-30087-0_11

11.1 Introduction

Many complex systems experience dynamic shock and vibration environments during their lifetime. Prior to subjecting the system to the true environments when it is fielded, the system will undergo a qualification program to ensure that the system and its components are rugged enough to survive the dynamic environments. During qualification, the system is exposed to simulated environments using laboratory equipment such as vibration shakers and drop tables. Qualification is also typically performed at the subassembly and component level during development.

Subassembly and component level testing is generally intended to replicate the environmental stresses encountered in the service environment such that the analyst can determine if the components would be able to survive the true environment without degradation in performance. The subassembly and component test inputs are typically derived by extracting acceleration data from the system in the service environment at the component attachment locations. The data may come from field tests, laboratory tests, computer simulations, or a combination of test and simulation data. The relevant data are gathered, and a specification is generated based on the maximum predicted response. The specification is then input to the component fixture, ideally at a similar proximity to the attachment locations that were collected in the service environment. The process used to develop component specifications is shown schematically in Fig. 11.1.

As noted in [1], "the damage potential of the real condition and the simulated environment should be the same." However, there are multiple factors present that introduce differences between the real condition and simulated environment. As seen in Fig. 11.1, the specification development process typically involves drawing an envelope over data, which inherently increases the amount of conservatism present in the specification, namely at the frequency ranges where the data contains local minimum. Coincidentally, those valleys at the attachment points line up with the natural frequencies of the component, which can cause an over test of the component. Additionally, the mechanical impedance, or resistance to motion, of the fixture does not typically match that of the system. Thus, even if the same input were applied to a similar location on the fixture and the system, the component would likely not respond the same in each assembly [2].

Furthermore, the general guidance for designing component test fixtures is to make them as rigid as possible [3]. While this criteria may be achievable for certain structures, it is sometimes impossible given the test constraints. If flexible fixture modes exist within the bandwidth of interest, they will almost certainly be excited to some degree. As noted in [4], the level of vibration on the fixture will only be maintained at the location of the control accelerometer. Also, the level of vibration on the fixture will be highly dependent on the location of the control accelerometer for flexible modes. As the number of attachment points on the component increases, it becomes increasingly difficult to maintain the desired component input.

The issues noted above are widely recognized by the structural dynamics community, but the challenge of overcoming the differences remains. Several techniques exist that can improve the component response, and they were listed by

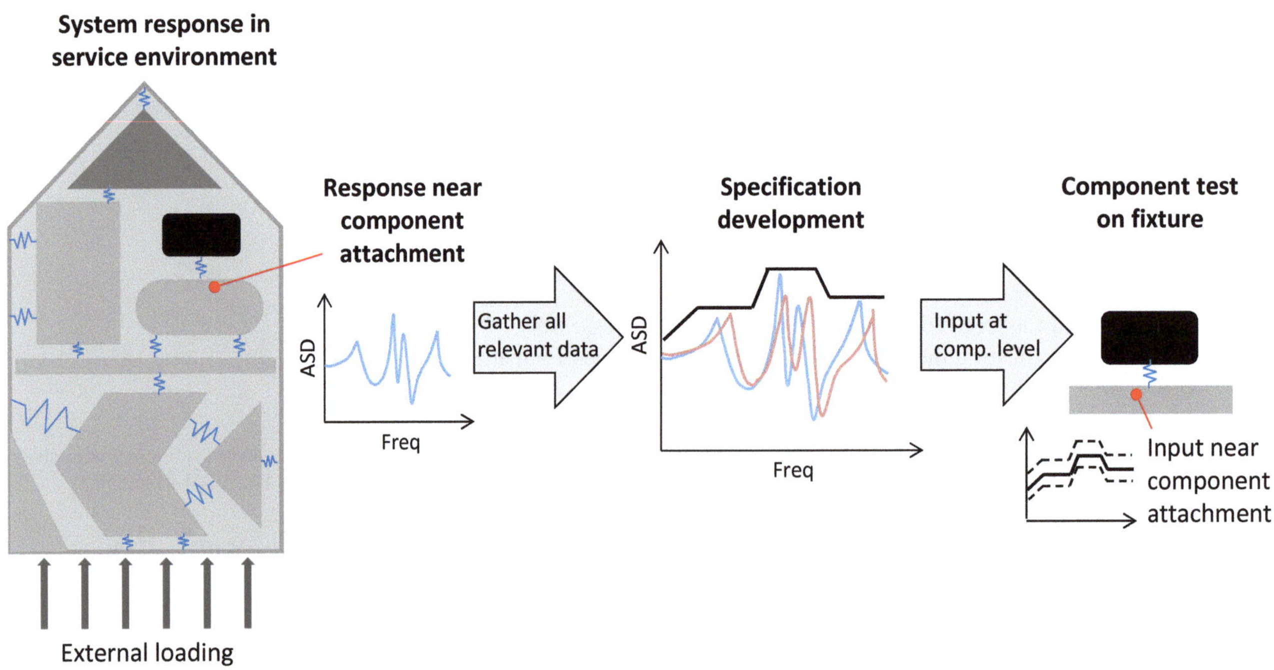

Fig. 11.1 Example of flow of data used to generate component vibration tests

Piersol et al. nearly 20 years ago [5]. The techniques include: force-controlled vibration testing, testing of the test article on its support structure, "notching" of the specified test levels, testing with multiple mode test fixtures, response-controlled testing, and analytical corrections for the interface impedance. NASA has published numerous handbooks on the application of force-limited vibration testing [6, 7], and Mil-Std 810 also endorses force control or acceleration limiting control strategies to avoid unrealistically severe vibration response [3]. Daborn et al. [8] have recently been successful in improving responses by attaching a test article to a more realistic fixture that matches the local impedance seen in the full system. This technique, dubbed IMMAT (Impedance Matched Multi-Axis Test), may be appropriate for certain systems but would become more challenging as the complexity of the system grows.

More recently, during an investigation focused on experimental dynamic substructuring using the transmission simulator method [9], a transformation was discovered [10] that may be extremely useful in solving the problem described above. Specifically, the transformation decouples a free system and defines it in terms of the component's fixed boundary modal response and the fixture's free modal response. Additionally, the transformed system includes mass and stiffness coupling terms that quantify the impedance between the component fixed boundary modes and fixture free modes. This transformation allows the user to quantify, in simple terms, the differences in impedance between the system configuration and component test configuration. Additionally, the theory can be used to determine the appropriate component test inputs to replicate the component fixed boundary modal response in the true environment.

11.2 Theory

The equations of motion (EOM) of a free system with an applied force can be written in modal space as

$$I\ddot{\overline{q}} + (2\zeta_{free}\omega_{free})\dot{\overline{q}} + \omega^2{}_{free}\overline{q} = \Phi^T\overline{a} = \overline{f}_{\overline{q}} \tag{11.1}$$

where ω_{free} are the natural frequencies of the free system, ζ_{free} are the critical damping ratios of the free system, and q are the modal displacements of the system. The vector of physical applied forces, a, is projected to a vector of modal forces, f_q, using the mode shapes, Φ, of the physical system. If the system responds linearly, damping is neglected, and f_q is comprised of sinusoidal inputs with frequency ω, Eq. (11.1) can be simplified as

$$\left[\omega^2{}_{free} - \omega^2 I\right]\overline{q} = \overline{f}_{\overline{q}} \tag{11.2}$$

Our goal is to derive a square transformation matrix, T, that partitions the modal degrees of freedom of the system into the fixed-boundary modal coordinates of the test article, p, and the free-free modal coordinates of the fixture, s. This is written as

$$\overline{q} = T\begin{Bmatrix} \overline{p} \\ \overline{s} \end{Bmatrix} \tag{11.3}$$

The transformation presented in Eq. (11.3) can be derived by constraining the motion of the fixture's boundaries in the system to match the free modal motion of the fixture, and is found to be

$$T = \begin{bmatrix} L_{fix}\Gamma & \Phi_b{}^+\Psi_b \end{bmatrix} \tag{11.4}$$

where Γ are the eigenvectors of the fixed boundary equations of motion (further detailed in [9]), Φ are the free modes of the system, Ψ are the free modes of the fixture, the subscript b denotes a subset of the boundary degrees of freedom (DOF) where measurements would be made, and

$$L_{fix} = null(\Psi_b{}^+\Phi_b) \tag{11.5}$$

If we premultiply Eq. (11.2) by the transpose of the transformation matrix, and insert Eq. (11.3) into Eq. (11.2), then we are left with an equation of the form

$$\left[\begin{bmatrix} \omega_{fix}^2 & K_{ps} \\ K_{ps}^T & K_{ss} \end{bmatrix} - \omega^2 \begin{bmatrix} I & M_{ps} \\ M_{ps}^T & M_{ss} \end{bmatrix}\right] \begin{Bmatrix} \overline{p} \\ \overline{s} \end{Bmatrix} = T^T \{\overline{f}_{\overline{q}}\} = \begin{Bmatrix} \overline{f}_{\overline{p}} \\ \overline{f}_{\overline{s}} \end{Bmatrix} \tag{11.6}$$

where the stiffness and mass matrices now contain non-zero coupling terms (K_{ps} and M_{ps}, respectively). Note that Eq. (11.6) is very similar to the equation presented in [10], but the applied forces and therefore modal forces, f, have not been set to zero. The forces are carried through the equations because the physical applications of these equations require an applied force. Additionally, Eq. (11.6) allows the user to understand how the applied modal force on the system translates to a modal force on p and s. If we are only applying the external forces to the fixture, then f_p will approach zero or at least be much smaller than f_s. If we consider the upper partition of Eq. (11.6), with f_p approximately equal to zero, the response of the p fixed boundary modes can be written in terms of the s free fixture modes as

$$\left[\omega_{fix}^2 - \omega^2 I\right]\overline{p} = \left[-K_{ps} + \omega^2 M_{ps}\right]\overline{s} \tag{11.7}$$

Equation (11.7) can be used to study how much each fixture mode, s, contributes to the response of each component mode, p. Additionally, if the component motion in the system is known, Eq. (11.7) can be used to calculate the fixture motion, s, required to excite the component appropriately. Once the necessary fixture motion is known, the required forces to excite the fixture modes can be equated using the lower partition of Eq. (11.6) as

$$\overline{f}_{\overline{s}} = \left[K_{ps}^T - \omega^2 M_{ps}^T\right]\overline{p} + \left[K_{ss} - \omega^2 M_{ss}\right]\overline{s} \tag{11.8}$$

Then, if both f_p and f_s are known (f_p assumed to be zero), the vector of modal forces can be premultiplied by the pseudoinverse of the transformation matrix transpose, as in Eq. (11.6), to calculate the modal forces in terms of the system q DOF. The required physical forces are then calculated by premultiplying the modal forces by the pseudoinverse of the mode shape matrix transpose, as in Eq. (11.1). Typically the physical forces could only be input at a discrete number of translation DOF from the mode shape matrix, so only those DOF would be included in the transformation.

Further details on the theory behind this work and development of the transformation can be found in [10].

11.3 Case Studies

Two simple models were generated to investigate the theory presented in Sect. 11.2. The model details are outlined in Sect. 11.3.1 and further described in Appendix 1. First, a sinusoidal excitation is generated with a frequency close to one of the subassembly natural frequencies in Sect. 11.3.2. This problem represents a situation where the field environment provides significant excitation to a mode of the system, and the component is then tested in a fixture that has different mode shapes and frequencies than the system. The response of the component in the two configurations is compared, and then a modified input is developed to cause the component response in the fixture to mimic the response in the system. Then, a sinusoidal excitation is generated with a frequency close to one of the fixture natural frequencies in Sect. 11.3.3. This problem represents a situation where the component responds a certain way in the system to the field environment, and then the component is tested on a fixture that has modes within the test bandwidth that get excited. Again, the response of the component in the two configurations is compared, and a modified input is developed to cause the component response in the fixture to mimic the response in the system.

11.3.1 Models Developed

For this work, all modeling was done in Matlab using planar beam elements with two DOF (translation and rotation) per node [11]. The component of interest was generated to be a 16-inch Aluminum beam. First, a system "truth model" containing four aluminum beams plus the component of interest was generated. Then, a model was generated to represent the component in a qualification configuration, where the component was secured to a relatively stiff steel beam. The systems are highlighted in Fig. 11.2, and the modes of the systems and components are detailed in Appendix 1. These models

Fig. 11.2 Component configurations in full system and on fixture

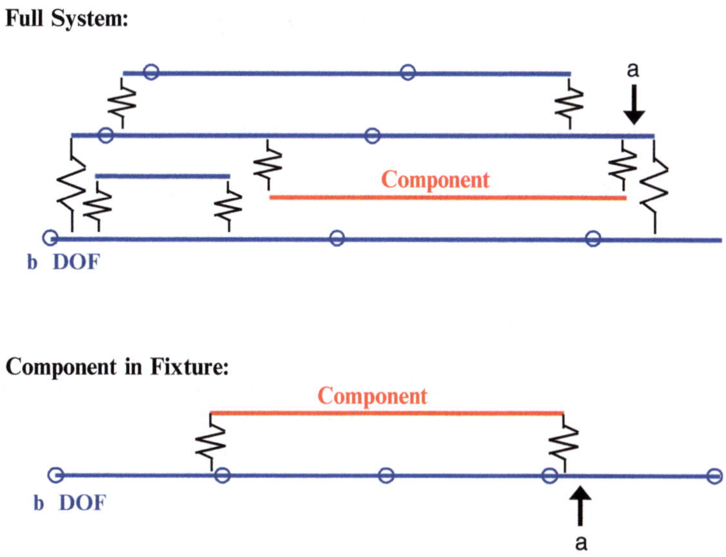

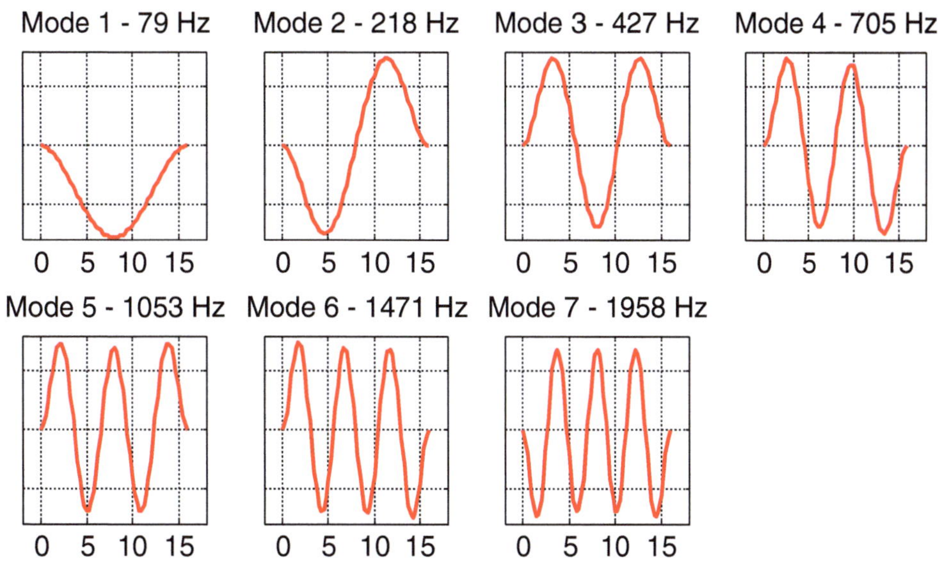

Fig. 11.3 Fixed boundary component response

represent a typical situation where the full system is relatively flexible and has modes dominated by several different components within the assembly, whereas the fixture is relatively stiff and has very few elastic modes within the bandwidth of interest. As noted in the figure, an applied force *a* was applied to each system in a similar location close to one of the component attachment locations. The bandwidth of interest was chosen to go up to 2000 Hz, a common bandwidth for vibration testing at Sandia. The full system contains 11 modes within 2000 Hz, and the blue subassembly alone contains 4 free modes within 2000 Hz. As shown in Fig. 11.2, 7 boundary DOF were chosen across the main beams in the full system. The component/fixture assembly contains 10 modes within 2000 Hz, and the blue fixture alone contains 3 free modes within 2000 Hz. Also shown in Fig. 11.2, five boundary DOF were chosen across the fixture. In each system, the number of boundary DOF was chosen to be approximately 1.5 times the number of fixture modes, as recommended by Mayes.

Both models were used to determine the fixed-boundary component modes. The calculated fixed-boundary modes were nearly identical between the two systems, and are shown in Fig. 11.3. The models were then put into the modal Craig Bampton form of Eq. (11.6) for the following analyses.

11.3.2 Response with Excitation Frequency Near System Mode

A sinusoidal force was input to each system at the locations identified in Fig. 11.2. For this study the forcing frequency was chosen to be 1125 Hz, close to one of the elastic modes of the subassembly detailed in Appendix 1.

The physical response of the complete system is shown in Fig. 11.4, and the response of the component in fixture is shown in Fig. 11.5. Note the difference in scaling of the y-axis, where the system responses are of much greater magnitude than the response of the component/fixture assembly. The differences in response are not too surprising. The excitation frequency is close to a system mode and not a fixture mode, so the system gets more significantly excited.

Both systems were put into the Modal Craig-Bampton form of Eq. (11.6). The desired fixed boundary motion of the component, p, was determined using the system model. As described in the Theory section, the desired component modal response was used to calculate the fixture motion, s, required to excite the component in the same manner as it is excited in the system. Once the fixture motion was known, the fixture modal forces, f_s, were used to calculate the system modal forces f_q. Then, the system input forces were calculated. The original input location was not used in this calculation because it would be difficult to accurately excite multiple fixture modes using the single DOF. The fixture input locations were chosen to be identical to the aforementioned boundary DOF, as those DOF would likely be accessible and measurable. The original applied force is compared to the newly calculated, modified applied force in Fig. 11.6. As noted in Fig. 11.6, the frequency of the forcing function was not changed, but the magnitudes of the modified forcing function are generally much larger than the original forcing function. This difference is expected because the component was not originally responding as much as the component response in the system, and therefore higher input levels were required to excite the component appropriately.

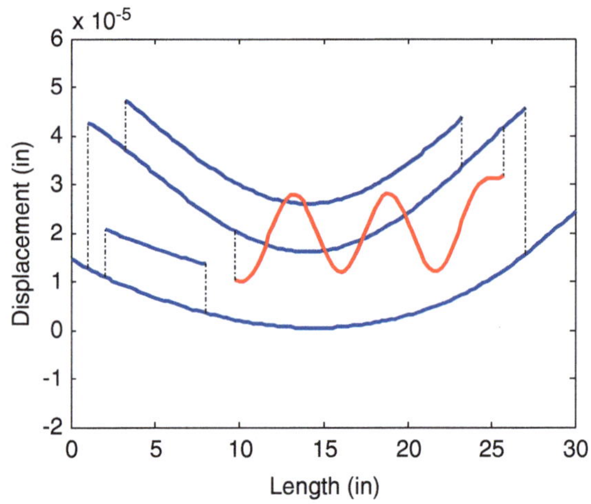

Fig. 11.4 Response of component in system due to excitation with frequency near a system mode

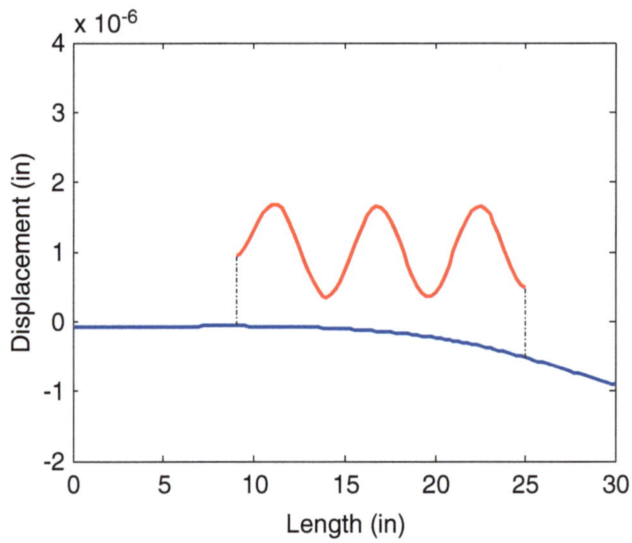

Fig. 11.5 Response of component in fixture due to excitation with frequency near a system mode

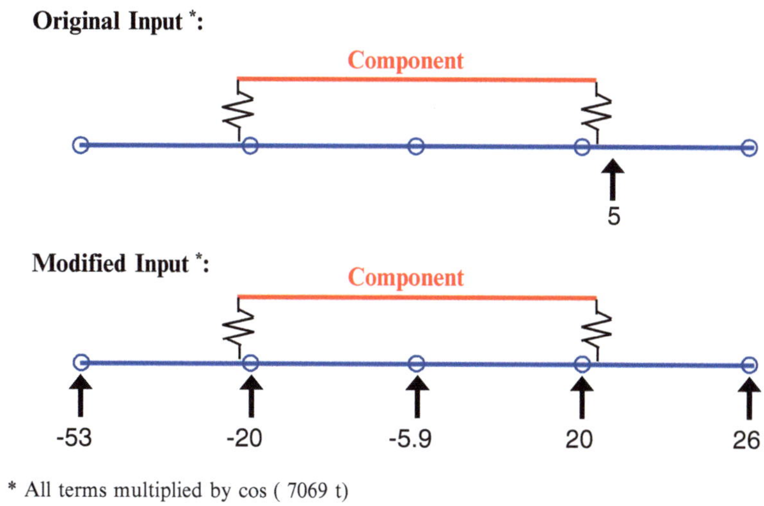

Fig. 11.6 Distribution of applied forces on fixture, original and modified, with excitation frequency near a system mode

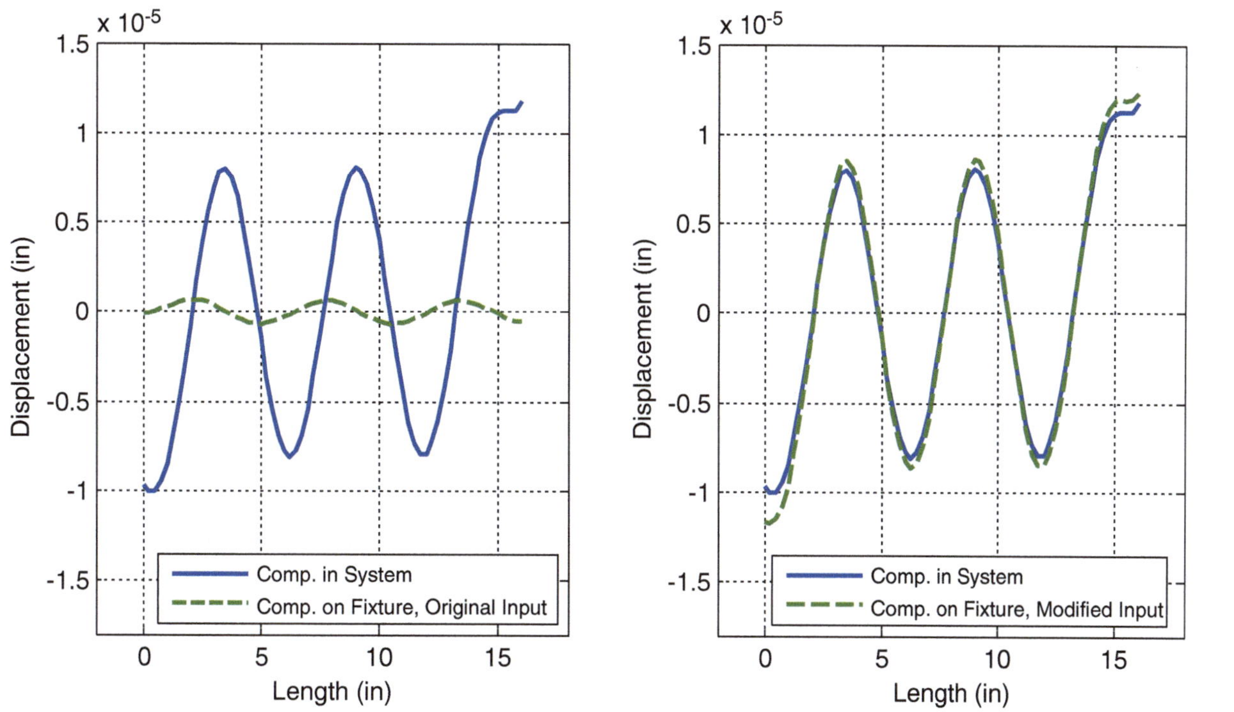

Fig. 11.7 Comparison of component response in system to component response on fixture with original and modified input, excitation frequency near a system mode

The component responses in the fixture due to both the original and modified inputs are compared against the "truth" response of the component in the system in Fig. 11.7. As seen, the component response is significantly improved using the modified input. The component response due to the modified input matches the "truth" component response much better than the component response due to the original input. In this case, using the original input on the fixture would cause an undertest to the component. Undertesting a component could lead to undetected failures occurring in a field environment, which is highly undesirable.

The Modal Assurance Criteria (MAC) were calculated between the component responses shown in Fig. 11.7 and are tabulated in Table 11.1. As seen, using the modified input resulted in a MAC of nearly 100 % between the component response on the fixture and the component response in the system, whereas the MAC was less than 1 % using the original input. Upon further examination of Fig. 11.7, it is unsurprising that the MAC from the original input is so low, as the

Table 11.1 MAC between component response in system and component response on fixture with original and modified input, excitation frequency near a system mode

	Component response on fixture, original input	Component response on fixture, modified input
Component response in system	00.49 %	99.85 %

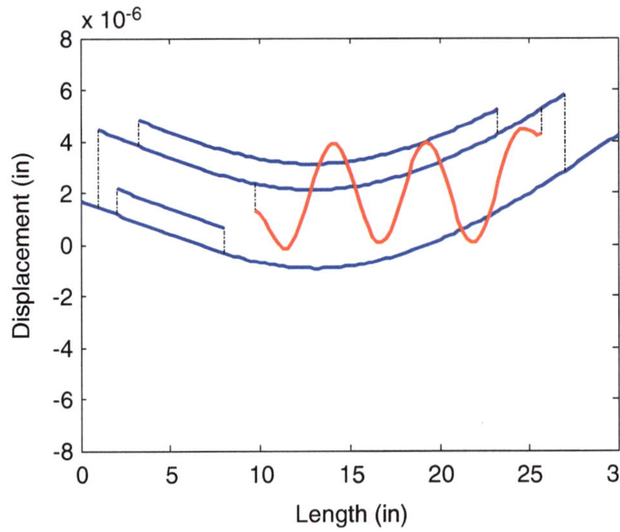

Fig. 11.8 Response of component in system due to excitation with frequency near a fixture mode

component response in the fixture was a maximum where the component response in the system was a node. The differences noted here would result in significantly different stress states and failure locations.

Additionally, the terms on the right hand side of Eq. (11.7) were investigated for each fixture/subassembly mode per each component mode. The terms identify how each fixture/subassembly mode is contributing to the component response. The terms are plotted in Appendix 1. As seen in the plots, the elastic modes of the subassembly contributed much more significantly to the component response than the subassembly rigid body modes. Additionally the 6th fixed boundary component mode was excited most in the full system configuration. When the component was excited in the fixture with the original input, the level of excitation provided to the component modes was much less than in the full system. Additionally the 5th fixed boundary component mode was excited most in the fixture with the original input, rather than the 6th. When the input to the fixture was modified, the response of each component mode matched more closely with the component response in the system. As seen, the first rigid body mode and the first elastic mode of the fixture were mainly used to excite the component correctly.

11.3.3 Response with Excitation Frequency Near Fixture Mode

A sinusoidal force with a different frequency was then input to each system at the locations identified in Fig. 11.2. For this study the forcing frequency was chosen to be 1325 Hz, close to one of the elastic modes of the fixture detailed in Appendix 1.

The physical response of the complete system is shown in Fig. 11.8, and the response of the component in the fixture is shown in Fig. 11.9. Note the difference in scaling of the y-axis, where this time the component/fixture assembly responds more than the system. The differences in response are not too surprising. The excitation frequency is close to a fixture mode and not a subassembly mode, so the component/fixture assembly gets more significantly excited than the system.

Both systems were put into the Modal Craig-Bampton form of Eq. (11.6). The desired fixed boundary motion of the component, p, was determined using the system model. As described in the Theory section, the desired component modal response was used to calculate the fixture motion, s, required to excite the component in the same manner as it is excited in the system. Once the fixture motion was known, the fixture modal forces, f_s, were used to calculate the system modal forces f_q. Then, the system input forces were calculated. The original input location was not used in this calculation because it would be difficult to accurately excite multiple fixture modes using the single DOF. The fixture input locations were chosen to be identical to the aforementioned boundary DOF, as those DOF would likely be accessible and measurable. The original

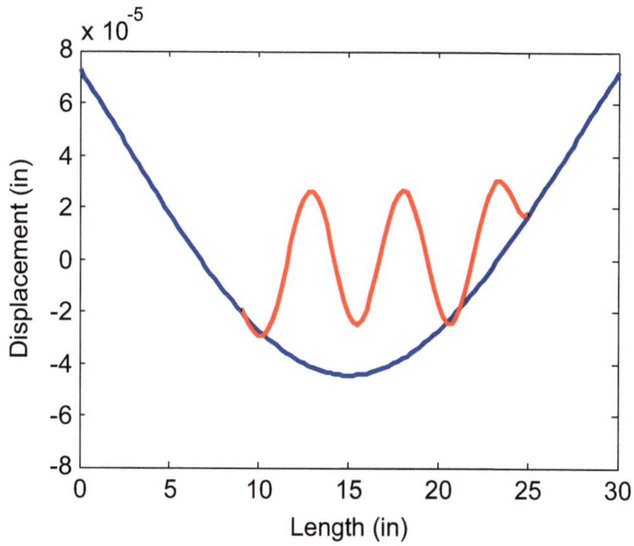

Fig. 11.9 Response of component in fixture due to excitation with frequency near a fixture mode

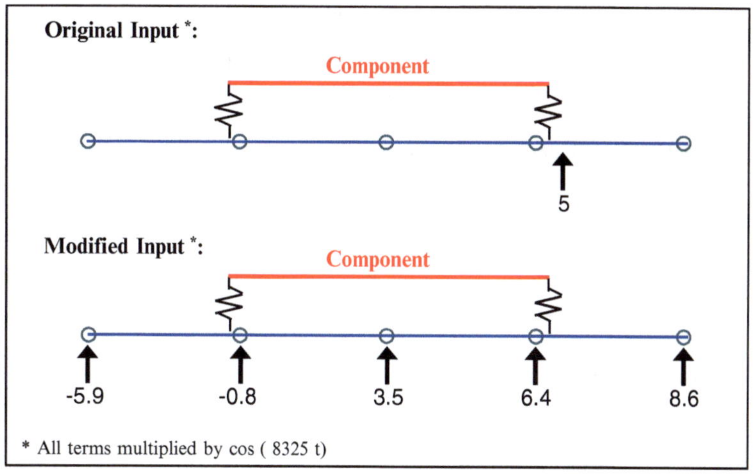

Fig. 11.10 Distribution of applied forces on fixture, original and modified, with excitation frequency near a fixture mode

applied force is compared to the newly calculated, modified applied force in Fig. 11.10. As noted in Fig. 11.10, the frequency of the forcing function was not changed.

The component responses in the fixture due to both the original and modified inputs are compared against the "truth" response of the component in the system in Fig. 11.11. As seen, the component response is significantly improved using the modified input. The component response due to the modified input matches the "truth" component response much better than the component response due to the original input. In this case, using the original input on the fixture would cause an overtest to the component. Overtesting a component could lead to false failures occurring in a qualification test that would not actually be seen in the field environment.

The Modal Assurance Criteria (MAC) were calculated between the component responses shown in Fig. 11.11 and are tabulated in Table 11.2. As seen, using the modified input resulted in a MAC of nearly 100 % between the component response on the fixture and the component response in the system, whereas the MAC was around 75 % using the original input.

Additionally, the terms on the right hand side of Eq. (11.7) were investigated for each fixture/subassembly mode per each component mode. The terms identify how each fixture/subassembly mode is contributing to the component response. The terms are plotted in Appendix 1. As seen in the plots, the 6th fixed boundary component mode is excited most in the full system configuration. When the component was excited in the fixture with the original input, the excitation was dominated by the first elastic mode of the fixture, which is expected because of the excitation frequency. When the input to the fixture was modified, the response of each component mode matched more closely with the component response in the system. The elastic fixture motion was brought down to be in family with the excitation provided by the rigid body modes.

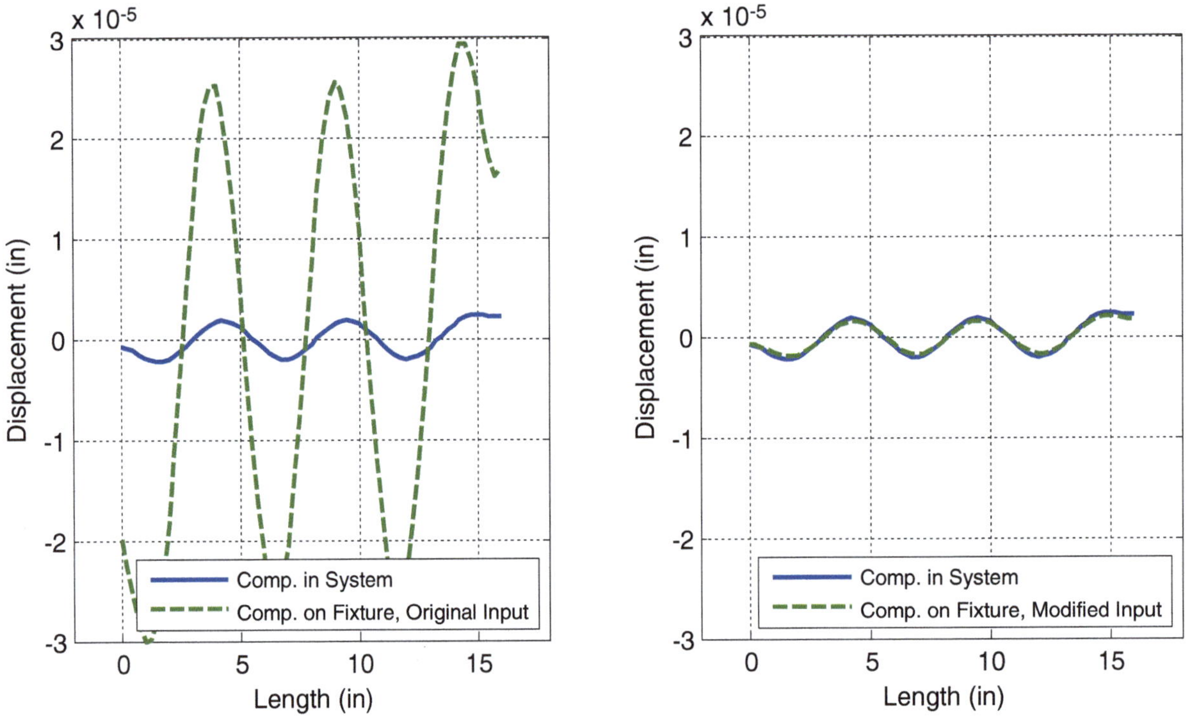

Fig. 11.11 Comparison of component response in system to component response on fixture with original and modified input, excitation frequency near a fixture mode

Table 11.2 MAC between component response in system and component response on fixture with original and modified input, excitation frequency near a fixture mode

	Component response on fixture, original input	Component response on fixture, modified input
Component response in system	75.59 %	99.80 %

11.4 Conclusions and Future Work

In this work, the modal Craig-Bampton theory presented in [10] was investigated. The method provided a better understanding of how the fixture/subassembly modes contributed to the response of a component due to an applied force. Two forcing functions were investigated, one with a frequency close to one of the elastic modes of the subassembly, and one with a frequency close to one of the elastic modes of the fixture. When the first force was applied, the component responded much greater in the system than in the fixture, as expected. When the second force was applied, the component responded much greater in the fixture than in the system, also as expected. In both cases, the modal Craig-Bampton theory was used to develop fixture inputs that excited the component notably more similar to the way it was excited in the full system. The appropriate fixture inputs were developed by using the rigid body and elastic modes of the fixture to excite the component. While the forces were only applied at discrete frequencies here, the problems exemplify the differences that could be encountered due to excitation in different frequency bands.

The work presented here demonstrated the differences that can be encountered when testing a component on a fixture that does not replicate the dynamics of the next level of assembly. Next, it is desired to solve similar problems for different inputs. Primarily, the problem should be solved across a spectrum rather than just at a few discrete frequencies. It is also desired to demonstrate the methodology for inputs other than harmonic force inputs, such as auto spectral densities or mechanical shocks. For non-stationary inputs, it would be necessary to describe the modal response of the component in terms of energy, as noted in [10].

Appendix 1: Additional Model Details

The free modes of the models up to 2000 Hz are shown in the following figures. The free modes of the truth system (including the component) are shown in Fig. 11.12, and the free modes of the subassembly without the component are shown in Fig. 11.13. The free modes of the component in the fixture are shown in Fig. 11.14, and the free modes of the fixture alone are shown in Fig. 11.15. The free modes of the component itself are shown in Fig. 11.16.

As noted in Sects. 11.3.2 and 11.3.3, the contribution of component response from each subassembly/fixture mode was calculated. The contributions from Sect. 11.3.2, where the input frequency was close to a subassembly mode, are shown in Fig. 11.17 for the subassembly, Fig. 11.18 for the fixture with the original input, and Fig. 11.19 for the fixture with the modified input. The contributions from Sect. 11.3.3, where the input frequency was close to a subassembly mode, are shown in Fig. 11.20 for the subassembly, Fig. 11.21 for the fixture with the original input, and Fig. 11.22 for the fixture with the modified input. Note that for both cases, the figure showing the subassembly contributions has the same y-scale as the figure showing the fixture contributions with the modified input, however the figure showing the fixture contributions with the original input has a significantly different y-scale.

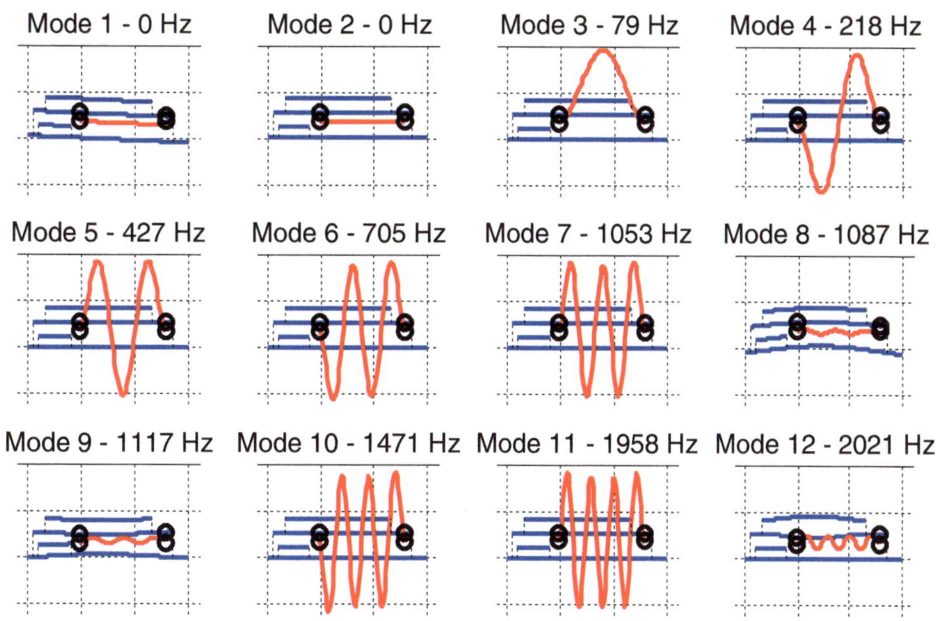

Fig. 11.12 Free modes of system (with component)

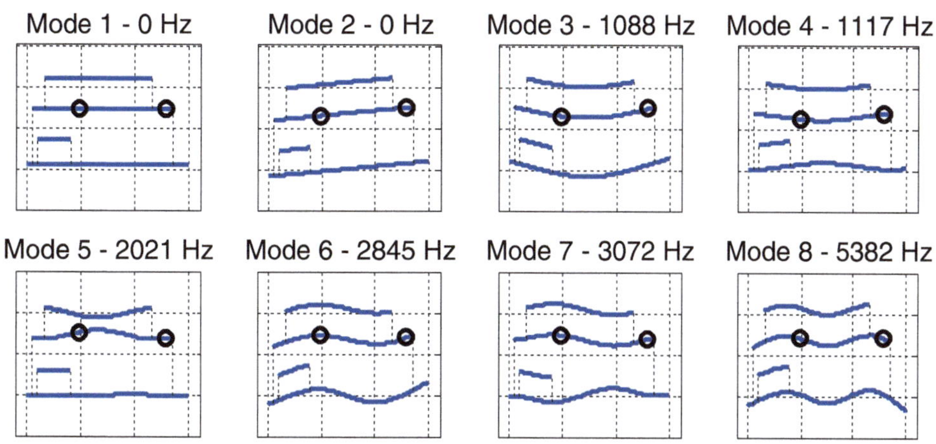

Fig. 11.13 Free modes of subassembly (without component)

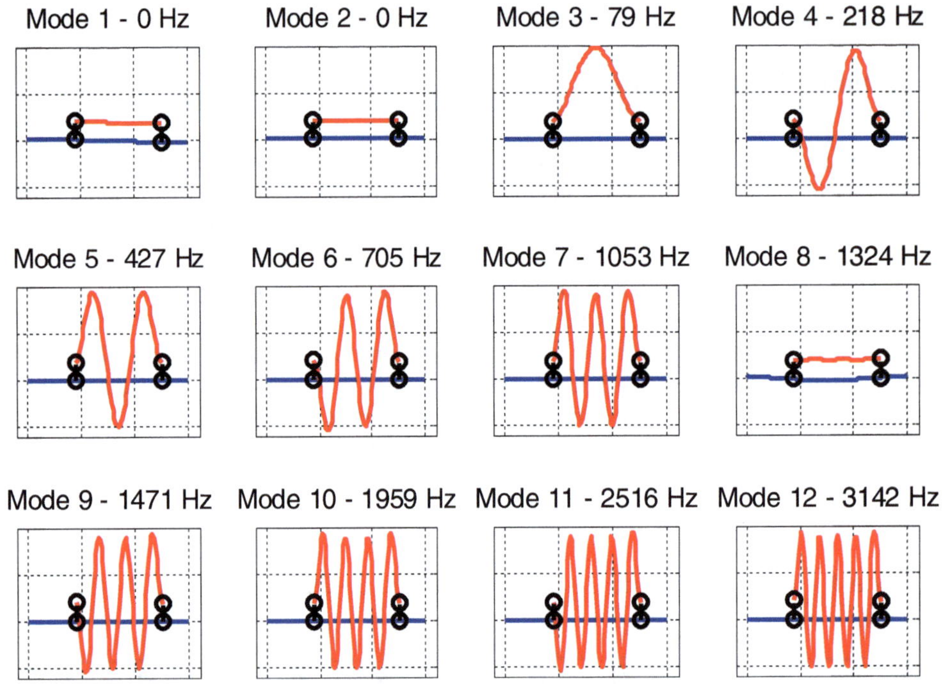

Fig. 11.14 Free modes of component in fixture

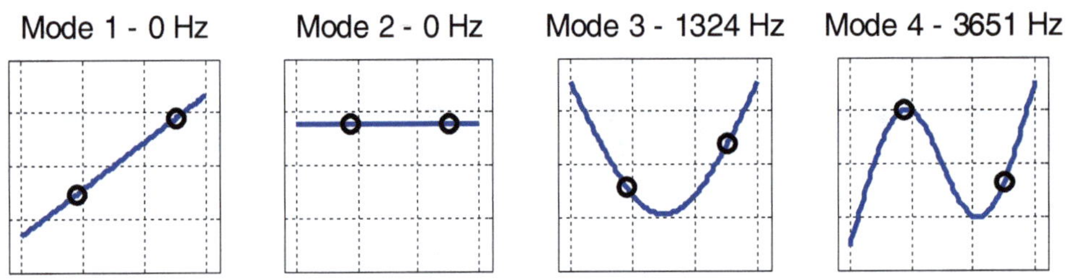

Fig. 11.15 Free modes of fixture

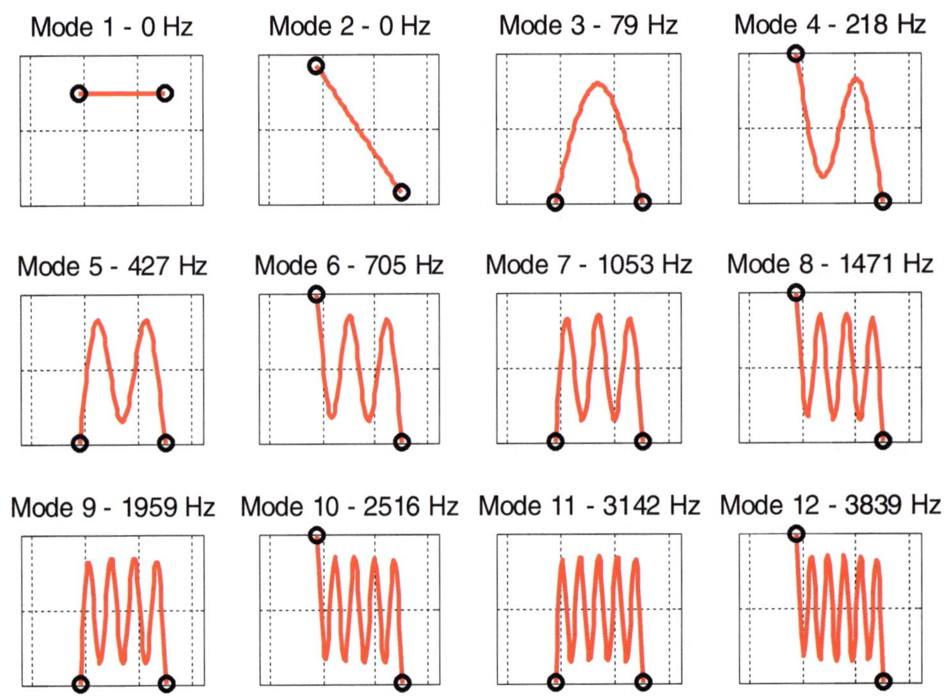

Fig. 11.16 Free modes of component

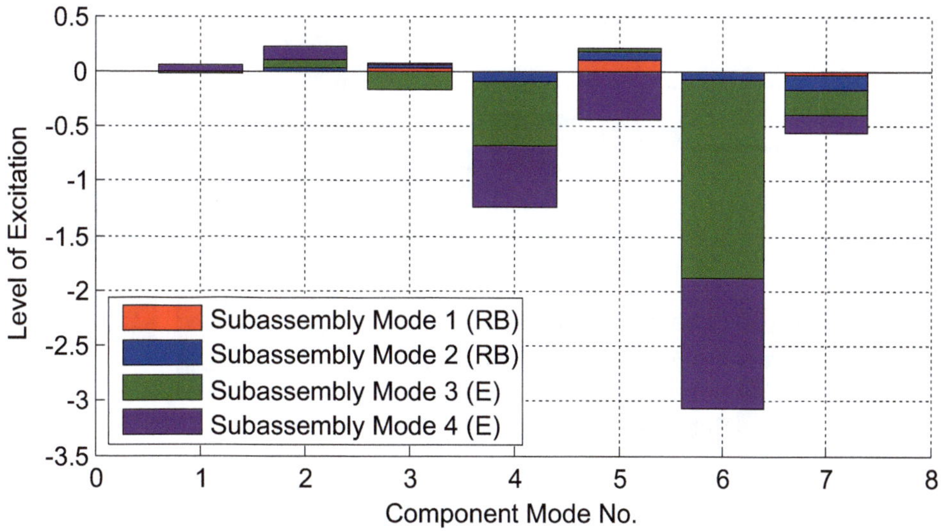

Fig. 11.17 Excitation of component from each subassembly mode with excitation frequency near a subassembly mode

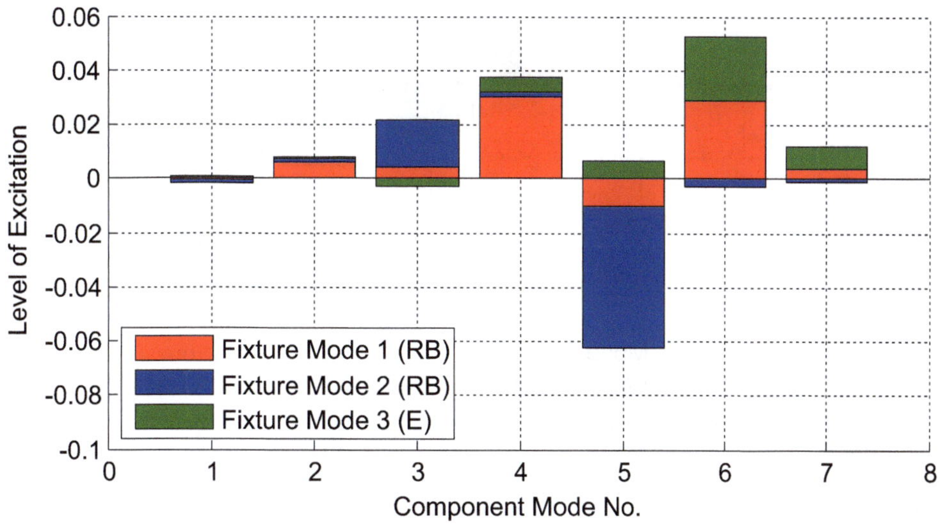

Fig. 11.18 Excitation of component from each fixture mode (original input) with excitation frequency near a subassembly mode

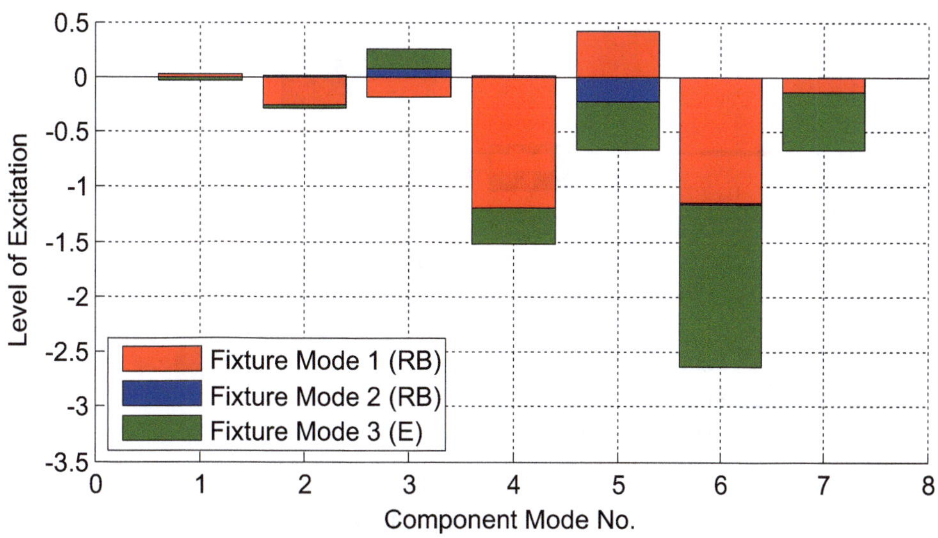

Fig. 11.19 Excitation of component from each fixture mode (mod. input) with excitation frequency near a subassembly mode

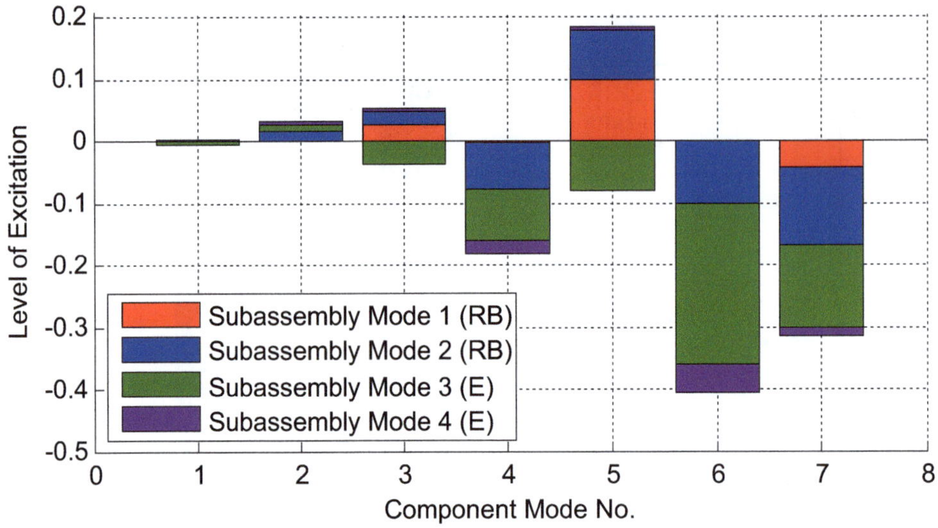

Fig. 11.20 Excitation of component from each subassembly mode with excitation frequency near a fixture mode

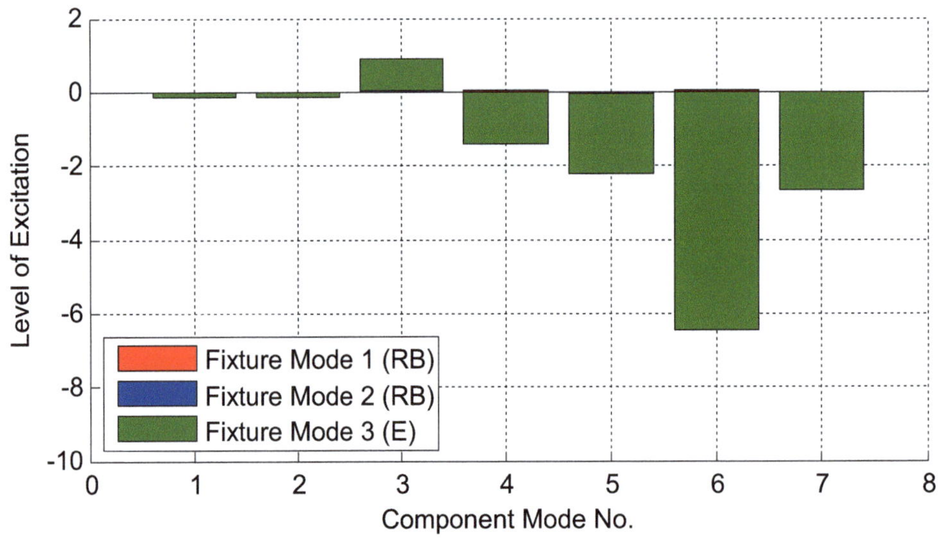

Fig. 11.21 Excitation of component from each fixture mode (original input) with excitation frequency near a fixture mode

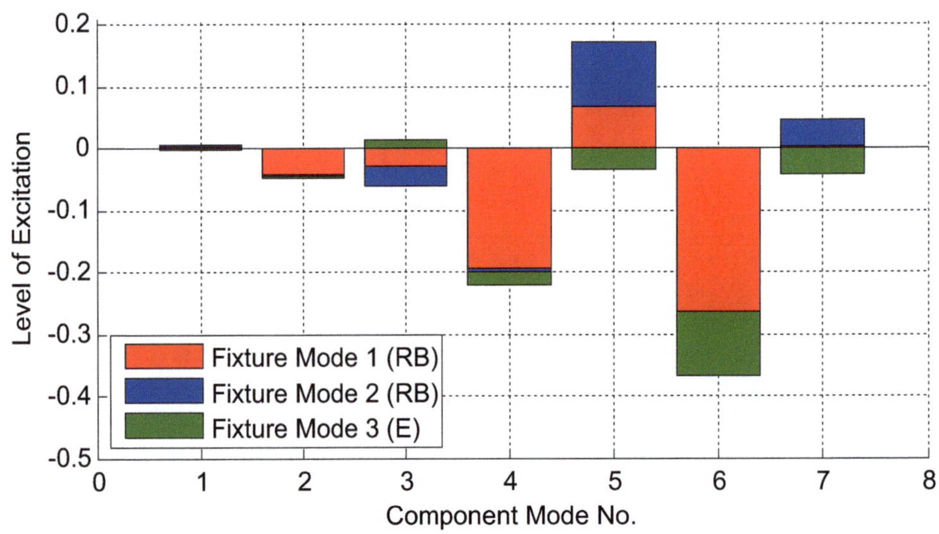

Fig. 11.22 Excitation of component from each fixture mode (mod. input) with excitation frequency near a fixture mode

References

1. Kumar, M., Rao, T.N., Jagadisan, K., Rao, K.J.: Tailoring of Vibration Test Specifications for a Flight Vehicle. Research Centre Imarat, Hyderabad
2. Gatscher, J.A., Kawiecki, G.: Comparison of mechanical impedance methods for vibration simulation. Shock Vib. **3**, 223–232 (1996)
3. Environmental Engineering Considerations and Laboratory Tests. Department of Defense Test Method Standard, MIL-STD-810G, 31 January 2008
4. Avitabile, P.: Why you can't ignore those vibration fixture resonances. Sound Vib. (1999)
5. Piersol, A.G., White, P.H., Wilby, J.F., Hipol, P.J., Wilbey, E.G.: Vibration test procedures for orbiter sidewall-mounted payloads. Astron Research and Engineering Report 7114-01, November 1988
6. Force Limited Vibration Testing. NASA Technical Handbook, NASA-HDBK-7004C, Revision C, 30 November 2012
7. Scharton, T.D.: Force limited vibration testing monograph. NASA Reference Publication RP-1403, May 1997.
8. Daborn, P.M., Roberts, C., Ewins, D.J., Ind, P.R.: Next-generation random vibration tests. In: Proceedings of the 32nd International Modal Analysis Conference, Orlando, FL, February 2014, Paper #92
9. Mayes, R.L.: A Craig-Bampton experimental dynamic substructure using the transmission simulator method. In: Proceedings of the 33rd International Modal Analysis Conference, Orlando, FL, February 2015, Paper #353
10. Mayes, R.L.: A modal Craig-Bampton substructure for experiments, analysis, control and specifications. In: Proceedings of the 33rd International Modal Analysis Conference, Orlando, FL, February 2015, Paper #352
11. Avitabile, P.: Review Finite Element Modeling. 22.515 Modal Analysis I Class Notes, University of Massachusetts Lowell, 19 June 2004

Chapter 12
Modal Updating of Tail of a Military Helicopter

Kurtuluş Ersoy, Mert Atasoy, and Cem Genç

Abstract Modal updating is a method to correct the dynamic behavior of FEM which contain uncertainties because of the boundary conditions, partial modeling, joint and contact stiffness values etc. Conducting structural dynamic analysis which represent different modal behavior than the real structure can produce considerable different analysis results like displacement, stress and fatigue life response in addition to the natural frequency and mode shape results. In this study, modal updating technique using natural frequency and mode shape differences as residual values is applied to the tail of a military helicopter. The finite element model by using beam and shell elements is created in Altair HYPERMESH® and modal analysis results are obtained using Altair RADIOSS® solver. Modal test, which is called ground vibration test (GVT) for the aerospace structures, is conducted by using LMS SCADAS® data acquisition system and modal test results are obtained using LMS TESTLAB® Polymax. Modal updating is conducted using LMS VIRTUAL LAB® together with NASTRAN SOL200® solver and point mass values assigned for some parts of the tail together with stiffness values for the fuselage-tail connection assigned by using CBUSH elements are updated.

Keywords Modal analysis • Modal updating • Model updating • Ground vibration test (GVT) • Experimental validation • Structural dynamics

12.1 Introduction

In general, finite element models can resemble different dynamic behavior than the real structures because of modeling uncertainties like boundary conditions, fastener and contact behaviors etc. Conducting modal test to control the resonance for all of the configurations, before and after the integration is not practical. Using finite element models which are not updated is not adequate to estimate the resonance risk, fatigue life of integration regions like helicopter tail or airplane wing or to get the correct displacement and stress values in different dynamic loading conditions. Therefore, it is very important to update finite element model according to the modal test results and to get accurate structural dynamic analysis results for different configurations.

Modal updating studies have been conducted for approximately 35 years. Baruch updated the stiffness and flexibility matrices [1]. Berman and Nagy developed methods to update mass and stiffness matrices [2]. Lin and Zhu made it possible to use damping matrices together with stiffness and mass matrices by using frequency response method [3]. Göge and Link are compared the method by which natural frequency and mode shape differences used as residuals with the method by which the natural frequency and frequency response function differences used as residuals. Both methods are able to update the mass and stiffness matrices, while the second one is also able to update the damping matrix. It is shown that both methods are produced reasonable results [4]. In this study, the method by which natural frequency and mode shape differences used as residuals is used.

12.2 Methodology

Modal updating procedure basically consists of three stages after obtaining the modal analysis and test results in the appropriate format. These steps are correlation analysis, parameter sensitivity analysis and optimization phase.

K. Ersoy (✉) • M. Atasoy • C. Genç
Radar & Electronic Warfare Business Sector Aircraft Integration Engineering Department, ASELSAN Inc., Ankara 06830, Turkey
e-mail: kersoy@aselsan.com.tr

A. Brandt, R. Singhal (eds.), *Shock & Vibration, Aircraft/Aerospace, Energy Harvesting, Acoustics & Optics, Volume 9*,
Conference Proceedings of the Society for Experimental Mechanics Series, DOI 10.1007/978-3-319-30087-0_12

In the correlation analysis part, the differences between the finite element model and test model are obtained. Correlation analysis part is completed in two stages, namely geometric correlation and dynamic correlation. Data acquisition points in modal test are paired with related nodes in the finite element model in geometric correlation phase. In the dynamic correlation phase, modal test and finite element model natural frequency and mod shape results are compared with each other. Mod shape correlation degree is determined according to Modal Assurance Criterion (MAC) function. MAC function is calculated as

$$\boldsymbol{MAC_{rs}} = \frac{\left\lfloor \{\varphi\}_r^T \{\varphi\}_s \right\rfloor^2}{\left\lfloor \{\varphi\}_r^T \{\varphi\}_r \right\rfloor \left\lfloor \{\varphi\}_s^T \{\varphi\}_s \right\rfloor} \tag{12.1}$$

MAC = 1 represents the perfect correlation whereas MAC = 0 is due to no correlation. For the practical purposes, for example for satellite projects, for main bending mode, MAC value is stated to be higher than 0.9 and frequency deviation is stated to be smaller than 3 %. For other structural modes, MAC value should be higher than 0.8 and frequency deviation should be smaller than 10 % [5].

Correlation phase is a very convenient method to compare two finite element analysis models which are created by using different techniques besides being the first stage of modal updating. Uhl et al. determined to update finite element model of a helicopter by making correlation analysis and it was shown that MAC values are increased by omitting some data acquisition points from modal test results [6]. Çınar evaluated the similarity of mod shape results for two different finite element models, which are created by shell and solid elements, for an aluminum honeycomb plate [7].

In the sensitivity analysis part, the sensitivity of modal frequencies and mode shapes to the finite element modeling parameters. Accordingly, the importance of the modeling parameters are compared with each other and parameter sequence in optimization part is determined. Sensitivity analysis is very useful especially in the finite element models which contain a lot of parameters since the duration of optimization analysis part is shortened.

In the optimization analysis part, the parameters determined in the sensitivity analysis part are defined as analysis input. Natural frequency and mode shape differences between finite element model and modal test model are defined as analysis output. By using optimization solvers, NASTRAN SOL200® etc., finite element model is updated to give results similar with modal test results.

12.3 Modal Analysis Details of Helicopter Tail Boom

This section discusses finite element modelling of the helicopter tail and modal analysis results for which modal updating applied.

Finite element model of the helicopter tail is created by using 1D beam elements for stringer and frame structures and 2D shell elements for skin as in Fig. 12.1 by using HYPERMESH®. Modelling strategy is determined according to the integration locations and approximately half of the helicopter tail is modelled in detail and for the remaining part, assumed mass values for which it was not possible to weigh are added by using point mass elements as in Fig. 12.2. Internal equipment are also added as point mass elements as in Fig. 12.3 and for all of the mass elements, mass moment of inertia values are also taken into account and they are used as parameters to update the finite element model.

Fuselage part of the helicopter is not modelled and fuselage-tail connection of the helicopter is modelled having translational and rotational stiffness values by using CBUSH elements as in Fig. 12.4. At the first stage, all of the stiffness values are entered high values as if they are rigid elements. They are assigned different property cards, PBUSH, by grouping them symmetrically as in Fig. 12.5 to increase the number of parameters in the modal updating stage and to get more realistic finite element model.

By using described finite element model, modal analysis results for the first vertical and lateral bending modes are as in Figs. 12.6 and 12.7, respectively by using RADIOSS® solver.

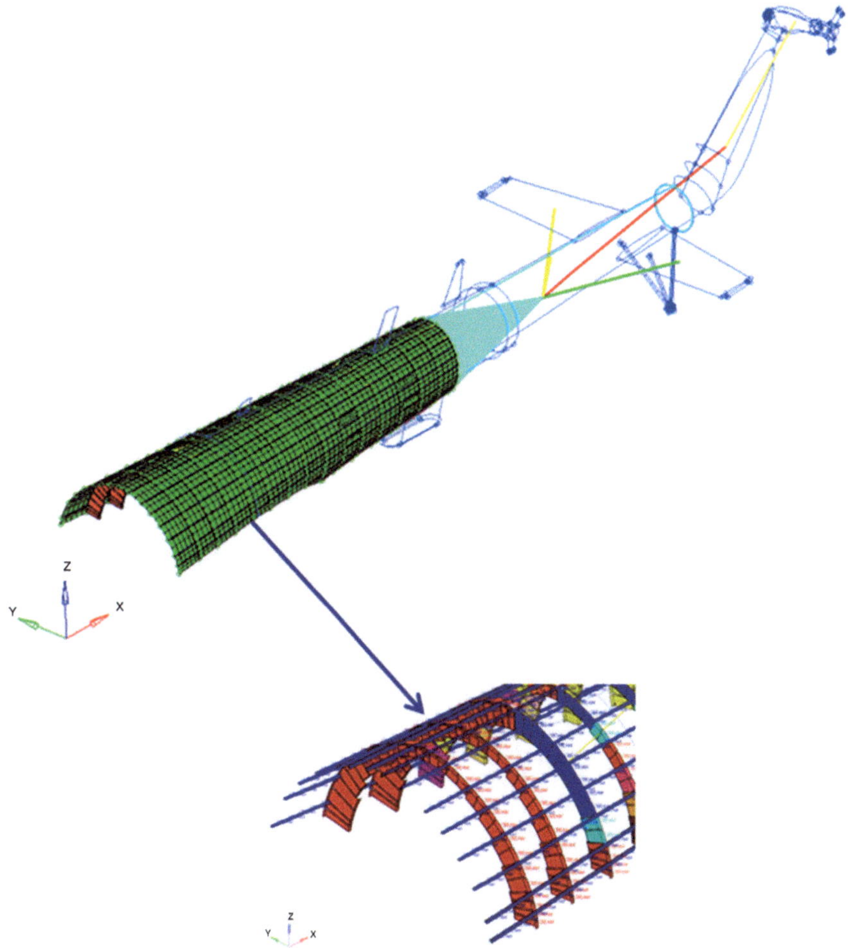

Fig. 12.1 General overview of finite element model of helicopter tail boom

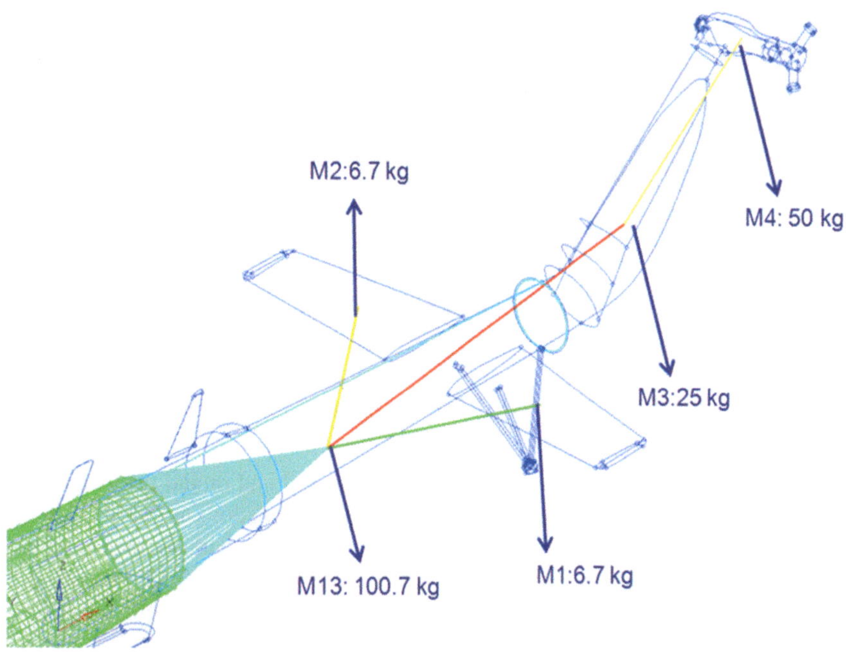

Fig. 12.2 Mass addition for non-modelled parts in finite element model of helicopter tail boom

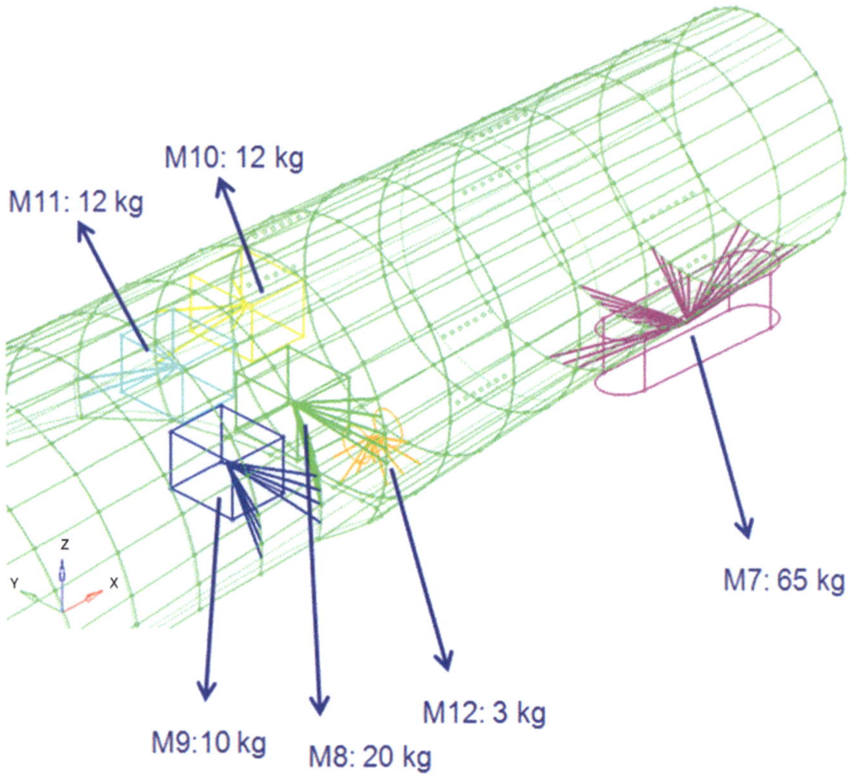

Fig. 12.3 Mass addition for electronic equipment in finite element model of helicopter tail boom

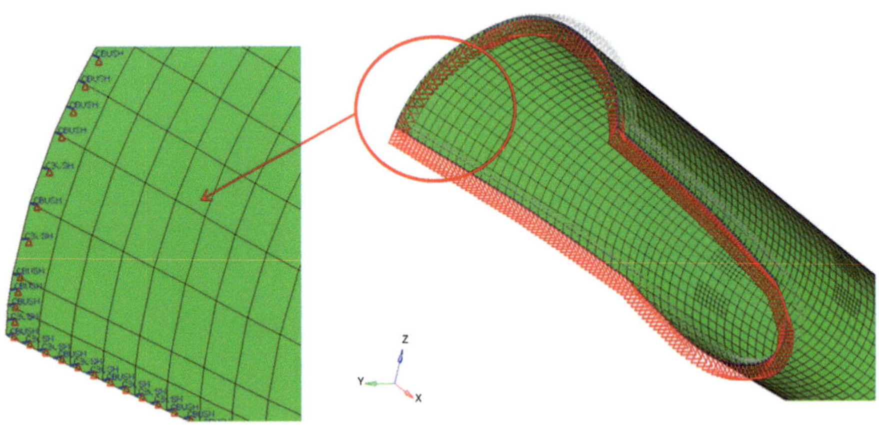

Fig. 12.4 Boundary condition for finite element model of helicopter tail boom

12.4 Modal Test Details of Helicopter Tail Boom

This section discusses modal test of the helicopter tail and modal test results used as reference in modal updating.

Accelerometers are located in the determined frame sections as in Fig. 12.8. The structure is excited by two modal shakers in vertical and lateral directions and acceleration and force data are measured by using LMS SCADAS® test system. Modal test model is created by using accelerometer locations as in Fig. 12.9 by using LMS TESTLAB®.

By using described modal test model, modal test results for the first vertical and lateral body bending modes are as in Figs. 12.10 and 12.11, respectively.

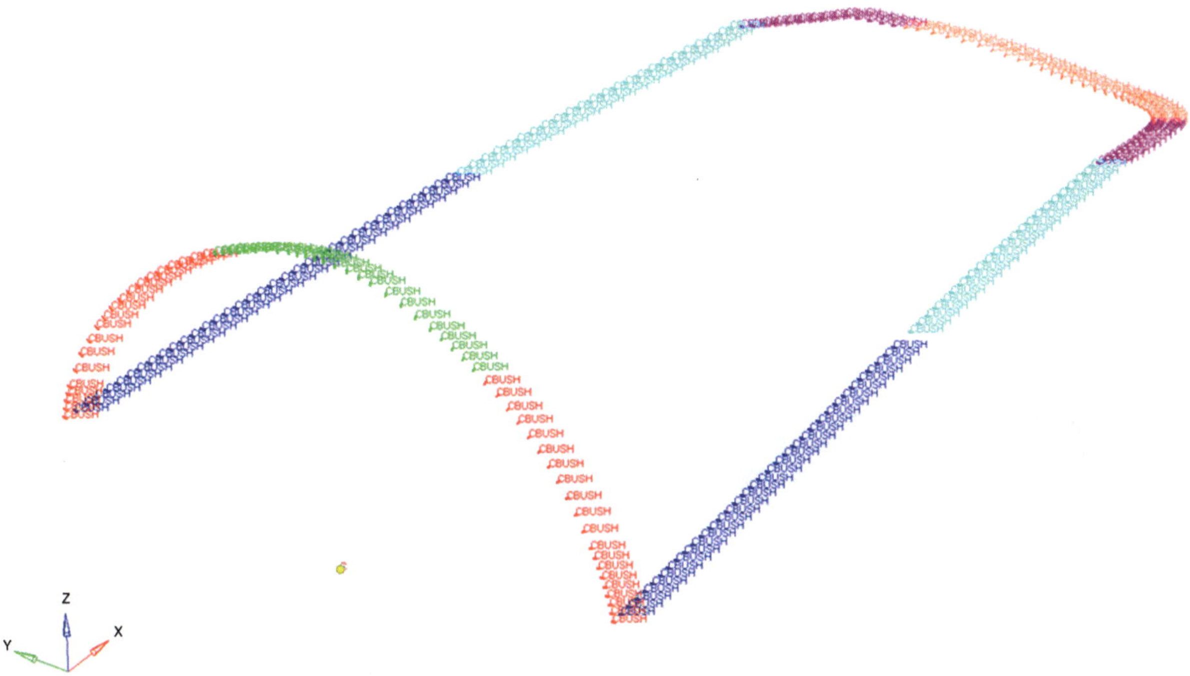

Fig. 12.5 CBUSH elements categorization for finite element model of helicopter tail boom

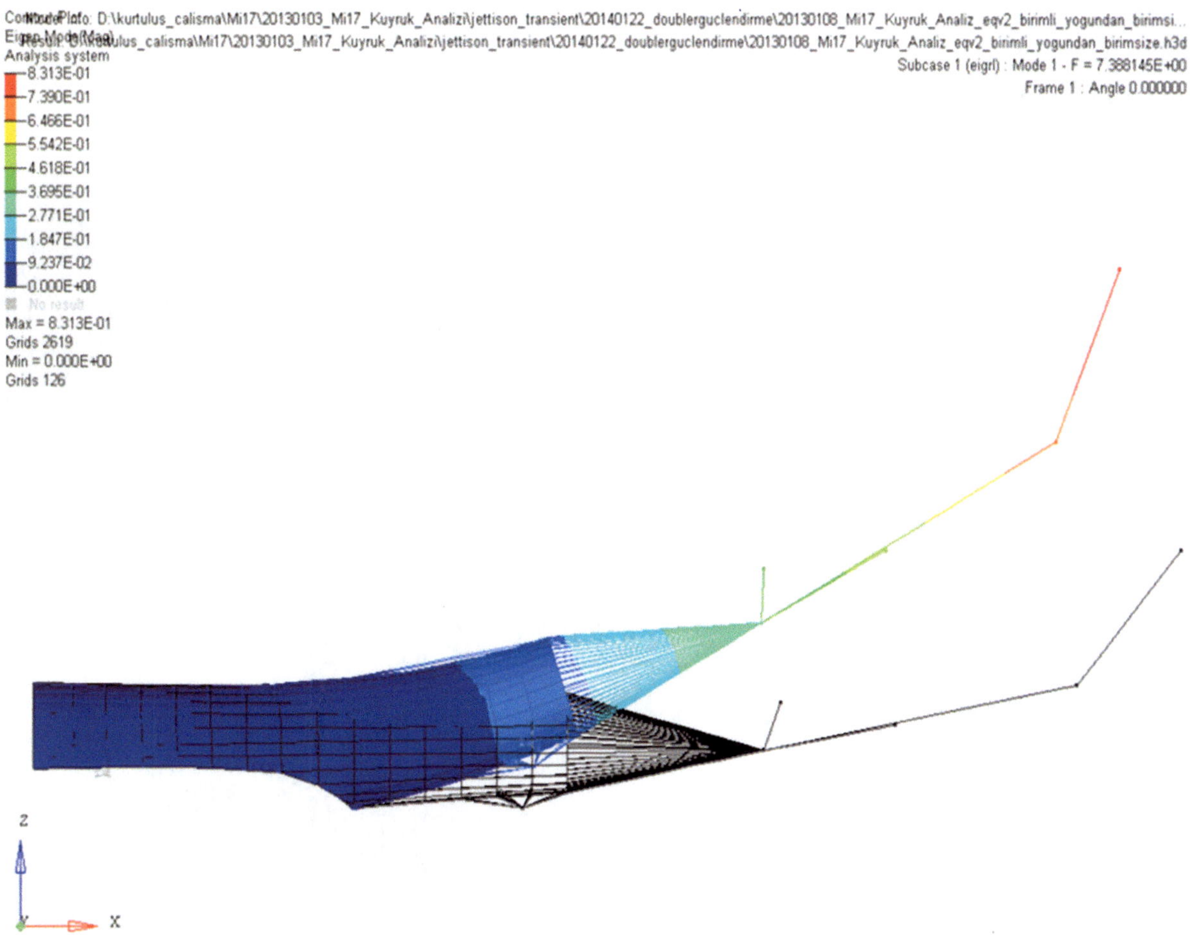

Fig. 12.6 Vertical bending mode result by finite element model of helicopter tail boom (7.39 Hz)

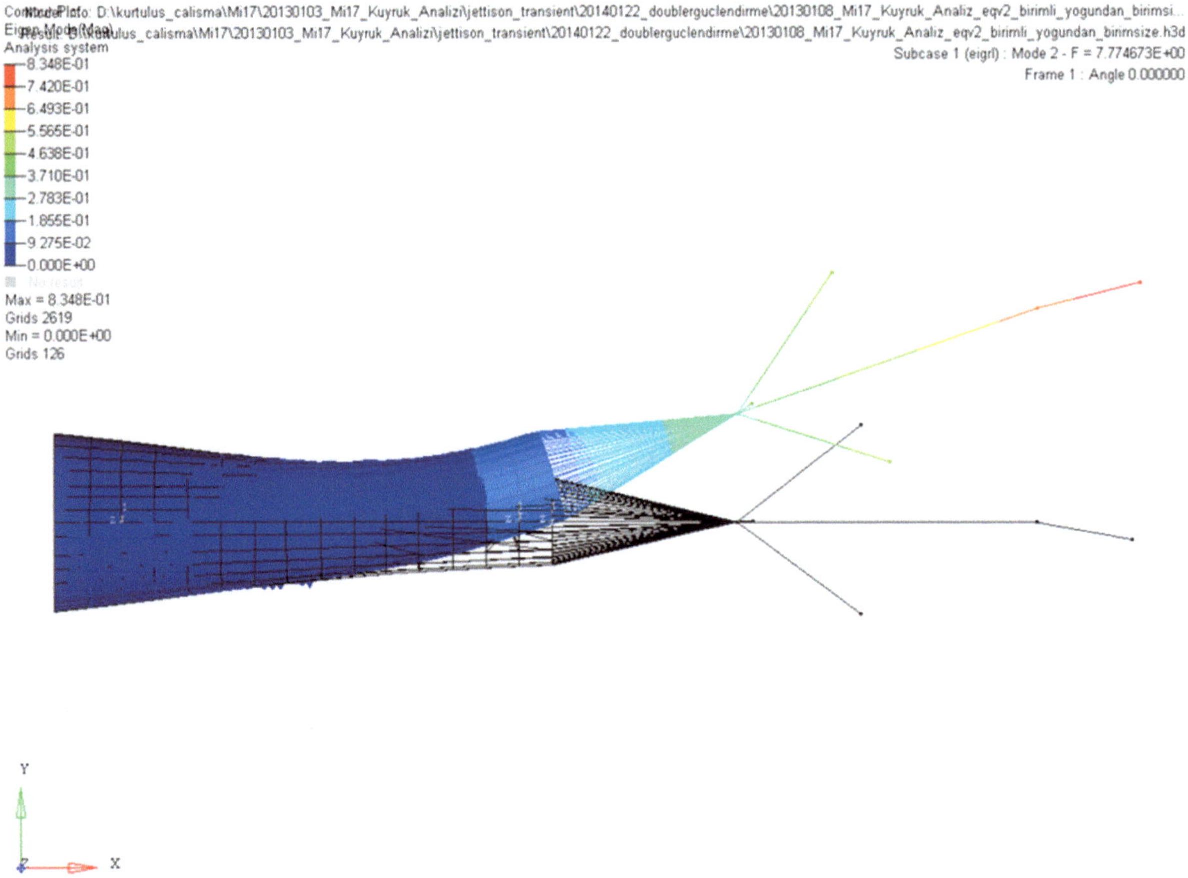

Fig. 12.7 Lateral bending mode result by finite element model of helicopter tail boom (7.77 Hz)

Fig. 12.8 Modal test setup for the helicopter tail boom

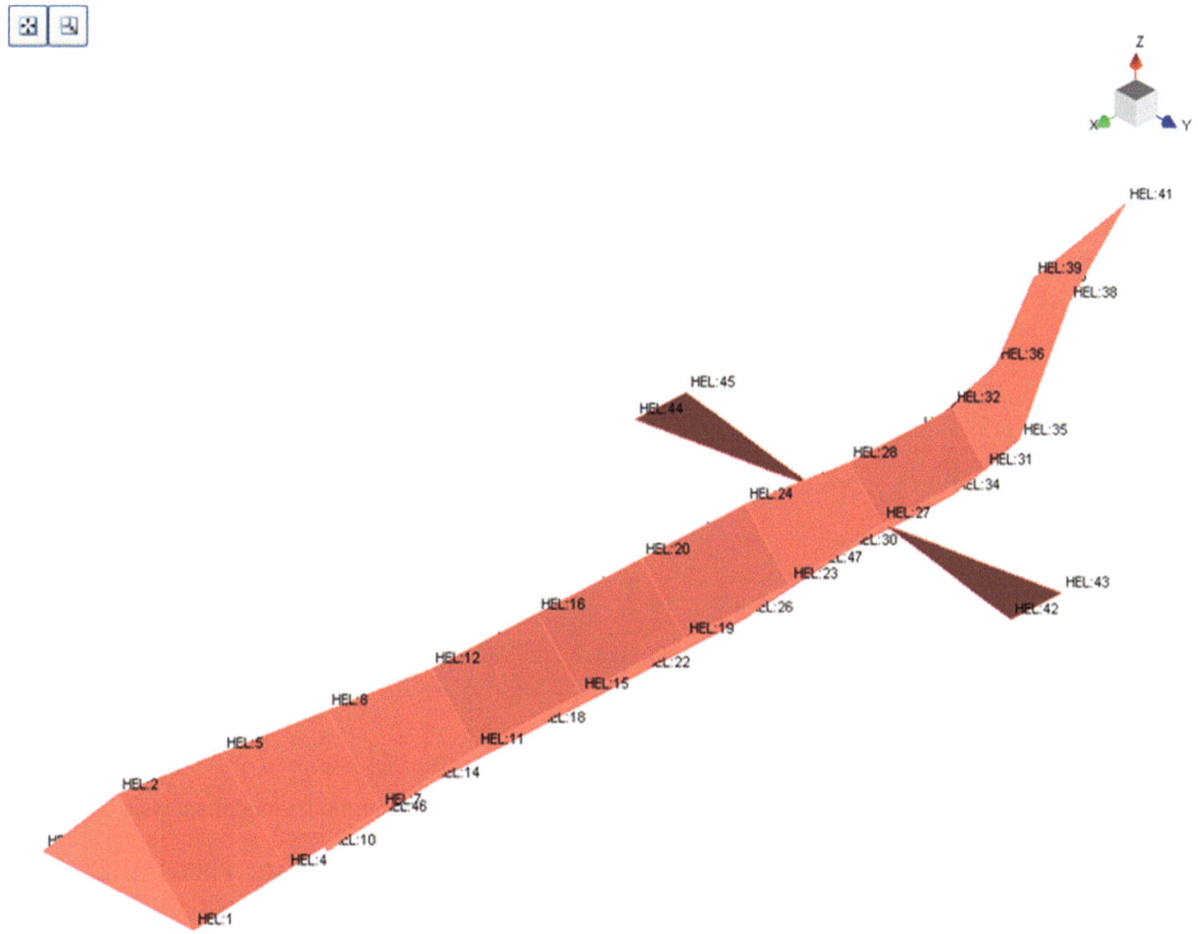

Fig. 12.9 Helicopter tail boom accelerometer locations

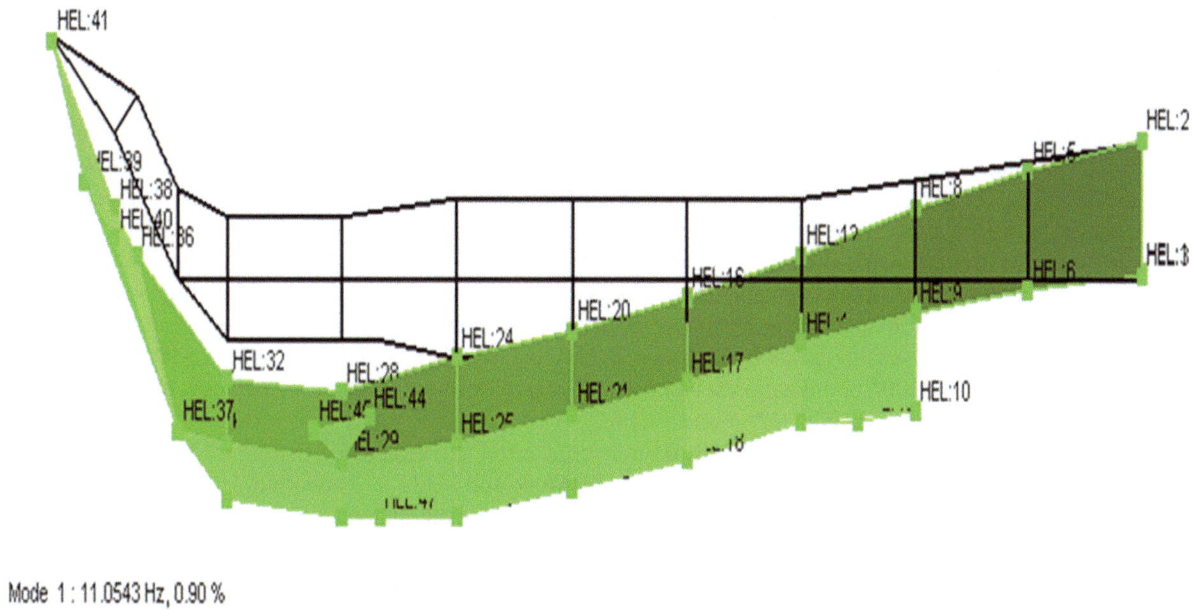

Fig. 12.10 Vertical bending mode of the helicopter tail boom (11.05 Hz)

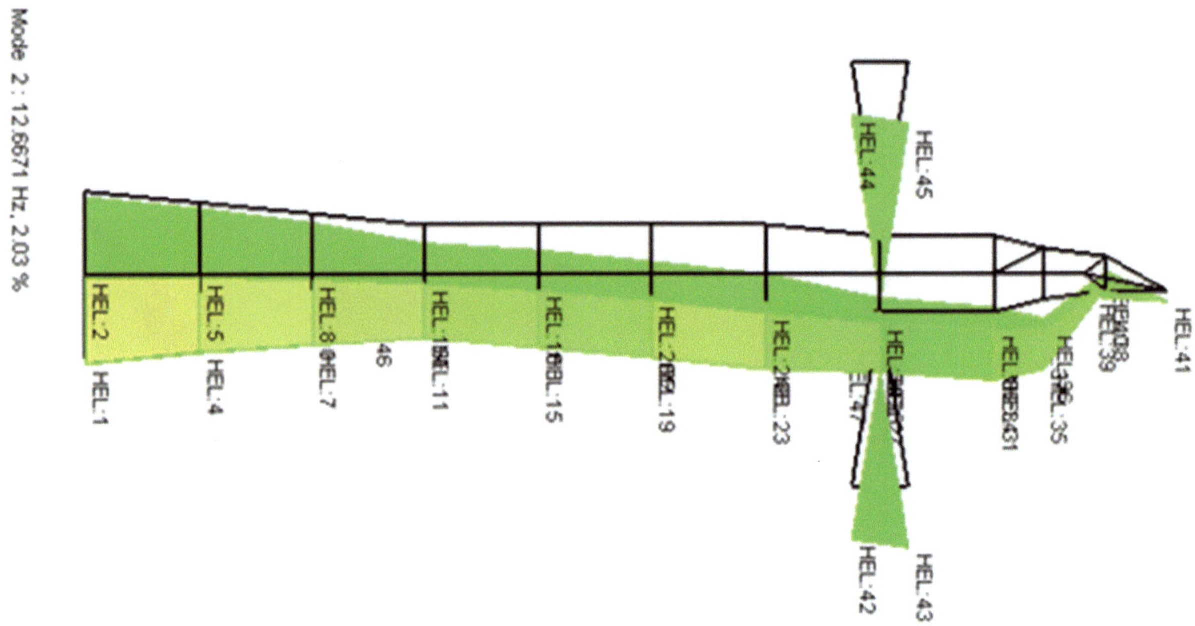

Fig. 12.11 Lateral bending mode of the helicopter tail boom (12.67 Hz)

12.5 Modal Updating

This section discusses modal updating of the helicopter tail by using finite element model modal analysis and modal test results discussed in the previous sections.

The first step in the modal updating is geometric correlation conducted by using LMS VIRTUAL LAB®. Modal test accelerometer locations are used as reference nodes and related finite element nodes are selected to be verification nodes as in Fig. 12.12.

Completing the geometric correlation, MAC values are obtained as in Fig. 12.13 by using LMS VIRTUAL LAB® and dynamic correlation results before the modal updating are summarized in Table 12.1.

Normalized sensitivity results, which are obtained by using LMS VIRTUAL LAB® together with NASTRAN SOL200® solver, for the most effective parameters for vertical and lateral body bending modes are given in Figs. 12.14 and 12.15, respectively. M4, M13 etc. represents mass values of un-modelled parts and CB6_K5, CB4_K5 etc. represents stiffness values for the boundary condition.

Investigating Table 12.1, the difference between vertical and lateral body bending modal frequencies for finite element model results (7.39 and 7.77 Hz) are lower than the modal test results (11.05 and 12.67 Hz). Since updating the mass values affects both modal frequencies similarly, it is determined to update the mass values firstly to decrease the difference between the lateral body bending modal frequency from finite element model, 7.77 Hz, and the lateral body bending modal frequency from test model, 12.67 Hz nearly to zero. After that, boundary condition stiffness parameters are updated. According to the sensitivity results, the stiffness parameters which are more sensitive for vertical modes are selected as in Fig. 12.16 in yellow color and they are used in the modal updating procedure according to their vertical to lateral mode sensitivity ratios. This updating strategy is applied since stiffness parameters are modeled as if they are rigid elements with high stiffness values at the first stage as mentioned before and it is meaningless to increase the stiffness parameters more.

Completing correlation and sensitivity analysis phases, modal updating is completed by using LMS VIRTUAL LAB® together with NASTRAN SOL200® optimization solver. By using mentioned modal updating strategy, mass parameters and stiffness parameters are updated as in Figs. 12.17 and 12.18.

*MAC_2_1 and Freq_Diff_2_1 represents Modal Assurance Criterion and Frequency Difference between finite element modal analysis and modal test results for Vertical Body Bending Mode and MAC_1_2 and Freq_Diff_1_2 represents Modal Assurance Criterion and Frequency Difference between finite element modal analysis and modal test results for Lateral Body Bending Mode.

Completing all of the modal updating steps, comparison of modal frequencies before and after the modal updating is given in Table 12.2. As it can be seen, after modal updating, both of the vertical and lateral bending results from finite element model is very similar with modal test results. In addition, MAC values for both modes are also increased about % 1–1.5.

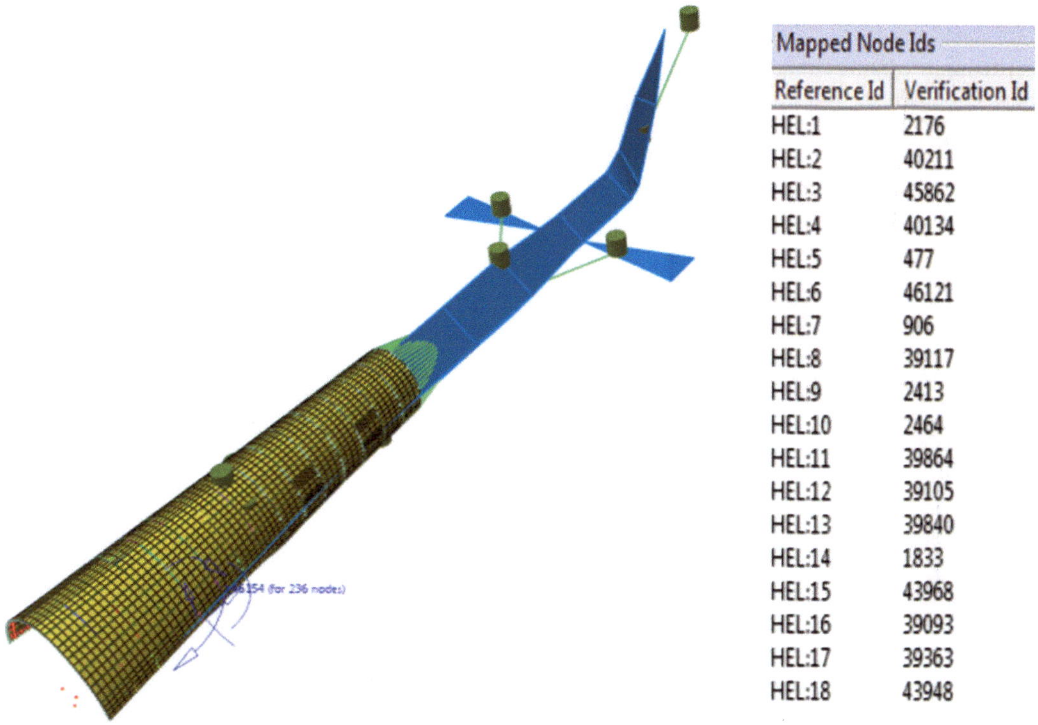

Fig. 12.12 Geometric correlation of finite element model and the test model

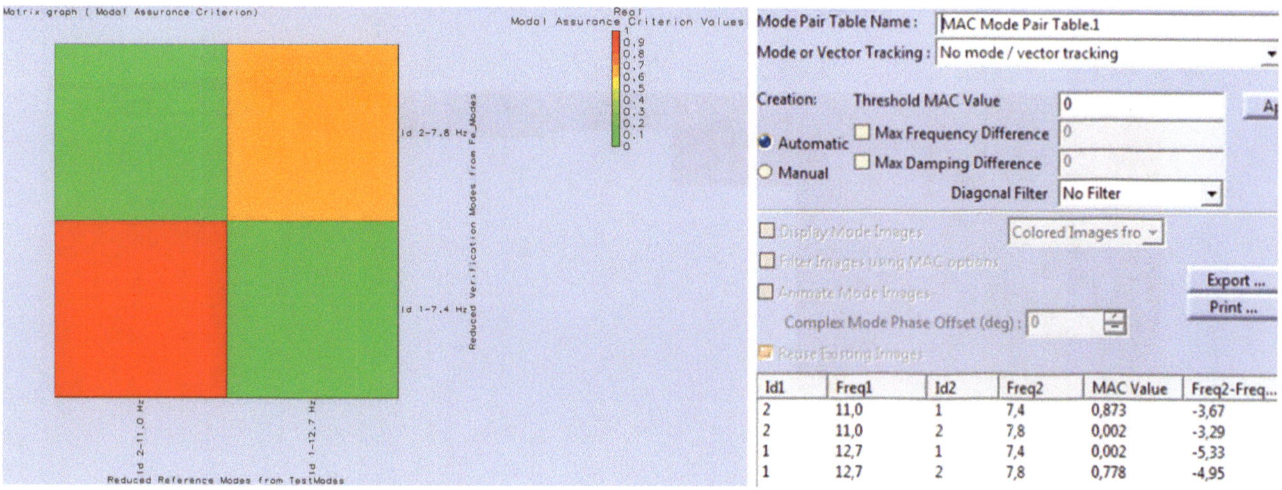

Fig. 12.13 MAC graph and MAC mode pair table

Table 12.1 Dynamic correlation results before the modal updating

	Modal test	Finite element model	Difference	MAC
Vertical bending mode	11.05 Hz	7.39 Hz	33.1 %	0.873
Lateral bending mode	12.67 Hz	7.77 Hz	38.6 %	0.778

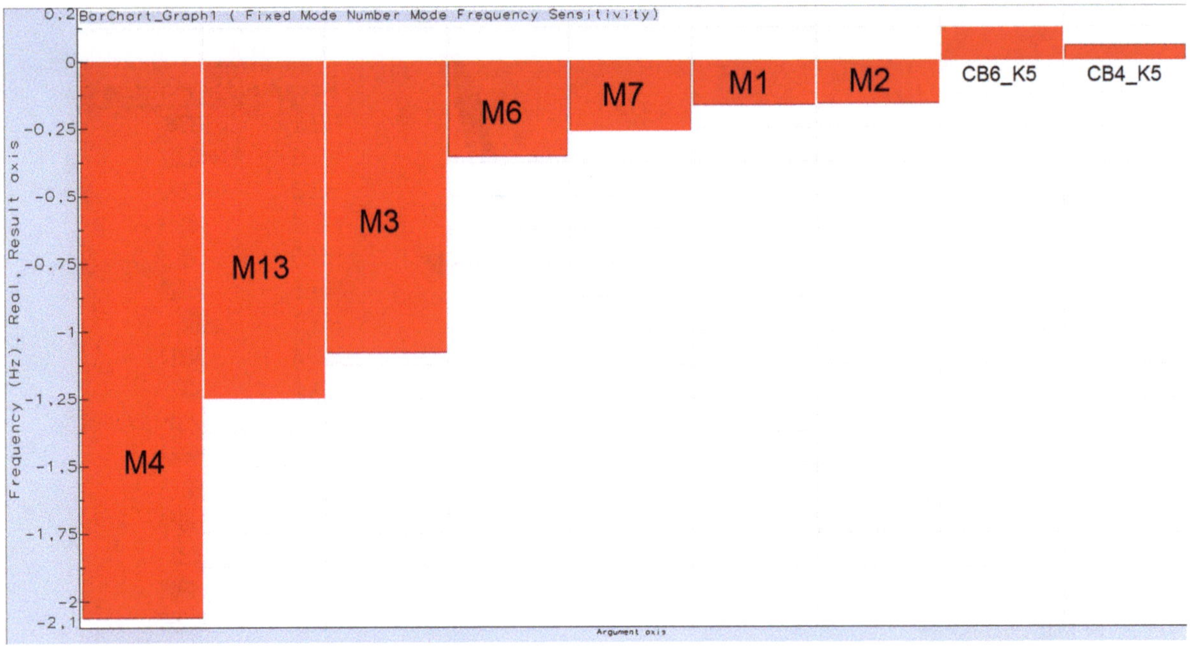

Fig. 12.14 Sensitivity results for vertical bending mode

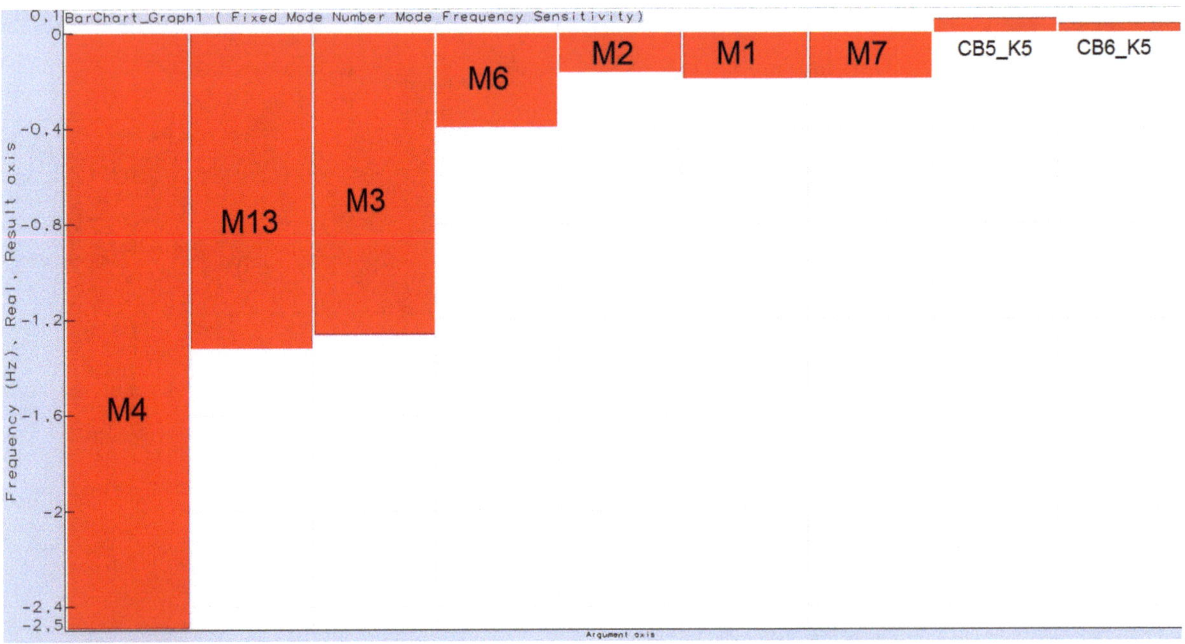

Fig. 12.15 Sensitivity results for lateral bending mode

Parameter	Vert. Mod -Sensitivity	Lat. Mod -Sensitivity	Vert. Sens./Lat. Sens.
Design Variable Group for Sensitivity (Bushings) - Nominal stiffness RX (K4).6_65	1.28E-05	2.86E-07	44.80
Design Variable Group for Sensitivity (Bushings) - Nominal stiffness RX (K4).6_66	2.40E-05	2.69E-07	89.43
Design Variable Group for Sensitivity (Bushings) - Nominal stiffness RX (K4).6_67	3.19E-04	8.83E-05	3.61
Design Variable Group for Sensitivity (Bushings) - Nominal stiffness RX (K4).6_68	0.001	0.001	1.00
Design Variable Group for Sensitivity (Bushings) - Nominal stiffness RX (K4).6_69	2.34E-04	0.002	0.12
Design Variable Group for Sensitivity (Bushings) - Nominal stiffness RX (K4).6_70	0.001	0.001	1.00
Design Variable Group for Sensitivity (Bushings) - Nominal stiffness RY (K5).9_65	0.004	1.42E-04	28.13
Design Variable Group for Sensitivity (Bushings) - Nominal stiffness RY (K5).9_66	0.005	3.86E-05	129.70
Design Variable Group for Sensitivity (Bushings) - Nominal stiffness RY (K5).9_67	0.007	0.002	3.50
Design Variable Group for Sensitivity (Bushings) - Nominal stiffness RY (K5).9_68	0.028	0.038	0.74
Design Variable Group for Sensitivity (Bushings) - Nominal stiffness RY (K5).9_69	0.01	0.097	0.10
Design Variable Group for Sensitivity (Bushings) - Nominal stiffness RY (K5).9_70	0.132	0.049	2.69
Design Variable Group for Sensitivity (Bushings) - Nominal stiffness RZ (K6).10_65	3.11E-04	1.47E-05	21.20
Design Variable Group for Sensitivity (Bushings) - Nominal stiffness RZ (K6).10_66	4.20E-05	6.94E-05	0.61
Design Variable Group for Sensitivity (Bushings) - Nominal stiffness RZ (K6).10_67	3.10E-04	7.81E-05	3.96
Design Variable Group for Sensitivity (Bushings) - Nominal stiffness RZ (K6).10_68	3.18E-04	0.001	0.32
Design Variable Group for Sensitivity (Bushings) - Nominal stiffness RZ (K6).10_69	0.003	0.001	3.00
Design Variable Group for Sensitivity (Bushings) - Nominal stiffness RZ (K6).10_70	0.014	0.007	2.00
Design Variable Group for Sensitivity (Bushings) - Nominal stiffness X (K1).1_65	1.23E-06	5.34E-08	22.95
Design Variable Group for Sensitivity (Bushings) - Nominal stiffness X (K1).1_66	1.63E-06	2.89E-08	56.50
Design Variable Group for Sensitivity (Bushings) - Nominal stiffness X (K1).1_67	1.29E-05	9.13E-07	14.17
Design Variable Group for Sensitivity (Bushings) - Nominal stiffness X (K1).1_68	1.34E-04	1.56E-05	8.55
Design Variable Group for Sensitivity (Bushings) - Nominal stiffness X (K1).1_69	4.38E-05	8.58E-06	5.10
Design Variable Group for Sensitivity (Bushings) - Nominal stiffness X (K1).1_70	2.67E-05	1.07E-05	2.50
Design Variable Group for Sensitivity (Bushings) - Nominal stiffness Y (K2).2_65	5.99E-06	3.03E-07	19.80
Design Variable Group for Sensitivity (Bushings) - Nominal stiffness Y (K2).2_66	7.94E-07	1.36E-06	0.58
Design Variable Group for Sensitivity (Bushings) - Nominal stiffness Y (K2).2_67	9.42E-05	9.81E-06	9.60
Design Variable Group for Sensitivity (Bushings) - Nominal stiffness Y (K2).2_68	0.001	6.62E-05	15.10
Design Variable Group for Sensitivity (Bushings) - Nominal stiffness Y (K2).2_69	0.001	2.66E-04	3.76
Design Variable Group for Sensitivity (Bushings) - Nominal stiffness Y (K2).2_70	8.62E-05	6.42E-05	1.34
Design Variable Group for Sensitivity (Bushings) - Nominal stiffness Z (K3).3_65	1.50E-04	7.21E-06	20.76
Design Variable Group for Sensitivity (Bushings) - Nominal stiffness Z (K3).3_66	2.98E-04	2.79E-06	106.81
Design Variable Group for Sensitivity (Bushings) - Nominal stiffness Z (K3).3_67	2.11E-04	6.35E-05	3.32
Design Variable Group for Sensitivity (Bushings) - Nominal stiffness Z (K3).3_68	0.001	0.002	0.50
Design Variable Group for Sensitivity (Bushings) - Nominal stiffness Z (K3).3_69	2.42E-04	0.004	0.06
Design Variable Group for Sensitivity (Bushings) - Nominal stiffness Z (K3).3_70	0.025	0.013	1.92

Fig. 12.16 Sensitivity result comparison for vertical and lateral bending modes

Exp	Iter	Mode	Goal	M2 (kg)	M3 (kg)	M4 (kg)	M12 (kg)	MAC_1_2	MAC_2_1	Freq_Diff_1_2	Freq_Diff_2_1	Status
1	0	START	3,79387e+007	6,7077	25	49,968	3,0019	0,775058	0,871038	4,94833	3,6678	SUCCEEDED
2	1	GRAD	3,79398e+007	6,71208	25	49,968	3,0019	0,775058	0,871038	4,94839	3,66786	SUCCEEDED
						…						
49	10	GRAD	1,31915e+007	3	12,5	25,03	3,00179	0,784568	0,874686	3,06771	1,9444	SUCCEEDED
50	10	GRAD	1,31701e+007	3	12,5	25	3,00359	0,784583	0,874691	3,06549	1,94238	SUCCEEDED

Exp	Iter	Mode	Goal	M6 (kg)	M10 (kg)	M11 (kg)	M13 (kg)	MAC_1_2	MAC_2_1	Freq_Diff_1_2	Freq_Diff_2_1	Status
1	0	START	1,31701e+007	22,551	11,9955	11,9955	100,707	0,784583	0,874691	3,06549	1,94238	SUCCEEDED
2	1	GRAD	1,31722e+007	22,5645	11,9955	11,9955	100,707	0,784583	0,874691	3,0657	1,94258	SUCCEEDED
						…						
29	6	GRAD	5,2509e+006	11,275	11,9906	12,0027	50,35	0,778995	0,872683	2,07657	0,96889	SUCCEEDED
30	6	GRAD	5,2575e+006	11,275	11,9906	11,9955	50,4104	0,779002	0,872685	2,07765	0,969975	SUCCEEDED

Fig. 12.17 Sample optimization steps by using mass parameters*

Exp	Iter	Mode	Goal	CB2_K1 (N_m)	CB2_K3 (N_m)	CB2_K4 (Nxm_rad)	CB2_K5 (Nxm_rad)	MAC_1_2	MAC_2_1	Freq_Diff_1_2	Freq_Diff_2_1
1	0	START	786920	1e+009	1e+009	1000	1000	0,784822	0,887421	0,0012598	0,887084
2	1	GRAD	786923	1,101e+012	1e+009	1000	1000	0,784822	0,887421	0,0012598	0,887086
						…					
9	2	GRAD	677884	1	1	1,1e+012	1	0,784726	0,888102	0,00101089	0,823337
10	2	GRAD	677858	1	1	1	1,1e+012	0,784726	0,888102	0,00100994	0,82332

Exp	Iter	Mode	Goal	CB1_K1 (N_m)	CB1_K2 (N_m)	CB1_K3 (N_m)	CB1_K4 (Nxm_rad)	MAC_1_X	MAC_2_X	Freq_Diff_1_X	Freq_Diff_2_X
1	0	START	677858	1e+009	1e+009	1e+009	1000	0,784726	0,888102	0,00100994	0,82332
2	1	GRAD	677861	2,1e+009	1e+009	1e+009	1000	0,784726	0,888102	0,00100994	0,823322
						…					
9	2	GRAD	587670	1	1	1,1e+009	1	0,784654	0,888634	0,000584602	0,766596
10	2	GRAD	421497	1	1	1	1,1e+009	0,784758	0,889729	0,000982285	0,649227

Fig. 12.18 Sample optimization steps by using stiffness parameters*

Table 12.2 Modal updating results

	Modal test	Finite element model	Updated finite element model
Vertical bending mode [Hz] (%Diff.)	11.05	7.39 (33.1 %)	11.30 (2.2 %)
Vertical bending mode—MAC with modal test	–	0.873	0.887
Lateral bending mode [Hz] (%Diff.)	12.67	7.77 (38.6 %)	12.67 (0 %)
Lateral bending mode—MAC with modal test	–	0.778	0.787

12.6 Conclusion

In an ideal modal updating procedure, the best method is to create finite element model of the structure for which every detail is included and to fine tune unknown parameters like fastener and weld stiffness. However, especially in aircraft integration projects, it is not always practical to model every part of the structure if the integration is not applied by the aircraft manufacturer. Instead, the integration locations are modelled in detail and other parts are included to get similar dynamic behaviours.

In this study, helicopter tail is modelled according to integration locations and for other parts, mass values are assumed and connected to the modelled part. Together with unknown mass values, stiffness parameters to represent fuselage-tail connection of the helicopter are used to update the finite element modal analysis results according to modal test results. Updated finite element model is used to get more reasonable comparison of fatigue life of the tail before and after the integration and to get correct displacement and stress values in dynamic loading conditions since the modal frequencies and their relation with excitation frequencies is the most important parameter in dynamic loading structural analysis. In addition, the resonance risk of the tail in different loading conditions or design alternatives is controlled. The details for these computations other than modal updating procedure are not mentioned in this paper.

Acknowledgements The authors would like to thank ASELSAN Inc. for useful support during the work.

References

1. Baruch, M.: Optimization procedure to correct stiffness and flexibility matrices using vibration tests. AIAA J. **16**, 1208–1210 (1978)
2. Berman, A., Nagy, E.J.: Improvement of a large analytical model using test data. AIAA J. **21**, 1168–1173 (1983)
3. Lin, R.M., Zhu, J.: Model updating of damped structures using FRF data. Mech. Syst. Signal Process. **20**, 2200–2218 (2006)
4. Göge, D., Link, M.: Assessment of computational model updating procedures with regard to model validation. Aerosp. Sci. Tech. **7**, 47–61 (2003)
5. ECSS Requirements & Standards Deviation, Space Engineering Modal Survey Assessment, ECSS Secretariat, pp. 15–490 (2008)
6. Uhl, T., Bochniak, W., Lisowski, W., Hermans, L., Auweraer, H.V., Malecki, J.: Fe-model correlation of a helicopter using ground test results. IMAC 17 **1**, 940–946 (1998)
7. Çınar, O.: Evaluation of effective elastic properties of honeycomb sandwich structures by optimization involving modal behaviour. Master Thesis, METU (2014)

Chapter 13
Modeling of High Frequency Shock Tests

Washington J. DeLima, Melanie N. Ambrose, and Richard Jones

Abstract An impact machine (air gun) is used to reproduce high frequency pyroshock tests. The complexity of the tests cause low repeatability and can result in lengthy test setup time and out-of-tolerance tests. In order to better understand the behavior of this test, a study using experimental and numerical simulation is performed to identify the test parameters that affect the repeatability.

Keywords Air gun • Pyroshock • Simulations • Characterization • Validation

13.1 Introduction

Components in aerospace systems can be exposed to pyroshock events during flight and therefore they need to be qualified for these events [1, 3]. Laboratory reproduction of pyroshock events are usually obtained by resonant plate machines. These machines simulate pyroshocks by launching a projectile toward the plate using an air gun at a specific pressure (gun pressure). The impact of the projectile with the plate excites the dominant bending mode of the plate. A resonant plate is designed to have a dominant bending mode at certain frequency. Pyroshock tests are specified as SRS (Shock response Function) of the acceleration time domain signal. Depending on the SRS specified and the device under test, the task to design a pyroshock test that matches the SRS can be very difficult and time consuming. Numerical simulation is a good alternative to identify the test parameters that affects the reliability of the test procedure. However, in order to increase the confidence level in the numerical model, it needs to be calibrated to reflect the actual test.

This paper presents an experimental characterization of air gun machine. Modal analysis and pyroshock tests were performed in a 3 kHz resonant plate. These series of experiments will be subsequently used in a validation procedure of a finite element method.

13.2 Experimental Test

13.2.1 Modal Test

An experimental modal test was performed in the 3 kHz resonant plate system. LMS TestLab Modal Test and TestLab Impulse Test were used to extract the modes of the system. Figure 13.1 shows the sum FRF of the measurement with the resonant frequencies of the system. The resonant frequency at 2808 Hz (Fig. 13.1) is the bending mode of the plate responsible for the knee frequency of the SRS from the measured pyrochock. In addition to the bending mode, the torsional mode at 1512 Hz and bending mode at 2483 Hz play important role in shaping the SRS curves associate with the pyroshock.

W.J. DeLima (✉) • M.N. Ambrose • R. Jones
Honeywell Federal Manufacturing & Technologies, Kansas City, MO, USA
e-mail: wdelima@kcp.com

A. Brandt, R. Singhal (eds.), *Shock & Vibration, Aircraft/Aerospace, Energy Harvesting, Acoustics & Optics, Volume 9*, Conference Proceedings of the Society for Experimental Mechanics Series, DOI 10.1007/978-3-319-30087-0_13

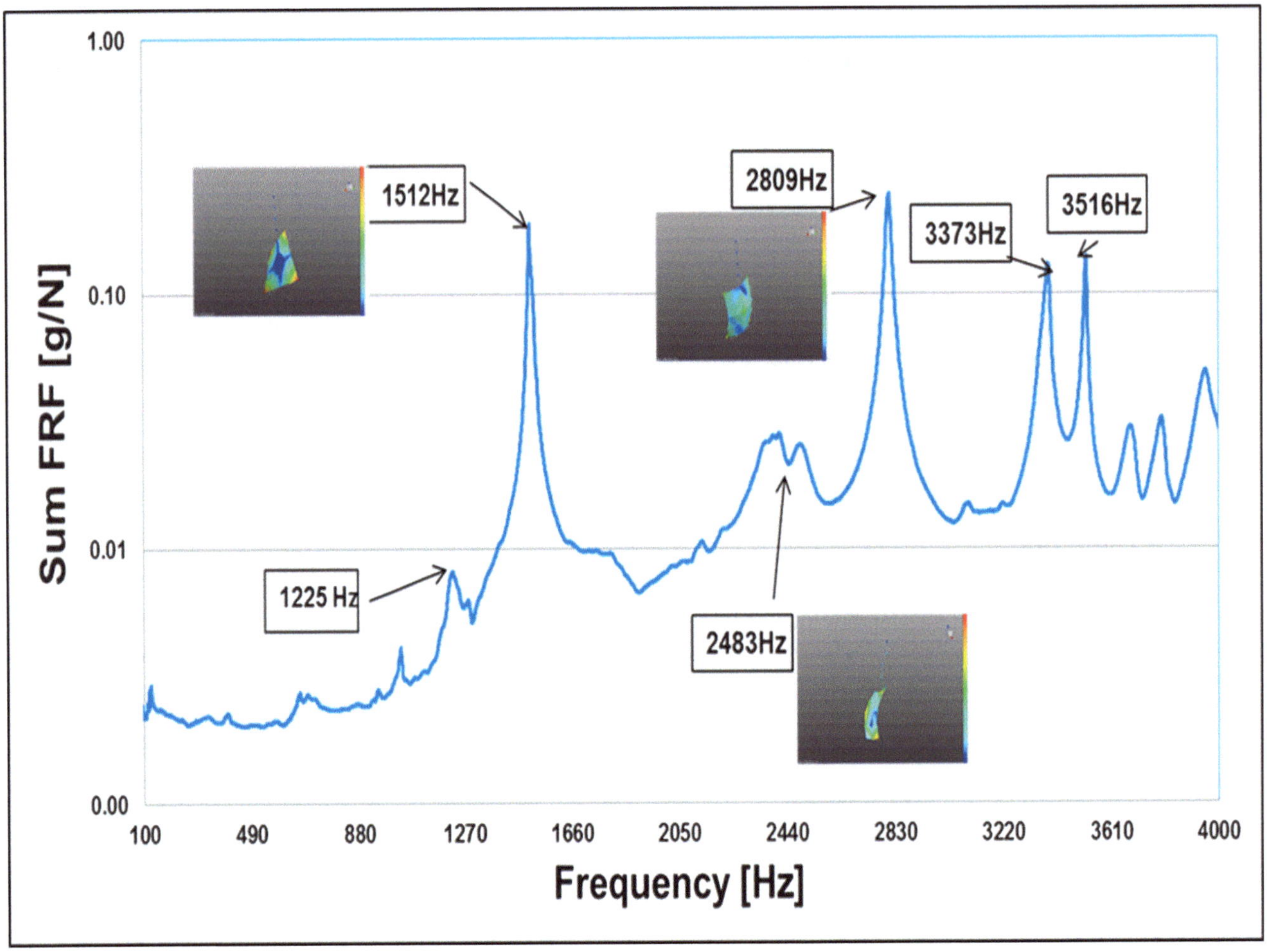

Fig. 13.1 Sum FRF of the measurement with the resonant frequencies of the system

Table 13.1 Test sequence

Test sequence	Plate 3 kHz	Air gun pressure PSI	Sequence number	Projectile Vel
1	A	65	1	63.751
2	A	65	2	63.717
3	A	65	3	63.712
4	B	65	1	63.717
5	B	65	2	63.737
6	B	65	3	63.69
7	C	65	1	63.712
8	C	65	2	63.751
9	C	65	3	63.751

13.2.2 Shock Test

A series of shock tests were performed with 3 kHz resonant plate to characterize the variation of the shock tests. Table 13.1 shows the test sequence. Three plates (plate A, B and C) were used. The air gun pressure was set to 65 psi and a 1/8″ thick felt was used at the point of impact. A 10 lb. projectile was used. In order to characterize the gun pressure variation the impact velocity of projectile was measured with a laser system. The average impact velocity of all nine tests was 63.72 fps with standard deviation of 2.3 %.

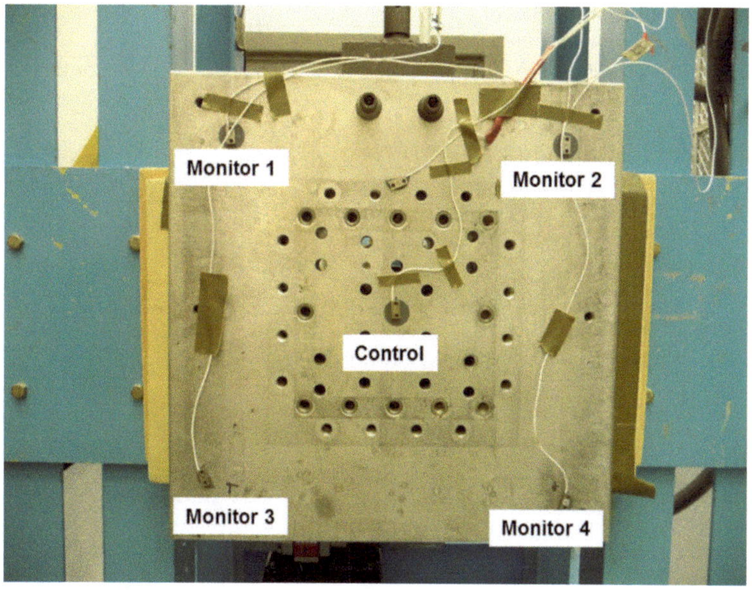

Fig. 13.2 3 kHz resonant plate A with accelerometers location

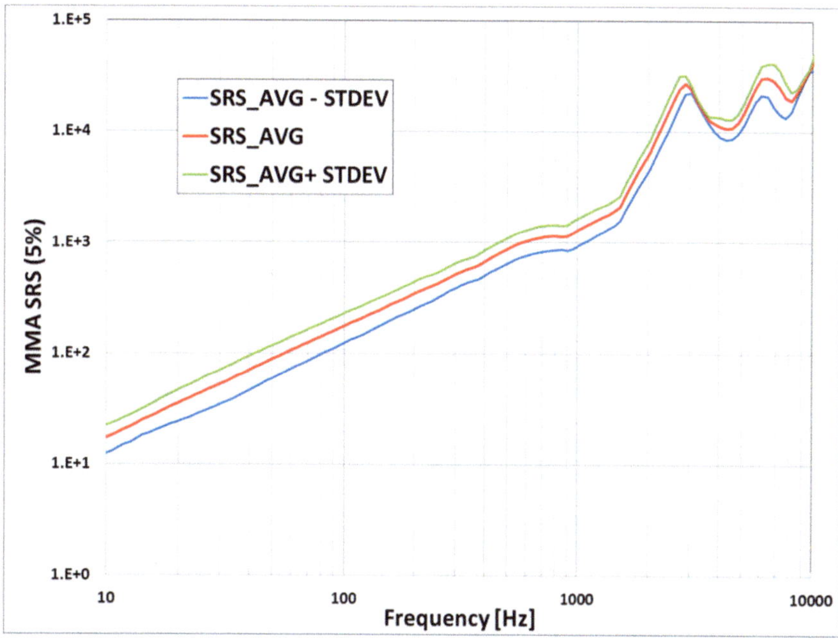

Fig. 13.3 Shock test—average SRS of all nine experiment

The acceleration was measured in five locations (Fig. 13.2) with piezoresitive accelerometer (Endevco 7270A-20 k). The control accelerometer is at the center of the plate. The SRS average of all measurements at the control location with its standard deviation is presented Fig. 13.3. The SRS curves for all accelerometers in one of the plates (plate A) are presented in Fig. 13.4. It is observed a dip in the SRS curve of the control accelerometer relative to the SRS of the monitor accelerometers around the torsional resonant mode (1512 Hz) of the plate. The reason for that is the control accelerometer location (center of the plate) is a nodal region of the torsional mode.

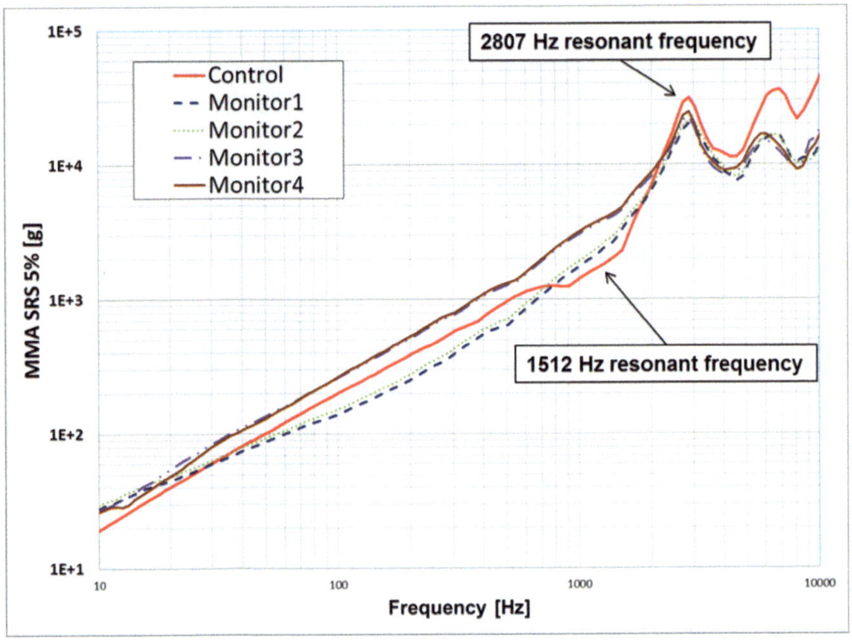

Fig. 13.4 Shock test—SRS of control and all four monitor accelerometers for plate A

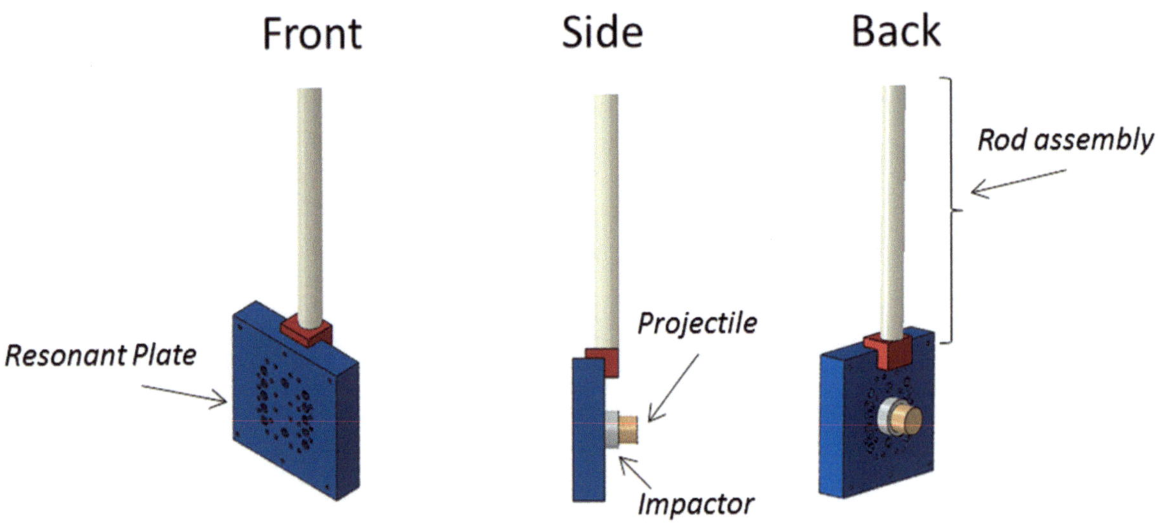

Fig. 13.5 Numerical modal setup, geometry

13.3 Numerical Simulations

13.3.1 Modal Analysis

Finite element models were created to better understand the dynamic behavior of the resonant plate testing. Abaqus 6.13-1 was used in the analysis. The models include the plate, mounting rod assembly, impactor and projectile (Fig. 13.5). All parts in the assembly are joined using tie constraints (infinite stiffness). The model was constructed with approximately 142 k hexahedral elements (Fig. 13.6). Two boundary conditions at the top of the mounting rod were considered including: free and fixed. A comparison of the experimental and numerical natural frequency and natural modes is presented in Table 13.2 and Fig. 13.7 respectively. There is a good agreement between both numerical boundary conditions and experimental results. The rest of the numerical analysis will be done with fixed boundary conditions.

Fig. 13.6 Numerical model setup, mesh

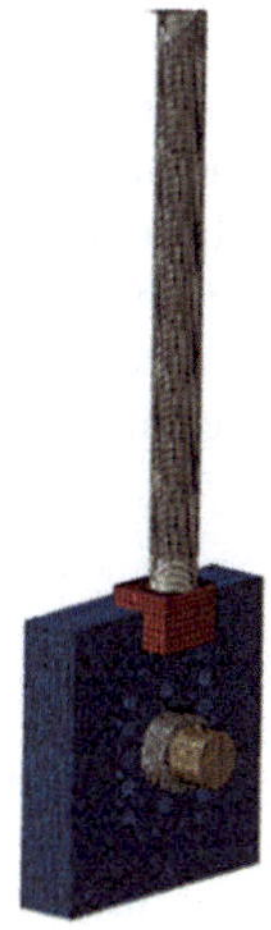

Table 13.2 Comparison of experimental and numerical natural frequency

Experimental	Fixed BC	Free BC
1227	1368	1373
1512	1478	1510
2483	2525	2584
2808	2792	2780
3373	3384	3327
3516	3489	3511

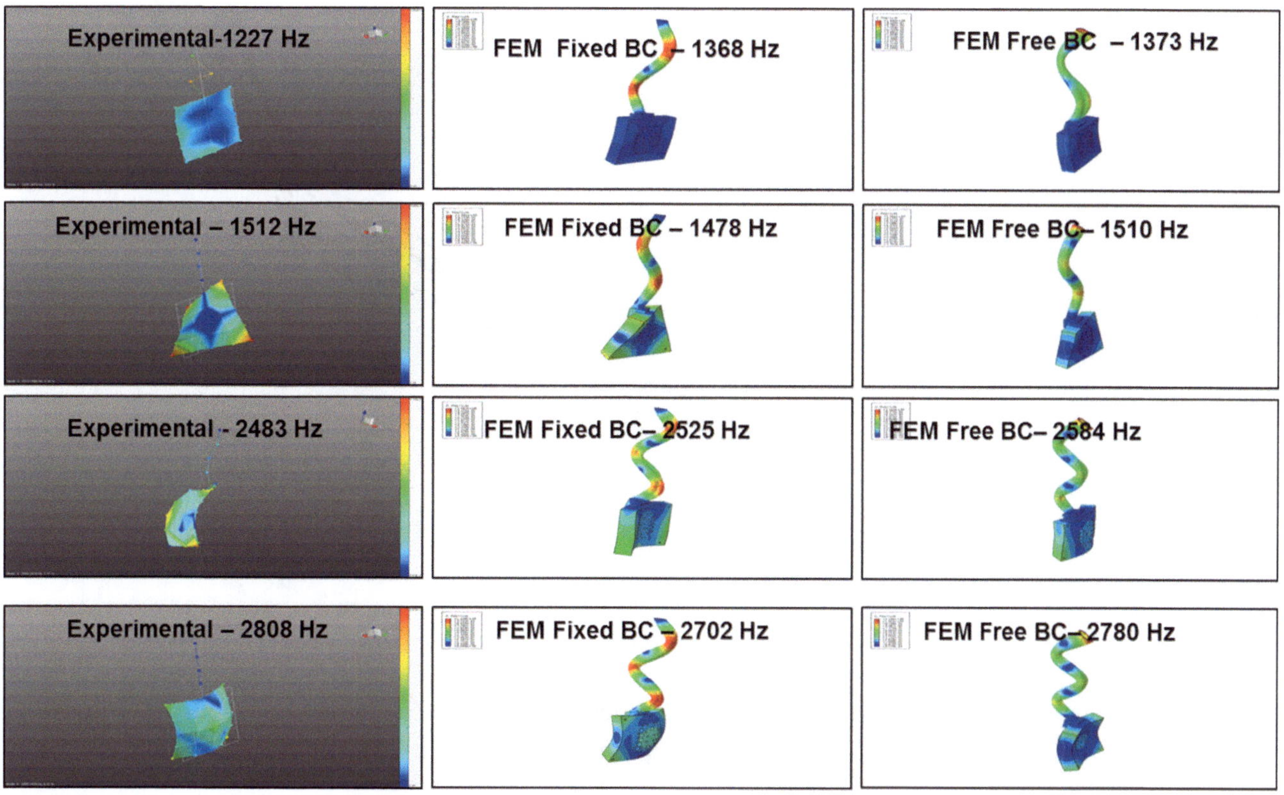

Fig. 13.7 Comparison of numerical and experimental natural modes

13.3.2 *Impact Simulation*

The same finite element model was used to evaluate the time domain response of the shock event. Abaqus Dynamic Explicit was used. The projectile was given a velocity of 63 fps and sent into contact with the impactor. Various Rayleigh material damping alpha and beta were calculated from the experimental data from various combination of the experimental damping ratio [2]. The 1/8″ felt was not modeled directly. Seven simulation runs were performed varying the material properties and the Rayleigh material damping (Table 13.3). The response at the center and the corner of the plate was captured to compare with the experimental data (Fig. 13.8).

Table 13.3 Simulations runs

	Plate				Impactor			
Simulation	E (PSI)	Density [lb s2/in4]	alpha	beta	E (PSI)	Density [lb s2/in4]	alpha	beta
A	10200000	0.0007324	51	1.8E-07	100000	0.0007324	0	0
B	10200000	0.0007324	51	1.8E-07	500000	0.0007324	0	0
C	10200000	0.0007324	1627	0.000005	5000000	0.0007324	0	0
D	10200000	0.0007324	114	2.2E-07	5000000	0.0007324	0	0
E	10200000	0.0007324	439	1.3E-07	5000000	0.0007324	0	0
F	10200000	0.0007324	51	1.8E-07	10200000	0.0007324	51	1.8E-07
G	10200000	0.0007324	0	0	10200000	0.0007324	0	0

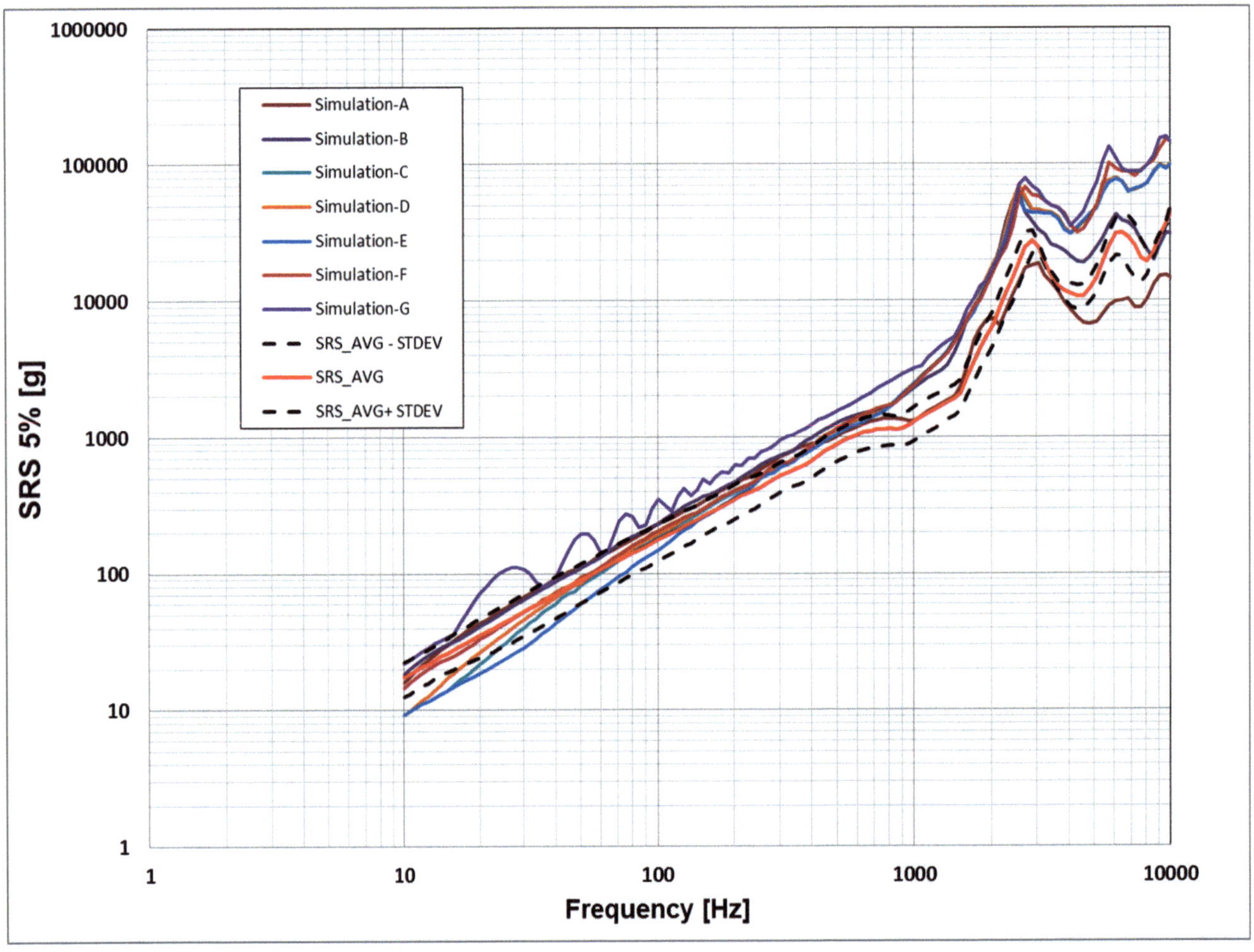

Fig. 13.8 Numerical explicit analysis: Rayleigh damping calculated from experimental data

It is observed that the variation in the Rayleigh damping of the plate improved the response of the simulation in the low frequencies as compared with no damping. However all the values of the Rayleigh material damping seem to give the same results. In order to reproduce the effect of the felt, the impactor was modeled with density of 0.283 lb/in3 and varied modulus of elasticity. The best results were obtained in simulations A.

13.4 Conclusion and Future Works

This paper presented an experimental characterization of air gun machine with 3 kHz resonant plate. These series of experiments (modal, pyroshock tests) were use in the characterization. A numerical model was created and validated using the experimental data. Various numerical and Rayleigh damping coefficients were evaluated in an attempt to calibrate the numerical model. In this first analysis in was concluded that the elastic properties of the impactor has great influence in the accuracy of the numerical results. Further analysis will investigate the sensitivity to the programmers used in between the impactor and the projectile.

References

1. Gute, G.D., MORRE, D.G.: Characterization of a resonant plate shock system using finite element analysis, SAND-95-2271C
2. Spears, R.E., Jensen, S.R.: Approach for Selection of Rayleigh damping parameters used for time history analysis. In: 2009 ASME Pressure Vessel and Piping Division Conference, July 2009
3. Davie, N.T., Bateman, V.I.: Pyroshock simulation for satellite components using a tunable resonant fixture – Phase 2, SAND93-2294 – UC-706

Chapter 14
A Novel Method to Correlate a Rocket Launcher Finite Element Model Using Experimental Modal Test Measurements and Identification Algorithms

Ronald N. Couch, Eliott J. Radcliffe, and Rickey A. Caldwell

Abstract The structural dynamics of a launcher system used to guide the initial trajectory of target test missiles is considered. Launcher induced body rates alter the trajectories of target vehicles at egress which can detrimentally affect the successful execution of flight test missions. As a result, flight mission analysts must have the capability to accurately reproduce the dynamic response of the launcher system to evaluate its effect on launch vehicle egress. Accurately reproducing this response requires a high fidelity finite element model that is correlated to modal survey test data. An experimental modal survey of a modified Nike-Hercules launch system is conducted and modal parameters are determined using conventional and output-only methodologies. A model correlation procedure is proposed and utilized that incorporates these experimentally determined modal parameters in conjunction with a set of correlation criteria to improve the accuracy of a high-fidelity launcher finite element model. After several iterations of the model correlation procedure, the model is able to accurately reproduce the dominant dynamic behaviors observed in the modal survey tests, especially at the key location of the launcher tip. This approach provides a basis for future launcher simulation and modeling efforts.

Keywords Modal survey • Finite element model correlation • Eigensystem realization algorithm

Nomenclature

DFT	Discrete Fourier transform
ERA	Eigensystem realization algorithm
FE	Finite element
FRF	Frequency response function
FRM	Frequency response matrix
IRF	Impulse response function
IRM	Impulse response matrix
MAC	Modal assurance criterion
POD	Proper orthogonal decomposition
POM	Proper orthogonal modes
POV	Proper orthogonal values

14.1 Introduction

The Missile Defense Agency (MDA), in conjunction with the U.S. Navy, oversees a robust flight test program to evaluate the nation's missile defense capabilities. These flight test missions are geared toward developing systems to defend U.S. and allied territory from limited ballistic missile attack. To achieve this goal, MDA oversees the development of threat representative targets.

Distribution Statement A. Approved for Public Release; Distribution is Unlimited.

R.N. Couch (✉) • E.J. Radcliffe • R.A. Caldwell
The Johns Hopkins University Applied Physics Laboratory, 11100 Johns Hopkins Road, Laurel, MD 20723, USA
e-mail: ronald.couch@jhuapl.edu

A. Brandt, R. Singhal (eds.), *Shock & Vibration, Aircraft/Aerospace, Energy Harvesting, Acoustics & Optics, Volume 9*,
Conference Proceedings of the Society for Experimental Mechanics Series, DOI 10.1007/978-3-319-30087-0_14

Launcher systems play a key role in successful flight test missions and can directly influence the behavior of target vehicles. Angular body rates imparted to the launch vehicle via the launcher structure during egress can significantly alter the vehicle's intended trajectory. These imparted body rates, or tip-off, arise from dynamic interaction of the launcher and launch vehicle at the moment of egress. For unguided targets, these tip-off rates represent off-nominal launch conditions that directly affect key trajectory parameters such as the apogee and down-range dispersion. Down-range dispersion is the statistical uncertainty bounding the predicted landing area during a flight test. The greater the magnitude of the downrange dispersion, the less accurate the target's aim point is likely to be. Furthermore, the magnitude of the downrange dispersion has range-safety implications, such as determining the size of the keep-out zone for maritime and commercial aviation traffic. For guided targets, egress induced tip-off rates represent motion that must be mitigated by the guidance system. As such, the magnitude of the tip-off rates influence design requirements for the guidance system. For these reasons, accurate egress simulations that can predict launch vehicle tip-off are required for the planning and certification of target flight test missions. Therefore, the contributions of the launcher to vehicle dynamics must be understood.

To accurately simulate and predict target launch scenarios involving the launcher system, it is necessary to develop a detailed understanding of the rigid body and elastic components of the tip-off rates imparted to the launch vehicle. This can be achieved by running monte-carlo trajectory simulations in conjunction with a finite element (FE) model of the coupled launcher-target vehicle structure. Relying on a FE model to predict egress conditions is standard practice; however, the FE model must suitably reproduce the dynamic behavior of the launcher to instill confidence in the fidelity of the launch simulation. Because of its implications at the moment of missile egress, it is especially important that the response of the launcher tip is accurately represented.

Although the FE method is a widely accepted and powerful analytical tool, it is difficult to accurately capture the dynamic response of a complex (i.e. "real") structure without test data. Therefore, the utility of an FE model as a dynamic tip-off simulation tool is limited unless it can be shown that it is grounded in test data. However, obtaining and appropriately applying the necessary test data for a model correlation effort is typically perceived as an expensive and unnecessary burden on an often tightly stretched program budget. Often, uncorrelated FE results are trusted implicitly and any modeling errors are absorbed by conservative design practices. Because of industry's growing reliance on FE model simulations and reduced emphasis on ground testing, a "low-impact" FE model updating procedure that does not adversely affect existing program resources must be incorporated into hardware development programs. The area of finite element model updating, or FE model correlation, is an important emerging field of study that includes many approaches of varying degrees of complexity [1, 2]. What is needed is a simple, portable procedure that correlates a finite element model of the launcher to a set of easily obtained test data for the purposes of accurate tip-off modeling.

Two modal identification methodologies are implemented and evaluated to support the correlation of the launcher FE model. The first is a conventional methodology known as the Eigensystem Realization Algorithm (ERA), where responses are referenced to a measured input force. The second is an output-only methodology, known as Proper Orthogonal Decomposition (POD) where modal parameters are derived without knowledge of the forcing.

The ERA method was originally developed at NASA Langley Research Center to characterize spacecraft structures [3] and has since become a common tool for modal identification [4]. The same research group has developed metrics for assessing quality of identified modal parameters [5], and applied these metrics to autonomously decipher real physical modes from computational artifacts [6]. Modifications of the Hankel matrix formulation of the standard ERA with data correlations (ERA/DC) have been shown with the promise of reducing noise or bias errors inherent to experimental modal tests [7]. A similar adaptation of the ERA/DC approach is also suitable for extracting modal parameters structures under natural excitation [8], though the current work follows the more conventional impulse response function (IRF) based approach, where the excitation force is a measured parameter.

POD is an algorithm that has recently been adapted from academic research [9] to an applied modal analysis tool. It is an output-only time-domain methodology that relies exclusively on the measured responses (i.e., accelerometer measurements) of a freely vibrating structure to extract modal parameters. A family of algorithms extends upon the basic POD methodology including, but not limited to, Mass-weighted Proper Orthogonal Decomposition, Complex Orthogonal Decomposition, Smooth Orthogonal Decomposition, and State Variable Modal De-composition [10, 11]. Each technique has inherent advantages and tradeoffs for a given application, but all rely on variations of the same fundamental principles.

In the current study, the primary areas of interest are the gross dynamics of the launcher structure and the specific response at the rail tip where the structure applies body rates to the launch vehicle at the moment of egress. A rigorous, iterative model correlation procedure using parameters obtained from modal test data and a set of model correlation criteria are proposed and implemented on a launcher FE model.

14.2 Launcher Structural Concerns

The system of interest is a modified Nike Hercules launcher, shown with an unguided target vehicle in Fig. 14.1. It is composed of three main substructures: a boom section, a base frame, and a pallet section. The boom section supports the target vehicle prior to and during launch and is essentially a cantilevered I-beam with a rail structure bolted to the top surface. Lugs attached to the launch vehicle provide a sliding contact interface with the rails during egress. The boom substructure is mounted to the base substructure via a hinge and a set of two, telescoping hydraulic arms. Boom elevation is controlled by extending the hydraulic arms until the desired launch angle is achieved. The nominal flight launch elevation is 70° upright from the stowed configuration, as indicated by the rotated (Y′Z′) frame relative to the inertial coordinate system in the YZ plane. The base substructure provides mechanical support for the boom and the electrical and hydraulic sub-assemblies that allow the launcher to be oriented to the desired elevation and heading. The base sub-structure is connected to a pallet structure via an azimuthal bearing. This pallet structure is intended to be stationary and serves as the primary interface with the "ground".

The launcher utilizes an over-slung rail configuration, where the launch vehicle egresses by sliding on rails mounted above the main structural member. The nature of the mechanical interface between the launch lug and rail introduces rigid body vehicle motion relative to the rail proportional to the launch angle. During launch, the lugs on the vehicle do not leave the rail simultaneously. The instant the first lug leaves the rail, gravitational forces induce a pitching moment on the launch vehicle that continues until the second launch lug departs from the rail. Because the launcher is intended to accommodate a wide variety of target vehicle sizes, the pitching moment imparted to longer, and heavier vehicles will be increased. This rigid body tip-off cannot be eliminated completely but can be mitigated somewhat by design changes to the launch lug and rail geometry. However, the necessary design changes are often impractical or cost-prohibitive for many flight test scenarios.

Elasticity in the base and boom substructures in addition to "looseness" in the structural joints also contribute to the magnitude of tip-off imparted to the launch vehicle. These sources of tip-off can be modeled but are difficult to predict on complex structures and must be determined empirically.

14.3 Modal Survey Testing

A series of modal surveys were conducted to characterize the dynamic response of the flight-configured launcher. The modal survey consists of striking the launcher with an instrumented sledgehammer at a set of prescribed impact locations distributed throughout the structure. The resulting excitation force and launcher dynamic response are measured using a network of accelerometers. The sensors are installed at stations along the rail, hydraulic lifts, launcher base, azimuthal bearing, and pallet.

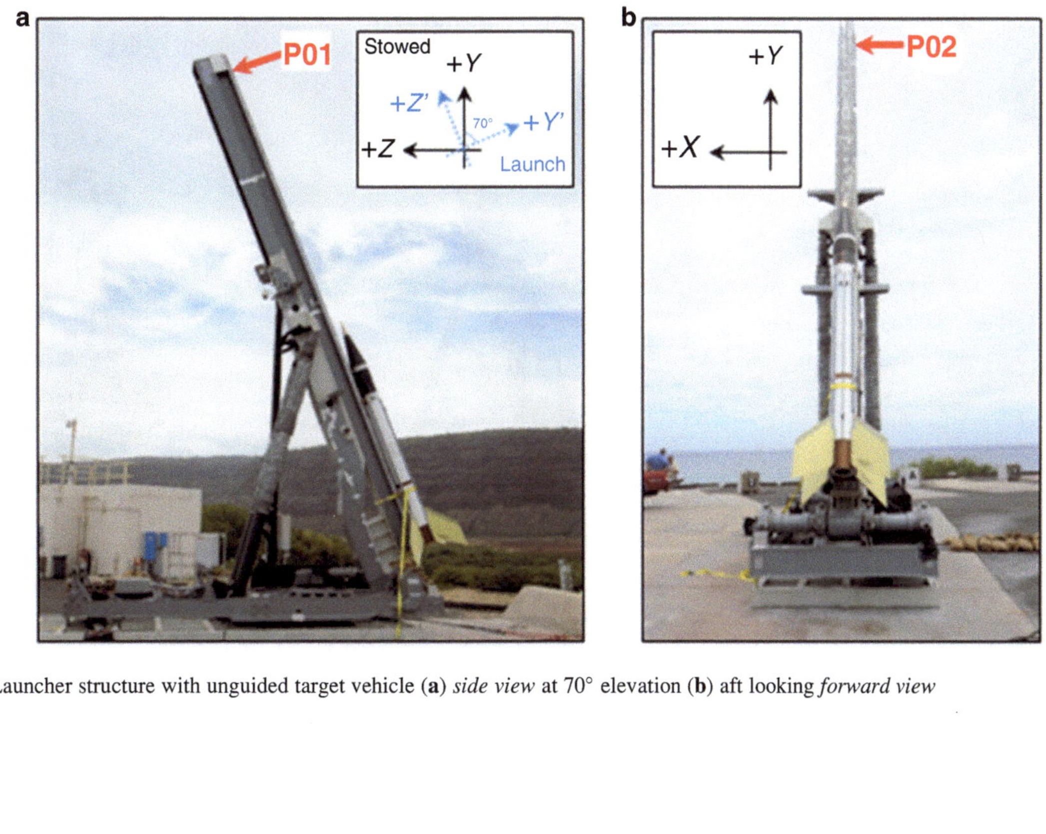

Fig. 14.1 Launcher structure with unguided target vehicle (**a**) *side view* at 70° elevation (**b**) aft looking *forward view*

The complete modal test includes logged datasets generated by hammer strikes at multiple locations along the rail and launcher base. Particular attention is given to the accelerometer located at the boom tip, because the dynamics encountered at this location significantly influence the magnitude and nature of the tip-off rates. Strike locations **P01** (Fig. 14.1a) and **P02** (Fig. 14.1b) are shown with the launcher in its nominal launch orientation. Location P01 is a strike downward in the $-Y'$ direction, and is intended to excite modes with displacements primarily in the YZ plane. Location **P02** is a horizontal strike in the +X-direction, which is intended to excite modes in the XY and XZ planes. These locations provide insight into the structural dynamics resulting from force excitation at the tip of the launch rail.

14.4 Launcher Finite Element Model and Analysis

The launcher FE model is developed using the Abaqus version 6.12 software. Figure 14.2 depicts the JHU/APL launcher FE model configured at the nominal launch elevation and stowed configurations. This FE model is suited for linear dynamic analysis such as the extraction of mode shapes, deflection, and internal forces. The model utilizes linear, room temperature material properties. It consists of 446,082 nodes and 387,347 elements. Mass properties of the FE model are based on the design mass properties of the actual system. The total mass of the model is within 30 kg of the reported design mass (i.e., within 1 %).

To assess the dynamic accuracy of the model, the simulated response of the launcher tip must be compared to the actual response measured during the modal survey. Modal transient analyses are simulated to represent each strike recorded during the modal survey tests. For each simulation, time histories of the acceleration, velocity, and displacement are extracted at 18 nodes corresponding to the accelerometer locations used during the modal survey tests. The precision of the simulated sensor and strike locations relative to the actual sensor and strike locations is a mesh-dependent parameter and a more refined mesh will increase precision. The target modes have characteristic shapes that invoke displacements of the entire launcher structure rather than localized displacements; therefore, their observability is largely insensitive to small deviations in the response locations along the launcher. Also, the physical location of the hammer strikes is somewhat low precision, controlled merely by how accurately a human being can swing a hammer to hit a specific location. In all cases, the selected sensor nodes and strike locations are within 2 in. of the placement of the actual sensors, which is more than adequate given the fidelity of the structural model.

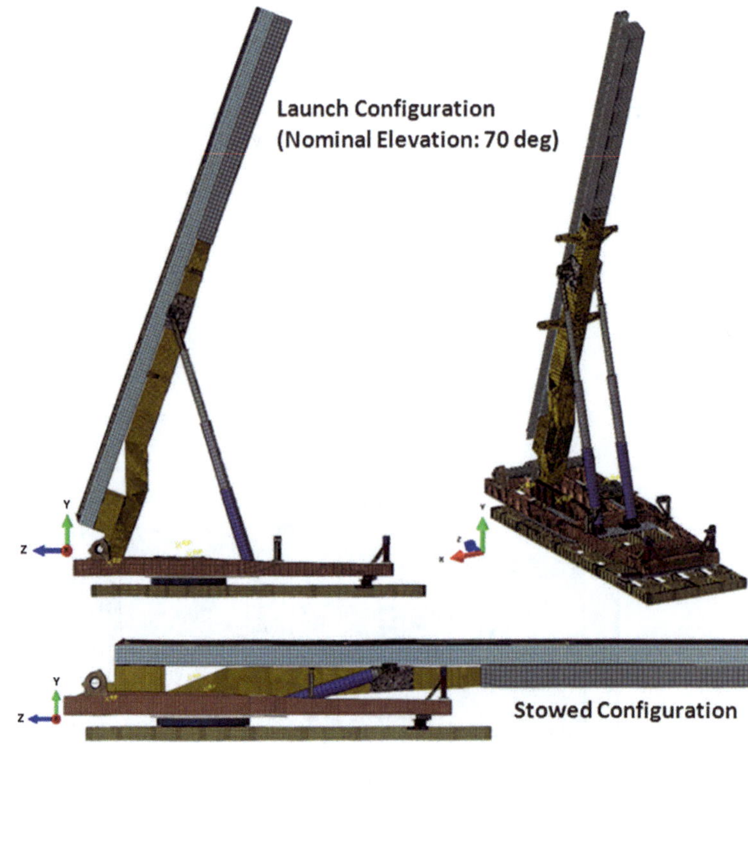

Fig. 14.2 JHU/APL launcher finite element model

For the FE modal transient simulation, it is necessary to include all the modes required to produce accurate predictions of the structure's dynamic behavior. For a large structure like the launcher, where the interest is in gross structural dynamics and not localized vibration phenomena, this usually means including all modes up to at least 100 Hz. To ensure that all the significant modes are determined, the eigenvalues and eigenvectors are calculated for all modes below 500 Hz, or in this case, the first 477 modes. Examination of those data shows that the significant launcher modes occur below 80 Hz (i.e. the first 30 modes) and over 90 % modal effective mass achieved when all 477 modes are included. Subsequent modal transient analyses incorporate these 477 modes.

Excitation of the model is achieved by assuming a large amplitude, short duration half-sine pulse forcing function. This forcing function is an idealized representation of the hammer force data collected during the modal survey tests. The amplitude of the forcing is arbitrarily set to 1000 lbf and the pulse width is determined from the average pulse width of the hammer strikes measured during the tests (i.e. 0.00407 s). Examination of the measured hammer force data shows that the peak amplitude of the applied force is typically between 2500 and 4000 lbf. Rather than precisely replicating each measured strike event, it is convenient to take advantage of the linear assumptions embedded in the FE modal transient solution. Thus, the simulated responses are scaled proportional to the measured force amplitude before comparing to the experimentally derived responses.

A uniform 1 % modal damping (relative to critical) is initially assumed for all modes, which ensures numerical stability and a reasonable initial estimate of response amplitudes. However, the modal damping of subsequent transient analyses is updated to achieve correlation between the FE model response and the modal survey test data. Figure 14.3 shows a simulated, X-direction, acceleration response of the launcher at the accelerometer node "S02" (see Fig. 14.3) to a strike near the rail tip in the −X direction. This simulated transient data is post-processed using the same algorithms utilized for the modal survey test data and used to assess how closely the FE model correlates to the test data.

The correlation procedure requires multiple sets of transient runs. The first set establishes a baseline, while subsequent sets incorporate model updates aimed at bringing the simulated results more in line with the test data. Several iterations of transient analyses are required to incorporate the effects of updating the FE model on the simulated dynamic response.

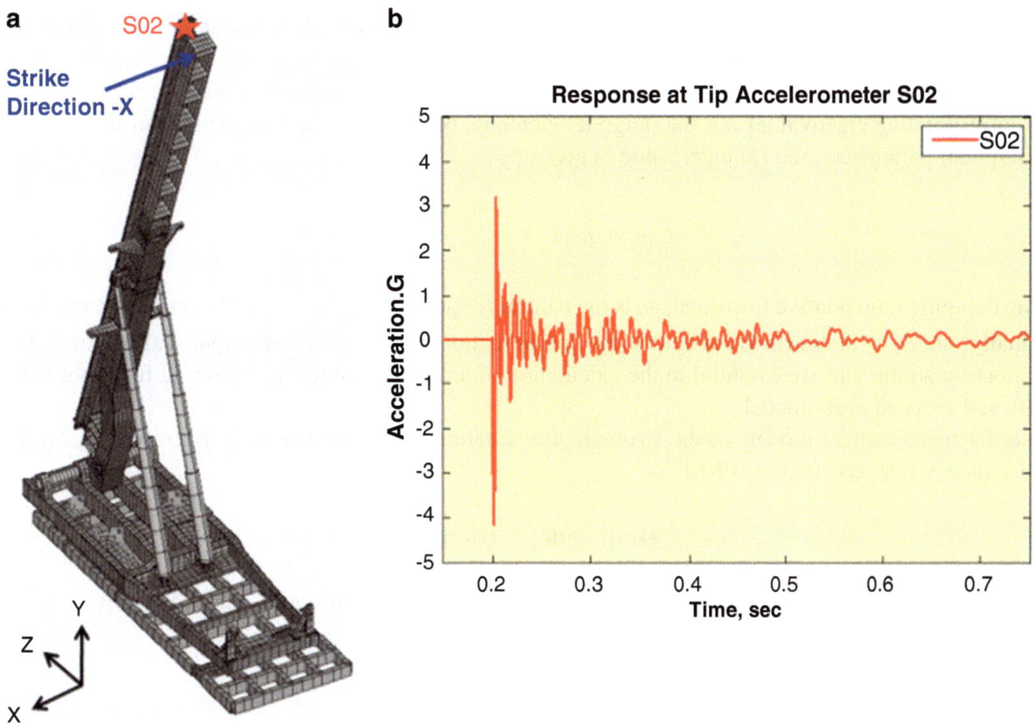

Fig. 14.3 (**a**) Simulated strike location P01 and (**b**) representative output at sensor node S02

14.5 Modal Identification Methodology

Dynamic accuracy of the FE model is achieved by anchoring a transient simulation data set to measurements obtained from a modal survey of the flight configured launcher. The data collected from these tests and generated from FE transient simulations are analyzed using modal identification methodologies to identify the vibration characteristics of the structure, including the dominant natural frequencies, damping factors, and mode shapes.

A conventional modal analysis technique that characterizes the structural response to specific excitation forces and locations, like ERA, is generally preferred for FE model correlation given a modal survey dataset where the input forcing is known. By comparison, output-only analysis offers similar capability under certain circumstances where it is impossible or otherwise inconvenient to measure input force, such as during launch or operational random vibration. For instance, output-only techniques may be applied in retrospect to datasets where launch platforms are instrumented for flight tests but where systematic modal surveys are not conducted. In the context of the current effort, The ERA-based analysis is reported here as the primary result and is used for follow-on FE model correlations. The POD analysis provides a secondary verification of certain critical high-energy modes and adds confidence to the results derived from the ERA analysis.

14.6 Eigensystem Realization Algorithm

Modal parameters are extracted from time domain impulse response matrices (IRMs) of the simulated or measured modal survey data using an ERA routine. The objective of ERA is to fit a minimum order realization to a modal dataset, which is used both for FE model correlation and for modeling the system response to arbitrary force excitations.

The ERA yields the system realization is in the form of a discrete-time state-space model. Although this form is useful for simulating the system response to an arbitrary forcing input, the goal of this work is to derive a conventional vibrational model in terms of fundamental modal parameters. This is provided by the end product of the ERA process, which includes an eigenvalue decomposition of the realized state matrices, and yields a modal "triplet" ($\mathbf{\Phi}$, $\mathbf{\Lambda}$, $\mathbf{L}$) that fully describes the system vibration characteristics. Each column of the $n \times 2q$ matrix $\mathbf{\Phi}$ corresponds to a system mode shape, as measured at the n accelerometer response nodes for each of the q retained modes. These are the experimental equivalent to the eigenvectors of a multi-degree-of-freedom analytical vibration system, or normal mode extraction of a dynamic FE model. The corresponding eigenvalues are the diagonal elements of the $2q \times 2q$ (diagonal) transient matrix $\mathbf{\Lambda}$. In terms of the system modal parameters, the *rth* eigenvalue is given by

$$\lambda_r = -\zeta_r \omega_r \pm j\omega_r \sqrt{1 - \zeta_r^2}, \; (r = 1, 2, \cdots, q) \tag{14.1}$$

where ζ_r is the damping ratio relative to critical, ω_r is the natural frequency, $\omega_r \sqrt{1 - \zeta_r^2}$ is the damped natural frequency, and $j = \sqrt{-1}$. Finally, the $2q \times m$ matrix $\mathbf{L}$ contains modal participation factors which are input scaling coefficients at each of the m force input locations, and are essential to the calculation of analytical frequency response functions (FRFs) and IRFs from the achieved reduced order model.

Reduced-order mathematical models of the structure are generated from the realized system ($\mathbf{\Phi}$, $\mathbf{\Lambda}$, $\mathbf{L}$). The analytical form of the frequency response matrix (FRM) is

$$\mathbf{H}(\omega) = \mathbf{\Phi} \left[j\omega \mathbf{I} - \mathbf{\Lambda} \right]^{-1} \mathbf{L} \,, \tag{14.2}$$

where $\mathbf{I}$ is the $2q \times 2q$ identity matrix and $(\cdot)^{-1}$ denotes the matrix inverse. Similarly, the IRM is given by

$$\mathbf{h}(t) = \mathbf{\Phi} e^{\mathbf{\Lambda} t} \mathbf{L} \,, \tag{14.3}$$

where $e^{\mathbf{\Lambda} t}$ is the matrix exponential of $\mathbf{\Lambda}$. Because $\mathbf{\Lambda}$ is diagonal, the exponential $e^{\mathbf{\Lambda} t}$ is a $2q \times 2q$ diagonal matrix with elements $e^{\lambda_r t}$, where each λ_r is defined by Eq. (14.1). The analytical forms defined by Eqs. (14.2) and (14.3) can be compared directly to the test data to assess the quality of fit. This comparison is used to minimize error, where multiple system realizations of increasing order are calculated until the identified parameters converge to a stable result. The realized system that meets a prescribed accuracy threshold with a minimal number of retained modes is reported as the final set of modal parameters.

14.7 Output-Only Techniques

Output-only techniques are used qualitatively to characterize modal properties from output measurements only, and without requiring a measurement of the force excitation. POD begins with the formulation of an ensemble matrix from a set of acceleration measurements, typically from a structure undergoing free vibration (i.e., impact testing). Each row of the matrix corresponds to an acceleration response signal and each column contains simultaneous measurements at discrete, uniformly sampled time intervals. A correlation matrix is formed by multiplying the ensemble matrix by its transpose.

Modal parameters are extracted from the correlation matrix by solving an eigenvalue problem, which yields the Proper Orthogonal Modes (POMs), $\boldsymbol{\nu}$, and Proper Orthogonal Values (POVs), $\boldsymbol{\sigma}$. For a sufficiently large number of sensor channels and acceleration samples, $\boldsymbol{\nu}$ converges to the normal mode shapes. The POVs, $\boldsymbol{\sigma}$, provide a measure of relative modal energy and are often used to distinguish between the highest energy modes of vibration for a structure versus lower energy "noise" modes.

The final stage of POD is to extract modal coordinates $\boldsymbol{\eta}$ from the original test measurements, which is achieved by pre-multiplying the ensemble matrix by the inverted POM matrix. Ideally, this matrix product isolates the contribution of each individual mode to the total acceleration signal. Under free vibration, a standard logarithmic decrement is used to determine modal frequencies and damping values. Thus, the output-only POD can be used in parallel to the conventional ERA process to derive the same modal parameters that are captured by the matrices $\boldsymbol{\Phi}$ and $\boldsymbol{\Lambda}$. The attendant scaling coefficients (matrix $\mathbf{L}$) are obviously not available without a known input force measurement; however, these terms are not always required for a structural modeling or FE model correlation effort.

Since POD provides much of the same modal information as the ERA analysis, it may be similarly applied to FE model correlation even under unconventional circumstances. For example, POD or other output-only techniques can be applied to in-service or legacy launcher acceleration datasets, given adequate instrumentation, even without execution of a planned modal survey. This trait allows for applications where POD may be applied in retrospect to improve existing simulation capability through the analysis of launch test datasets. Although a systematic modal survey is preferred, the lack of one does not necessarily prohibit modal identification.

14.8 Modal Identification Results

Figure 14.4 shows example FRFs derived from the pre-launch modal survey with the launcher configured in the nominal launch mode. Peaks in the FRFs are marked and used for initial estimates of the system natural frequencies. The response accelerometers represented in these plots are located at the tip of the rail and are aligned with the axes of the forcing inputs P01 (Fig. 14.4a) and P02 (Fig. 14.4b) (see Fig. 14.1). These responses are representative of the forcing "drive-point," and have good observability of most major structural modes of this particular launcher.

Shown in Fig. 14.5 are example mode shapes corresponding to modes with resonant frequencies at 5.6, 10, and 17 Hz. The un-deformed structure is shown in black, and the deformed structural components are shaded cyan. The modes shown in Fig. 14.5a, c are the first and second bending modes in the YZ plane, shown from the side view as in Fig. 14.1a. Figure 14.5b is the second bending mode in the XY and XZ planes, shown from the aft-looking forward view as in Fig. 14.1b. Other identified modes include a variety of characteristic geometries, including rail structure rocking, bending, twisting, and surging. Certain modes of the launcher exhibit traveling wave, or asynchronous vibration, which are characteristic of complex modal eigenvectors and likely non-proportional damping in the real structure. This behavior tends to be difficult to reproduce in a FE simulation, where proportionally damped linear normal modes are often assumed.

Mode shapes are reported in Table 14.1 for each FRF peak shown in Fig. 14.4, totaling 14 observable resonances of the tested configuration below 100 Hz. Additional resonances exist within this frequency range, but are omitted due to low observability at the rail tip. The first column assigns a number for each mode and the second gives a description of that mode shape. The third and fourth columns list initial modal frequency estimates based on marking the FRF peaks (e.g., Fig. 14.4. Sample FRFs for (a) a vertical impact on the rail, and (b) a horizontal impact on the rail), where the third column defines the input location at which the mode is observable and the fourth is frequency. The modal identification results using the ERA analysis are provided in columns five and six in Table 14.1. The subset of frequency and damping values identified using POD analysis are listed in the seventh and eighth columns and provide secondary validation for certain high-energy modes.

Primary modal parameter estimates are generated using the conventional, ERA-based approach using inputs P01 and P02, and a selection of response accelerometers with high measurement fidelity during modal testing. This is indicated by good signal coherence in the target $<$100 Hz bandwidth. Although additional inputs and responses may be used, these choices of

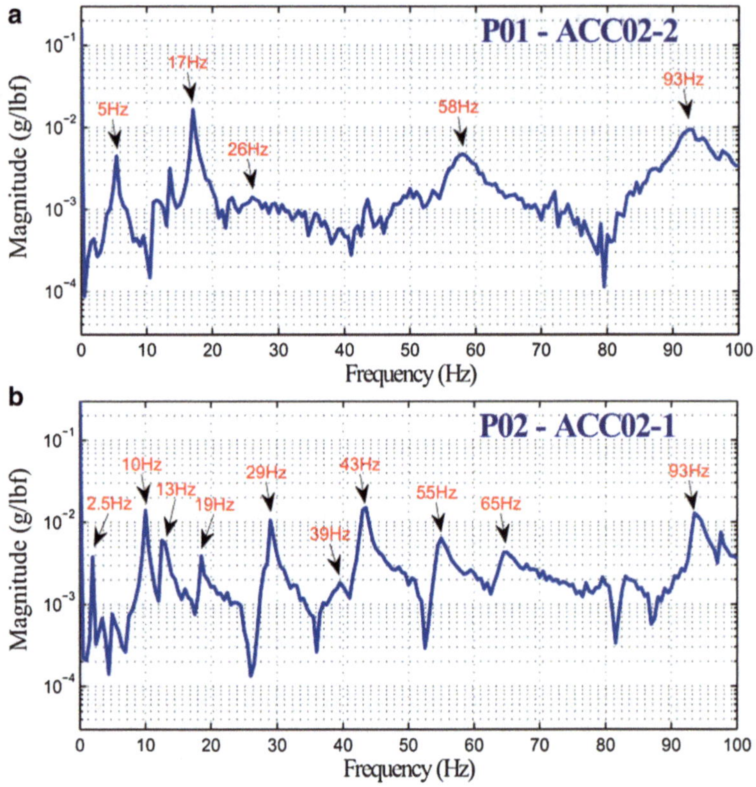

Fig. 14.4 Sample FRFs for (**a**) a vertical impact on the rail, and (**b**) a horizontal impact on the rail

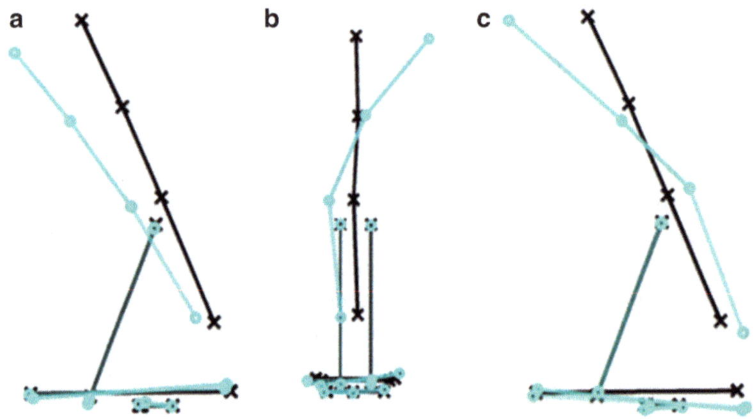

Fig. 14.5 Example mode shapes at (**a**) 5.6 Hz, (**b**) 10 Hz, and (**c**) 17 Hz

inputs and outputs provide the best combination of mode observability and good signal-to-noise. As a check on the effectiveness of ERA for identification, the ERA-fitted response is plotted against the original test data for a representative FRF (Fig. 14.6a) and the corresponding IRF (Fig. 14.6b). The quality of fit for both representations is very good, and the model is of sufficient order to capture the dominant peaks seen in the FRFs (see Fig. 14.4).

POD is used to extract a subset of the major rocking and bending modes that are observable at sensors along the launcher rail, and omitting sensors that are installed elsewhere on the launcher structure. The POD results are intended to independently validate the ERA results and demonstrate a novel methodology in an applied setting. Example results for two modes in the Y direction are shown in Fig. 14.7.

Figure 14.7 shows approximate mode shapes defined by the POMs at approximately 5 and 17 Hz. The high-energy bending mode at 17 Hz confirms the relative phasing and amplitude observed for this mode using the ERA methodology. By comparison, POD extracts the in-phase displacement of the 5 Hz mode, but the extracted shape has more curvature than is assumed for the rocking geometry of that mode (i.e., effectively rigid). The discrepancy at 5 Hz may be an artifact caused by

Table 14.1 Modal resonant frequency and damping

Mode	Description	FRF peak		ERA		POD	
		Input	Frequency (Hz)	Frequency (Hz)	Damping (%)	Frequency (Hz)	Damping (%)
1	Rocking rail in XY/XZ plane	P02	2.5	2.36	10	–	–
2	Rocking rail in YZ plane	P01	5	5.55	5	5.0	8.3
3	1st bending rail in XY/XZ plane	P02	10	10.0	3.4	–	–
4	Bending/Traveling rail in XY/XZ plane	P02	13	13.1	2.5	–	–
5	1st bending rail in YZ plane	P01	17	17.0	1.0	17.2	1.4
6	2nd bending rail in XY/XZ plane	P02	19	18.4	1.7	–	–
7	Vertical bending/surging	P01	26	27.7	12.7	–	–
8	3rd bending rail in XY/XZ plane	P02	29	28.9	1.2	–	–
9	Bending/Travelling rail—all axes	P02	39	39.2	4.9	–	–
10	Bending rail—all axes	P02	43	43.1	0.9	43.3	1.6
11	Bending rail, twisting base—all axes	P02	55	54.7	1.2	54.7	2.0
12	2nd bending rail in YZ plane	P01	58	57.8	3.7	–	–
13	4th bending rail in XY/XZ plane	P02	65	64.1	1.5	–	–
14	Twisting about rail Z axis	P01/P02	93	93.3	2.6	–	–

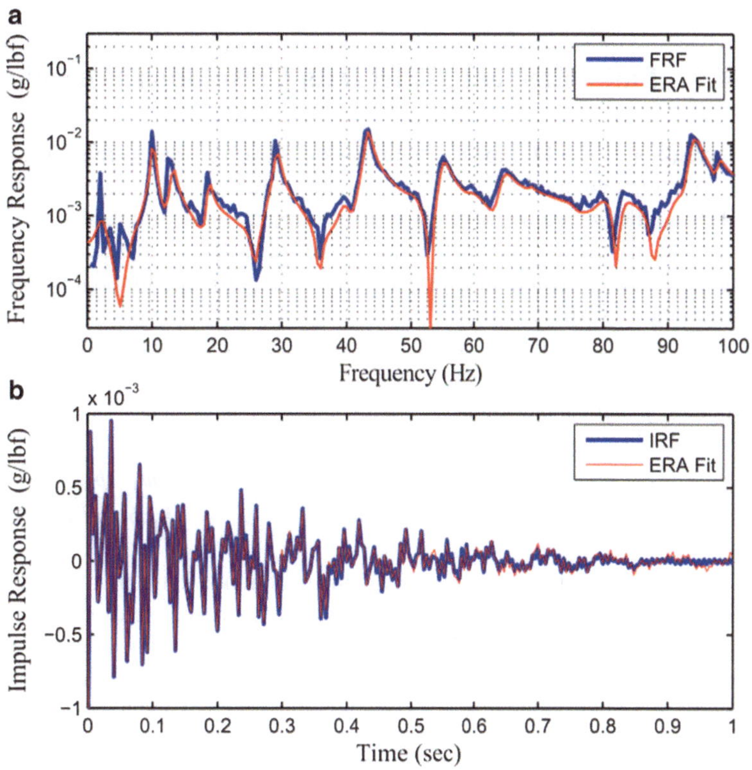

Fig. 14.6 Example ERA results for a representative (**a**) FRF and (**b**) IRF

noise or the limited sensor count along the rail. Figure 14.7b shows the modal coordinates of the 5- and 17-Hz modes, where it is clear that POD effectively isolates the time-domain transients for each mode. POVs are shown in Figure 14.7c, which indicates that the 17 Hz mode has significantly higher energy than the 5 Hz mode in the acceleration measurements. Discrete Fourier Transforms (DFTs) of the modal coordinates are plotted in Figure 14.7d, where the dominant peak resides at the natural frequency of each mode.

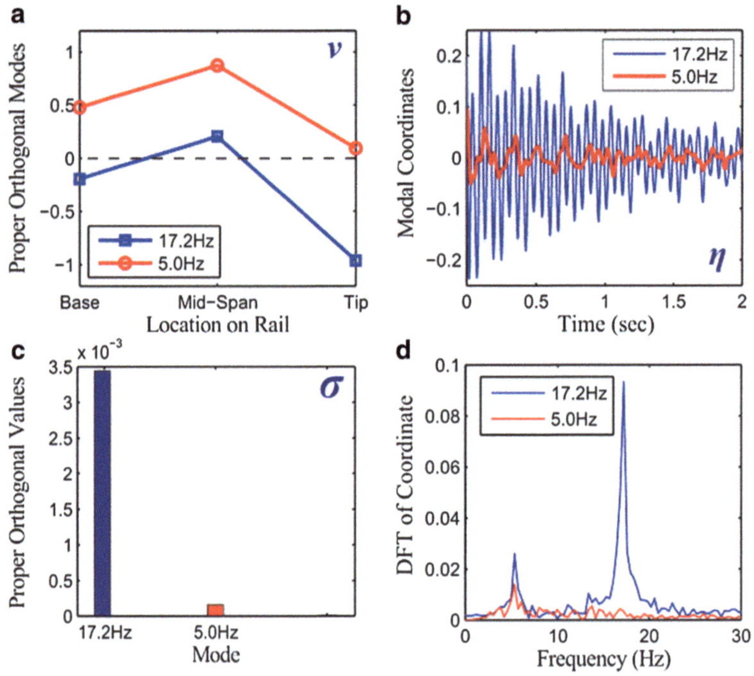

Fig. 14.7 Sample POD results

14.9 Finite Element Model Data Correlation Procedure

Correlation of the FE model is a multi-step, iterative process requiring data obtained from both experimental and numerical simulation sources. Modal parameters are extracted from the modal survey data and FE model simulation using both the ERA and POD methodologies. These experimentally determined and simulated modal parameters are compared relative to a set of FE model correlation criteria developed to assess the fidelity of the simulation relative to the measured data. The FE model is then updated and compared to the experimental modal response data in an iterative fashion until the correlation criteria are satisfied.

A common technique used to correlate an experimentally derived dataset to a simulation-derived dataset is the computation of Modal Assurance Criterion (MAC) values [12]. The MAC value calculation compares two mode shape vectors, $\boldsymbol{\phi}_1$ and $\boldsymbol{\phi}_2$, using the equation

$$\frac{|\boldsymbol{\phi}_1 \cdot \boldsymbol{\phi}_2|^2}{\boldsymbol{\phi}_1^H \boldsymbol{\phi}_1 \boldsymbol{\phi}_2^H \boldsymbol{\phi}_2}, \tag{14.4}$$

where the mode shape vectors correspond to columns in the mode shape matrix, $\boldsymbol{\Phi}$. The superscript $(\cdot)^H$ denotes the complex Hermitian and $(\cdot)$ denotes the dot product. In this case, one mode shape is derived from experimental data and the other is derived from simulated data from a FE model. A MAC value approaches unity when there is a good correlation between the mode shape and approaches zero when there is a poor correlation. Ideally, two arbitrary modes are either perfectly correlated and have a MAC value of 1 or completely uncorrelated and have a MAC value of 0 due to the linearly independent nature of mode shapes. In practice, identified modes are neither perfectly correlated nor completely uncorrelated due to inherent test limitations associated with a system having a finite number of available response measurements and sub-optimal signal-to-noise characteristics. Also, a large MAC value for two modes does not necessarily imply good frequency correlation. It is possible for two modes to have near-unity MAC numbers but occur at significantly different frequencies. An assessment of how well each mode predicted by the FE model correlates to the experimental data is determined by three criteria. These criteria are MAC number, frequency error, and rail tip frequency response magnitude.

First, MAC values are generated for each mode by comparing the simulated data to the experimental data according to Equation (14.4). In this case, a MAC number of greater than 0.75 indicates that a particular mode shape in the model has been reasonably correlated to the test data. Second, the percent difference between the measured modal frequency and predicted frequency, or frequency error, is determined. Good frequency correlation exists when the frequency error is less

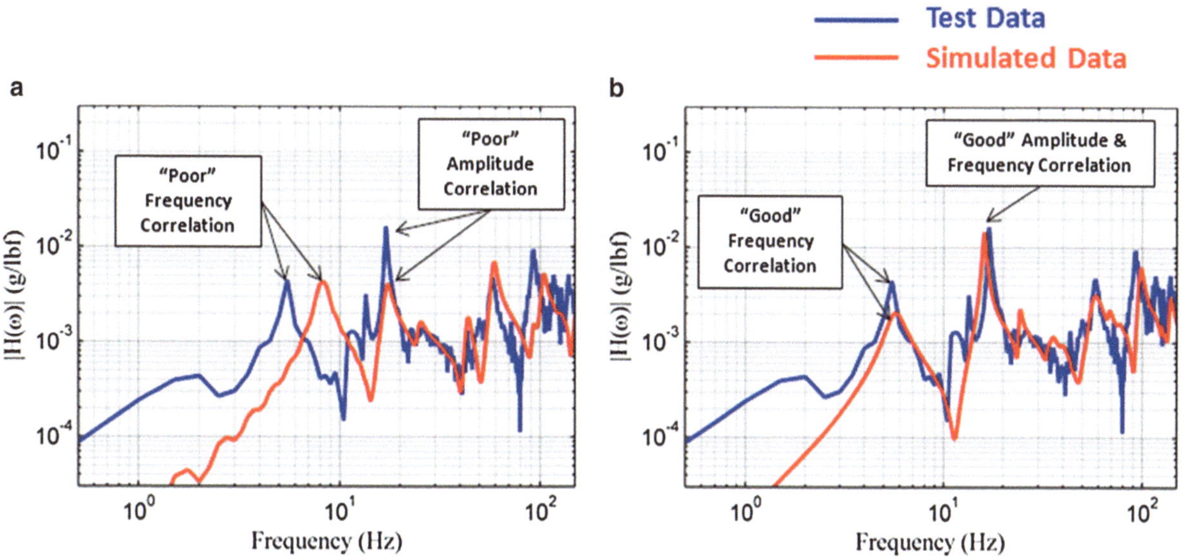

Fig. 14.8 Frequency response magnitude at the launcher tip of (**a**) poorly correlated model and (**b**) a well-correlated model

than or equal to 5 %. Third, the magnitudes of the launcher tip frequency responses of the simulated and experimentally determined data are compared to assess how closely the two match one another in terms of the location and amplitude of the dominant peaks. This assessment is mostly qualitative, because the model is not capable of capturing all of the real dynamic behaviors of the structure and other spurious behaviors introduced by non-physical phenomenon (i.e., noise) that are evident in an experimentally determined dataset. However, if the dominant modes identified from the experimental data are well represented in the model (i.e., MAC numbers near unity), the model does not introduce significant spurious modes, and the appropriate levels of damping are included in the FE model, then the frequency response of the FE model simulated data should be very close to the frequency response derived from experiment. This indicates that in the area of interest, the model is a good approximation of the real system.

An underlying assumption with this modal correlation procedure is that launch vehicle tip-off rates are largely a function of the launcher rail tip dynamics. It is possible that there may be additional locations other than the rail tip where it will be necessary for a dynamic launcher FE model to accurately simulate the experimental frequency response. If other such locations are identified, the frequency response of these areas will be considered as criterion in the modal correlation procedure. Figure 14.8 shows the frequency response at the launcher tip for both a poor and well-correlated model.

14.10 Modal Correlation Results

After several cycles of updating and comparing the FE model results to the experimentally derived parameters, overall correlation can be achieved. Table 14.2 shows the resonance frequencies derived from the experimental and simulated datasets, the MAC numbers, and percent frequency error for each mode. The simulated data are generated from the most recent iteration of the correlated launcher FE model.

The results in Table 14.2 show that of the 22 experimentally derived mode shapes, 17 of the modes derived from the FE model can be correlated to test data (i.e., MAC number greater than 0.75). The launcher model does not correlate the fundamental 2.36-Hz rail mode, although that is likely the result of problems with noise in the experimental dataset in the "near-DC" frequency range and not model error. It is encouraging that the frequency error associated with this mode appears to be very small. Of the 17 correlated modes, 10 show frequency errors at or below 5 %. Also, of the 22 experimentally derived modes, only 2 could not be correlated to modes derived from the FE model.

As detailed previously, after an analysis of the baseline frequency response is completed, the model is updated and then re-evaluated until the modal correlation criteria are met. In the context of the current work, the launcher has undergone four such iterations, and the rail tip frequency response of the most recent model is shown in Fig. 14.9.

Figure 14.9a shows the frequency response in the rotated Y direction to a strike at P01. The dominant peaks in the experimental data occur at 5.5, 13, 17, 58, and 93 Hz, and the simulated response aligns well with the actual response.

Table 14.2 JHU/APL launcher FE model modal assurance criterion table

Mode	Test freq. (Hz)	FE model freq. (Hz)	MAC no.	Freq. error (%)
1[a]	2.36	2.46	0.1[b]	−4.2
2[a]	5.55	5.7	0.95	−2.7
3[a]	9.99	9.65	0.93	3.4
4	13.07	14.64	0.94	−12.0
5[a]	17.04	16.12	0.99	5.4
6	18.37	20.71	0.66	−12.7
7[a]	27.65	34.81	0.85	−25.9
8[a]	28.94	33.38	0.84	−15.3
9	34.22	38.16	0.82	−11.5
10	39.23	–	–	–
11	43.13	42.67	0.94	1.1
12	50.98	55.01	0.78	−7.9
13	54.7	–	–	–
14[a]	57.83	67.75	0.7	−17.2
15[a]	64.12	63.7	0.78	0.7
16	93.29	80.82	0.83	13.4
17	97.6	97.24	0.76	0.4
18	99.22	105.92	0.78	−6.8
19	115.3	117.14	0.87	−1.6
20	115.7	119.83	0.82	−3.6
21	126.9	125.21	0.81	1.3
22	137.7	136.25	0.86	1.1

[a]Significant FE model mode (i.e., modal effective mass greater than 5 %)
[b]Mode is observable but not well correlated due to noise in the test data

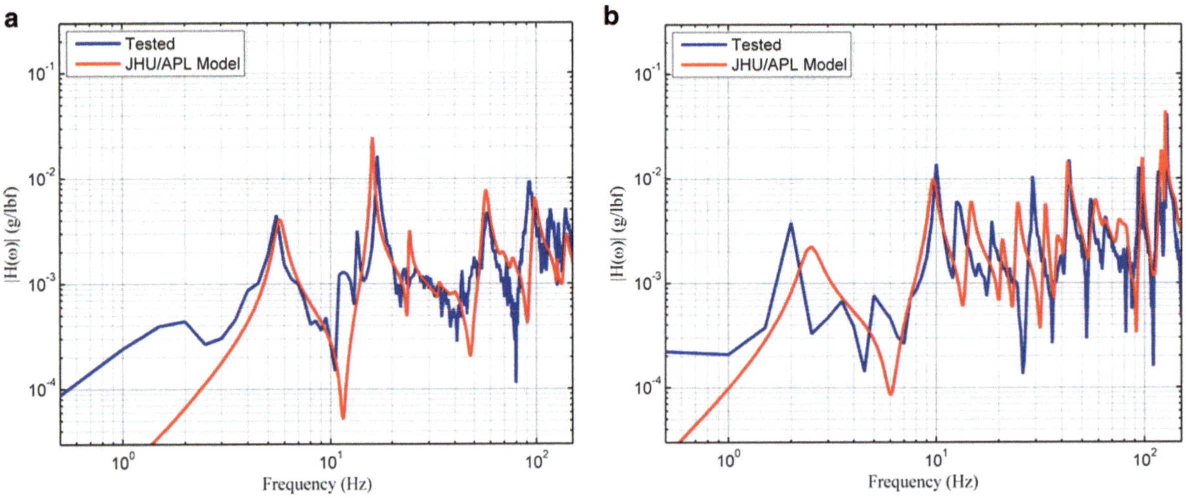

Fig. 14.9 Comparison of the experimental and simulated rail tip frequency response magnitude of the launcher FE model (**a**) in the rotated Y direction due to a strike at the P01 and (**b**) in the global X direction due to a strike at P02

Of those five dominant peaks, inspection of Table 14.2 shows that only the 5.5-Hz and 17-Hz modes are correlated with model-derived peaks as 5.7-Hz and 16-Hz respectively. Although there appears to be good agreement in the frequency response at 58-Hz and 93-Hz (57-Hz and 98-Hz in the model data), these mode shapes are not correlated (i.e. MAC numbers below 0.75) and are therefore not the same mode. Overall, the amplitudes of the simulated peaks are consistent with the experimental data indicating that the model uses appropriate levels of modal damping, and the FE model produces a good prediction of the rotated Y-direction frequency response at the rail tip.

The comparison of the transverse frequency (X direction) response is also shown in Fig. 14.9b. Due to the high modal density in the experimental data, it will be difficult to completely recreate this response in a FE model. The dominant peaks of the transverse response occur at 10, 13, 29, 43, 93, 116, and 127 Hz. The 2.36-Hz mode is visible in the dataset but it is contaminated by noise as evidenced by the "chopped" nature of the experimental frequency response below 10 Hz. As a result of this contamination, this mode is ultimately poorly correlated. A quick comparison of the two frequency response curves could lead to the erroneous conclusion that the model does not accurately predict the response in this direction. However, inspection of the data in Table 14.2 shows that all of the dominant modes in this direction are correlated to modes in the simulated data (9.65, 14.6, 33, 43, 81, 119, and 125 Hz, respectively). In addition, although some shifts in the peaks are evident, the overall levels of frequency and amplitude error are low. Only the peaks at 14.6, 33, and 81 Hz in the simulated data have frequency error over 5 %, indicating that the simulated frequency response in this direction aligns well with the experimental data. Finally, the amplitudes of the dominant peaks in the simulated data appear to correlate well with amplitudes in the experimental data, implying that the assumptions for modal damping are sufficient.

The modes corresponding to peaks at 58 and 93 Hz in the rotated Y-direction frequency response (Fig. 14.9) illustrate why it is important to use multiple, independent criteria for assessing correlation between model and experimental data. A model can produce a good approximation of the frequency response, implying that the dynamics in one direction, at one, localized area are well represented. However, the characteristic mode shapes can be very different. Aligned peaks in the frequency response do not necessarily imply agreement in their corresponding mode shapes. Likewise, good agreement in MAC values does not guarantee that the frequency response of the simulated data will match the frequency response in the experimental data. Frequency response and MAC values are complementary metrics that help to hasten the process of developing an experimentally anchored FE model.

14.11 Conclusions

The FE model correlation procedure presented here has produced a launcher model that provides an accurate representation of the dynamic response at the launcher tip. Most of the modes evident in the model have acceptable correlation to experimental data, and the simulated frequency response at the rail tip in both the vertical, rotated Y direction and the transverse, X direction are very close to the measured data. Nevertheless, there still exists room for improvement. Although many of the simulated modes are acceptably correlated to experimental data, more rigorous interpretations of correlation (i.e., MAC number of greater than 0.75) may require extensive adjustments to the model. Also, among the correlated modes, there are still quite a few that have frequency errors over 5 %. This is especially true for the modes below 50 Hz that dominate the magnitude of the dynamic response at the rail tip. An effort should be made to better align these modes in the frequency domain.

The model has difficulty reproducing the fundamental 2.36-Hz mode. This is likely due to measurement limitations in the experimental modal survey data at low frequency, and not necessarily due to error in either FE model. In future model correlation tasks, effort will be made to improve the quality of the modal survey data to specifically target the fundamental low-frequency modes. For instance, sensor architectures will be chosen with a suitable combination of high sensitivity and low noise floor in order to optimize the observability of the low-frequency resonance. Furthermore, increasing sensor count, particularly along the rail to enhance spatial resolution, will tend to yield more accurate modal parameter estimates and improved correlation for modes of increasing order and complexity.

The modal correlation procedure proposed here is intended to produce a rigorous, comprehensive assessment of a model's ability to accurately recreate the major dynamic behaviors of a system and is strongly rooted in both industry and academia. Once the desired level of correlation in the launcher model is achieved, the model can be used to evaluate the effect of launching larger, more energetic target vehicles. Thus, the utility of the correlated launcher FE model using this procedure is vastly improved and provides MDA with the capability to produce accurate target vehicle launch simulations for multiple target programs.

Acknowledgments The authors would like to thank Dr. Brian Feeny for his guidance on Output Only modal identification methodologies and Mrs. Heather Borowski for her work on the launcher finite element model.

References

1. Mottershead, J.E., Friswell, M.I.: Model updating in structural dynamics: a survey. J. Sound Vib. **167**(2), 347–375 (1993)
2. Zarate, B.A., Caicedo, J.M.: Finite element model updating: multiple alternatives. Eng. Struct. **30**, 3724–3730 (2008)
3. Juang, J., Pappa, R.S.: An eigensystem realization algorithm for modal parameter identification and model reduction. AIAA J. Guid. Control Dyn. **8**(5), 620–627 (1985)
4. Maia, S.: Theoretical and Experimental Modal Analysis Research Studies Press LTD (1997)

5. Pappa, R.S., Elliott, K.B., Schenk, A.: Consistent-mode indicator for the eigensystem realization algorithm. J. Guid. Control Dyn. **16**(5), 852–858 (1993)
6. Pappa, R.S., James III, G.H., Zimmerman, D.C.: Autonomous modal identification of the space shuttle tail rudder. J. Spacecr. Rocket. **35**(2), 163–169 (1998)
7. Juang, J.-n., Cooper, J.E., Wright, J.R.: An eigensystem realization algorithm using data correlations (ERA/DC) for modal parameter identification. Control-Theory Adv. Tech. **4**(1), 5–14 (1988)
8. Chiang, D.-Y., Lin, C.-S.: Identification of modal parameters from ambient vibration data using eigensystem realization algorithm with correlation technique. Mech. Sci. Tech. **24**(12), 2377–2382 (2010)
9. Feeny, B.F.: On proper orthogonal co-ordinates as indicators of modal activity. J. Sound Vib. **255**(5), 805–817 (2002)
10. Feeny, B.F., Caldwell, R.A.: Reduced mass-weighted proper orthogonal decomposition of an experimental non-uniform beam. In: Proceedings of the 2010 Inverse Problems Symposium Conference, 2010
11. Feeny, B.F., Caldwell, R.A.: Output only modal analysis of a non-uniform beam experiment by using decomposition methods. In: ASME 2011 International Design Engineering Technical Conferences and Computers and Information in Engineering Conference, 2011
12. Allemang, R.J.: The modal assurance criterion - twenty years of use and abuse. J. Sound Vib. **37**, 14–21 (2003)

Chapter 15
Numerical Studies on the Reduced Order Modeling of Frictionless Joint Contact Interfaces

M. Breitfuss and H.J. Holl

Abstract This contribution focuses on the consideration of contact interface stresses within reduced order models of mechanical systems comprising frictionless lap joints. The reduced order models investigated are based on two types of extensions to the trial vector basis utilized in the fixed interface reduction method following Craig and Bampton.

After a short introduction the problem formulation, the motivation for investigating a frictionless contact interface and the according discretization for obtaining a numerical model are given. The zero thickness element based numerical model for the contact stresses within the frictionless joint contact interface is mentioned as well. This is followed by a recapitulation of the model order reduction process utilizing a Galerkin projection and referring to the reduction basis according to Craig and Bampton. The methods for obtaining the above mentioned extensions to this basis are outlined as well. Finally some meaningful results obtained from static and dynamic loading of the reduced order models are discussed. The respective numerical model represents a cantilever beam consisting of two solid metal components.

The methods are evaluated in terms of the required number of additional trial functions necessary for obtaining a satisfying approximation of the acting contact stresses. It turns out that one method allows for a lower number of trial functions while keeping accuracy at an very acceptable level.

Keywords Bolted connection • Contact interface • Model order reduction • Proper orthogonal decomposition • Zero thickness element

15.1 Introduction

The model order reduction of linear dynamic systems is state of the art. A well known approach for this kind of problems is the fixed interface reduction method following Craig and Bampton [3]. In contrast the model order reduction of systems involving nonlinearities, either local or globally distributed, is still under investigation. The reduced order modelling of mechanical structures comprising a joint, like a bolted connection, is such a demanding problem due to the nonlinear relations involved within the contact interface.

The method according to Craig and Bampton utilizes a set of trial vectors, which are used as approximation of the system displacement fields and as subspace where the resulting equation of motion gets projected on. An approach, which extends this set of trial vectors with an additional type of trial vectors for the local displacements within the contact interface, was proposed by [8]. Subsequent contributions [9], [10] and [11] further refined this approach.

The suggested set of trial vectors already lead to very promising results during own investigations of the authors. Within this contribution the convergence of the computed contact stresses is investigated. The explanation starts with the problem formulation and the discretization to obtain a numerical model. This is followed by a brief review of the Galerkin projection based model reduction approach. In this section the computation of the additional trial vector set according to [11] is given and followed by an approach, which is intended to deliver an "optimal" set of additional trial vectors. These two sets of trial vectors are evaluated in terms of accuracy of the contact stresses obtained from the reduced order model. Finally some conclusions are drawn from the numerical studies carried out.

M. Breitfuss (✉) • H.J. Holl
Johannes Kepler University Linz, Altenbergerstr. 69, 4040 Linz, Austria
e-mail: markus.breitfuss@jku.at

A. Brandt, R. Singhal (eds.), *Shock & Vibration, Aircraft/Aerospace, Energy Harvesting, Acoustics & Optics, Volume 9*, Conference Proceedings of the Society for Experimental Mechanics Series, DOI 10.1007/978-3-319-30087-0_15

15.2 Problem Formulation

15.2.1 Governing Equations

The body Ω with the surface Γ comprising a contact interface Γ_{jc} as depicted in Fig. 15.1a is investigated. As boundary conditions the surface Γ_d is constrained and the stresses $\boldsymbol{t}_b$ on the surface Γ_t are prescribed. The surfaces Γ_d and Γ_t are disjunct surface regions of Γ. The dynamic equilibrium of a physical particle within the domain Ω at the position denoted by $\boldsymbol{x}$ is given by

$$\rho\ddot{\boldsymbol{u}} - div\boldsymbol{S} - \boldsymbol{k} = 0 \quad \forall \boldsymbol{x} \in \Omega \tag{15.1a}$$

with the boundary conditions

$$\boldsymbol{u} = 0 \quad \forall \boldsymbol{x} \in \Gamma_d \quad \text{and} \quad \boldsymbol{S}\boldsymbol{n} = \boldsymbol{t}_b \quad \forall \boldsymbol{x} \in \Gamma_t \tag{15.1b}$$

and the initial conditions for time $t = t_0 = 0$

$$^0\boldsymbol{u} = \boldsymbol{u}_0 \quad \forall \boldsymbol{x} \in {}^0\Omega \quad \text{and} \quad {}^0\dot{\boldsymbol{u}} = \boldsymbol{v}_0 \quad \forall \boldsymbol{x} \in {}^0\Omega \tag{15.1c}$$

where ρ denotes the mass density, $\boldsymbol{u}$ the displacement, $\dot{\boldsymbol{u}} = \boldsymbol{v}$ the velocity, $\ddot{\boldsymbol{u}} = \boldsymbol{a}$ the acceleration, $\boldsymbol{S}$ the Cauchy stress tensor, $\boldsymbol{k}$ the imposed force density and $\boldsymbol{n}$ the surface normal pointing outwards. The left superscript $^0(.)$ is supposed to emphasize that a certain quantity (.) is evaluated at time t_0 or on the bodies initial configuration $^0\Omega = \Omega \times t_0$ respectively.

The body comprises a contact interface Γ_{jc} depicted in Fig. 15.1b. Contact stresses

$$\boldsymbol{S}^{\pm}\boldsymbol{n}^{\pm} = \boldsymbol{t}_{jc}^{\pm} \qquad \forall \boldsymbol{x} \in \Gamma_{jc} \tag{15.2}$$

will occur where $\boldsymbol{t}_{jc}^{\pm}$ denotes the acting joint contact stresses. The superscript $^{\pm}$ is used to denote that a certain quantity is related to the respective side of the contact interface.

To define the relations within the contact interface Γ_{jc} all physical particles P^- on the contact surface Γ_{jc}^- have to be related with the physical particles P^+ on Γ_{jc}^+. Once determined for the initial configuration all contact pairs $\{P^-, P^+\}$ remain associated due to the assumption of small relative displacement within the contact interface.

Utilizing the interface normal direction $^0\boldsymbol{n}_{jc} = {}^0\boldsymbol{n}^-$ one can define the gap function

$$g_n = [\boldsymbol{x}^+ - \boldsymbol{x}^-]^{T\,0}\boldsymbol{n}_{jc} \qquad \forall \boldsymbol{x} \in \Gamma_{jc} \tag{15.3}$$

which expresses the contact pair distance in contact interface normal direction. As the contact in surface normal direction is of main interest for the upcoming explanations the kinematics in contact surface tangential direction are not further discussed.

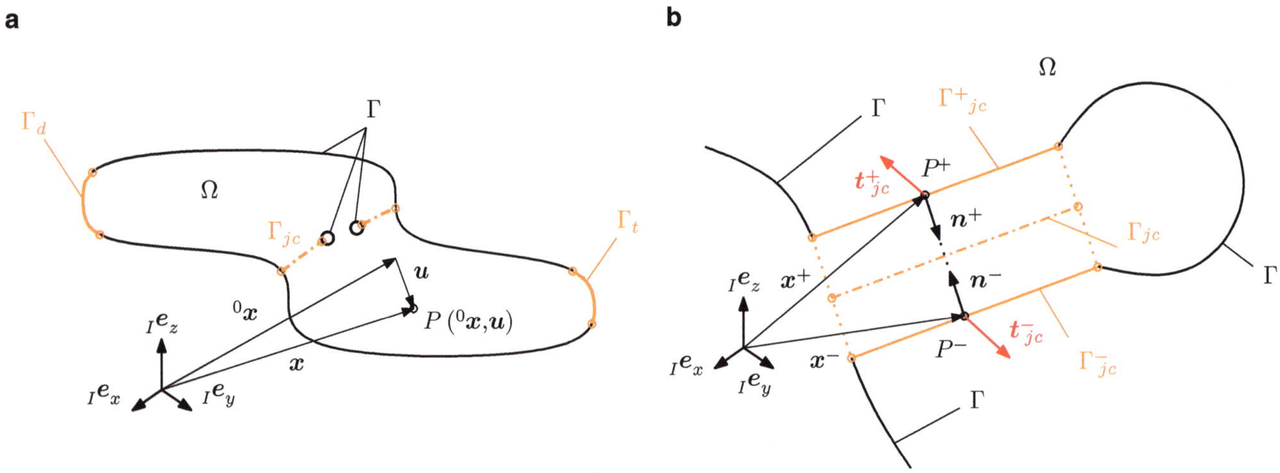

Fig. 15.1 Body Ω comprising a contact interface in actual configuration at time t (**a**) and contact stresses within the contact interface Γ_{jc} for the contact pair $\{P^-, P^+\}$ (**b**)

The occurring contact stresses for a contact pair $\{P^-,P^+\}$, arbitrarily selected from the currently contacting ($\boldsymbol{t}_{jc} \neq 0$) sub domain of the contact interface Γ_{jc} are, according to Newton's 3rd Law, equal in magnitude but acting in opposing directions

$$\boldsymbol{t}_{jc} = \boldsymbol{t}_{jc}^- = -\boldsymbol{t}_{jc}^+ \qquad \forall \boldsymbol{x} \in \Gamma_{jc}. \tag{15.4}$$

The contact interface stress vector can be separated into two components

$$\boldsymbol{t}_{jc} = \boldsymbol{t}_n + \boldsymbol{t}_t \tag{15.5}$$

where $\boldsymbol{t}_n$ is defined to point in contact interface normal direction and the remaining part $\boldsymbol{t}_t$ is oriented tangential to the contact surface.

The latter component is neglected as this contribution focuses on the computation of the stress component in normal direction. The energy dissipation due to friction within the contact interface will not be investigated. However it is required to suppress the tangential relative movement of the contact partners, at least to obtain a correct global behaviour of the body. This can be achieved by a permanent tangential coupling of the contact partners within a small area of the contact interface. This small area is assumed to permanently comprise a sufficiently high contact stress normal component due to the bolt pretension. This finally leads to a contact stress vector

$$\boldsymbol{t}_{jc} = \boldsymbol{t}_n = t_n \boldsymbol{n}_{jc}. \tag{15.6}$$

where t_n denotes the scalar valued contact stress component in interface normal direction.

Utilizing the gap function g_n and the contact stress component t_n one can finally formulate the nonlinear contact conditions of the frictionless contact interface Γ_{jc}

$$g_n \geq 0, \quad t_n \leq 0, \quad g_n t_n = 0 \qquad \forall \boldsymbol{x} \in \Gamma_{jc}. \tag{15.7}$$

15.2.2 Discretization

For the whole body domain Ω a linear constitutive law without rate dependencies is assumed and the occurring displacements $\boldsymbol{u}$ are expected to be small compared to the body dimensions. From the virtual displacements denoted by $\delta\boldsymbol{u}$, which are presumed to be small in relation to the body dimension as well, one can derive the virtual strain tensor $\delta\boldsymbol{\epsilon}$. Utilizing the principle of virtual work one obtains the weak formulation of the dynamic equilibrium

$$\delta W_\Omega + \delta W_{\Gamma_{jc}} = 0 \tag{15.8a}$$

where the virtual work of the body domain Ω reads

$$\delta W_\Omega = \int_\Omega \rho \ddot{\boldsymbol{u}} \cdot \delta\boldsymbol{u} \; d\Omega + \int_\Omega \boldsymbol{S} : \delta\boldsymbol{\epsilon} \; d\Omega - \int_{\Gamma_t} \boldsymbol{S}\boldsymbol{n} \cdot \delta\boldsymbol{u} \; d\Gamma - \int_\Omega \boldsymbol{k} \cdot \delta\boldsymbol{u} \; d\Omega \tag{15.8b}$$

and the virtual work of the contact interface domain Γ_{jc} is given by

$$\delta W_{\Gamma_{jc}} = -\int_{\Gamma_{jc}^-} \boldsymbol{t}_{jc}^- \cdot \delta\boldsymbol{u}^- d\Gamma - \int_{\Gamma_{jc}^+} \boldsymbol{t}_{jc}^+ \cdot \delta\boldsymbol{u}^+ d\Gamma. \tag{15.8c}$$

Both quantities, Eqs. (15.8b) and (15.8c), are treated separately.

15.2.2.1 Body Domain

The body domain ${}^0\Omega$ is discretized with the finite element discretization $\widehat{\Omega}$ comprising the node locations $\widehat{\boldsymbol{x}}$ and the elements $\widehat{\Omega}^{(e)} \subset \widehat{\Omega}$. The discretization of the surface regions Γ_d and Γ_t are denoted by $\widehat{\Gamma}_d$ and $\widehat{\Gamma}_t$ respectively. Based on the interpolation of the displacement field within the domain of each element e, given by

$$u^{(e)} \approx N^{(e)}\,\widehat{u}^{(e)}, \tag{15.9}$$

where $N^{(e)}$ holds the shape functions of the respective element formulation and $\widehat{\boldsymbol{u}}(e)$ denotes the nodal degrees of freedom of all nodes connected to the element, one can evaluate the integrals in Eq. (15.8b) using a numerical integration scheme. It is assumed that the prescribed displacement boundary conditions are directly incorporated into the principle of virtual work, see e.g. [1] or [12] for a detailed description. Performing these tasks leads to the linear, undamped equation of motion without contact consideration

$$\boldsymbol{M}\ddot{\widehat{\boldsymbol{u}}} + \boldsymbol{K}\widehat{\boldsymbol{u}} = \boldsymbol{f}_e \tag{15.10}$$

where $\boldsymbol{M}$ denotes the mass matrix and $\boldsymbol{K}$ denotes the stiffness matrix of the constrained system. Displacements and accelerations of each node are collected in $\widehat{\boldsymbol{u}}$ and $\ddot{\widehat{\boldsymbol{u}}}$ respectively. The vector of external forces $\boldsymbol{f}_e$ considers both the prescribed stresses $\boldsymbol{t}_b$ and the imposed force density $\boldsymbol{k}$.

15.2.2.2 Contact Interface Domain

The contact interface domain is discretized utilizing zero thickness elements which implement a penalty approach for regularization of the contact interface conditions Eq. (15.7). In this case the finite element discretization $\widehat{\Omega}$ is required to comprise a matching mesh of all element surfaces adjacent to the contact interface. Then the discretization $\widehat{\Gamma}_{jc}$ of the joint contact interface Γ_{jc} is obtained by a composition of zero thickness elements $\widehat{\Gamma}_{jc}^{(e)}$ where one surface connects to a single element belonging to $\widehat{\Gamma}_{jc}^{-}$ and the other surface to a single element part of $\widehat{\Gamma}_{jc}^{+}$.

Formulations for obtaining a zero thickness element related contribution to the stiffness matrix $\Delta\boldsymbol{K}_{jc} \equiv \Delta\boldsymbol{K}_{jc}(\widehat{\boldsymbol{u}})$ can be found in literature, e.g [4] or [6]. In case of reduced order modeling, e.g. using a Galerkin projection based approach, the authors suggest a formulation which contributes to the nodal force vector on the right hand side of Eq. (15.10). Such an approach is outlined in [2] where the formulation of a 3 (6) node zero thickness element is provided as well. For easier reference the contact stress equivalent nodal force vector of the zero thickness element is given in a slightly modified writing

$$\boldsymbol{f}_{jc}^{(e)} = \int\limits_{\widehat{\Gamma}_{jc}^{(e)}} [\boldsymbol{N}^{(e)} - \boldsymbol{N}^{(e)}]^T\, \boldsymbol{t}_{jc}^{(e)}\, d\widehat{\Gamma}. \tag{15.11}$$

$\boldsymbol{N}^{(e)}$ denotes the matrix of element shape functions and $\widehat{\Gamma}_{jc}^{(e)}$ denotes the domain of the respective zero thickness element surface. The contact stress component is computed via the "constitutive" relation

$$t_n = \begin{cases} 0 & \text{if} \quad g_n \geq 0, \\ c_n g_n & \text{if} \quad g_n < 0 \end{cases}. \tag{15.12}$$

with the penalty parameter c_n. The resulting nonlinear equation of motion with contact consideration is given by

$$\boldsymbol{M}\ddot{\widehat{u}} + \boldsymbol{K}\widehat{u} = f_e + f_{jc}. \tag{15.13}$$

where $\boldsymbol{f}_{jc}$ denotes the contact stress equivalent nodal force vector representing all zero thickness elements. The quantity $\boldsymbol{f}_{jc}$ is formally obtained utilizing the linear assembly Operator A on the nodal force vectors of each zero thickness element

$$\boldsymbol{f}_{jc} = \underset{\widehat{\Gamma}_{jc}}{\mathrm{A}}\, \boldsymbol{f}_{jc}^{(e)}. \tag{15.14}$$

15.3 Model Order Reduction

15.3.1 Basics

It is assumed that an approximation of the nodal displacements $\widehat{\boldsymbol{u}}$ in Eq. (15.13) is given by the Ritz ansatz

$$\widehat{\boldsymbol{u}} \approx \boldsymbol{\Phi q} \tag{15.15}$$

with the mode matrix $\boldsymbol{\Phi}$ which columnwise holds the displacement trial vectors $\boldsymbol{\varphi}_i$ and the according generalized coordinates $\boldsymbol{q}$. If the number of generalized coordinates $\boldsymbol{q}$ is significantly smaller than the dimension of $\widehat{\boldsymbol{u}}$ the mode matrix $\boldsymbol{\Phi}$ can be utilized as reduction basis to obtain a reduced order model.

Substituting the ansatz Eq. (15.15) into Eq. (15.13) and performing a Galerking projection by premultiplying the resulting equation with $\boldsymbol{\Phi}^T$ finally leads to the reduced equation of motion

$$\boldsymbol{M}_{red}\ddot{\boldsymbol{q}} + \boldsymbol{K}_{red}\boldsymbol{q} = \boldsymbol{f}_{red} + \boldsymbol{\Phi}^T\boldsymbol{f}_{jc} \tag{15.16}$$

where $\boldsymbol{M}_{red} = \boldsymbol{\Phi}^T\boldsymbol{M\Phi}$ denotes the reduced mass matrix, $\boldsymbol{K}_{red} = \boldsymbol{\Phi}^T\boldsymbol{K\Phi}$ denotes the reduced stiffness matrix and $\boldsymbol{f}_{red} = \boldsymbol{\Phi}^T\boldsymbol{f}_e$ denotes the projected vector of external forces.

An reduced order model is intended to approximate a high dimensional solution space by a lower dimensional solution space with acceptable tradeoff in accuracy. One has to keep in mind that the solution obtained from such a reduced order model is constrained to the low dimensional solution space spanned by the trial vectors $\boldsymbol{\Phi}$. Therefore a proper trial vector basis is crucial for obtaining a satisfactory reduced order model.

For this sake the vector of nodal displacements is partitioned

$$\widehat{\boldsymbol{u}} = \begin{bmatrix} \widehat{\boldsymbol{u}}_i \\ \widehat{\boldsymbol{u}}_b \end{bmatrix} \tag{15.17}$$

where $\widehat{\boldsymbol{u}}_i$ denotes the displacement of the inner degrees of freedom and $\widehat{\boldsymbol{u}}_b$ denotes the displacement of the n_b boundary degrees of freedom within the discretized surfaces $\widehat{\Gamma}_d$ and $\widehat{\Gamma}_t$. Appropriate trial vectors for the investigation of the vibrational behavior of a linear mechanical system (up to a certain frequency limit) are suggested in [3]. With the help of these trial vectors one can construct a mode basis

$$\boldsymbol{\Phi}_{classic} = [\boldsymbol{\Psi}_c\boldsymbol{\Phi}_n] \tag{15.18}$$

which holds n_b constraint modes $\boldsymbol{\Psi}_c$ and n_n fixed interface normal modes $\boldsymbol{\Phi}_n$. The computation of these two types of trial vectors is described in detail in [3].

Unless a very high number of trial vectors is considered this mode basis does not guarantee to accurately capture the local displacements within the contact interface. As consequence the computed contact stresses might not be sufficiently accurate as well. To overcome this limitation [8] proposed to extend an existing mode base, like the above mentioned classic mode basis following Craig and Bampton, by a set of so called joint interface modes $\boldsymbol{\Phi}_{jim}$. These modes account for the local displacements due to the contact interface. Furthermore Newtons 3rd law is explicitly considered within their formulation.

Methods for the computation are given in [8], [9] and [11], the resulting extended transformation matrix reads

$$\boldsymbol{\Phi} = [\boldsymbol{\Psi}_c \;\; \boldsymbol{\Phi}_n \;\; \boldsymbol{\Phi}_{jim}]. \tag{15.19}$$

15.3.2 Brief Review on Proper Orthogonal Decomposition

The Proper Orthogonal Decomposition (POD) is a method for extracting the essential information out of a given collection of k snapshots $\boldsymbol{y}_j$. This essential information is expressed by a low dimensional set of $\ell < k$ basis vectors $\boldsymbol{v}_i$ which approximate the collection of snapshots in an optimal sense, see [7]. These POD basis vectors are a solution to the minimization problem

$$\min_{v_1,\ldots,v_\ell} \sum_{j=1}^{k} \left\| y_j - \sum_{i=1}^{\ell} \left\langle y_j, v_i \right\rangle_W v_i \right\|^2 \tag{15.20a}$$

so that

$$\left\langle v_i, v_j \right\rangle_W = \delta_{ij} \qquad \text{for} \quad i,j = 1, \ldots, \ell \tag{15.20b}$$

where

$$\langle \boldsymbol{a}, \boldsymbol{b} \rangle_W = \langle \boldsymbol{a}, \boldsymbol{W}\boldsymbol{b} \rangle = \langle \boldsymbol{W}\boldsymbol{a}, \boldsymbol{b} \rangle = a^T \boldsymbol{W}\boldsymbol{b} \tag{15.21}$$

denotes the weighted inner product of two vectors with a positive definite weighting matrix $\boldsymbol{W}$. A basis $\boldsymbol{V} = [\boldsymbol{v}_1, \ldots, \boldsymbol{v}_\ell]$, which is a solution to the above minimization problem, can be obtained by solving the symmetric eigenvalue problem

$$\boldsymbol{Y}^T \boldsymbol{W} \boldsymbol{Y} \bar{\boldsymbol{v}}_i = \lambda_i \boldsymbol{v}_i \qquad \text{for} \quad i = 1, \ldots, \ell \tag{15.22a}$$

where $\boldsymbol{Y} = [\boldsymbol{y}_1, \ldots, \boldsymbol{y}_k]$. The final POD basis vectors are obtained by evaluating

$$\boldsymbol{v}_i = \frac{1}{\sqrt{\lambda_i}} \boldsymbol{Y} \bar{\boldsymbol{v}}_i. \qquad \text{for} \quad i = 1, \ldots, \ell \tag{15.22b}$$

This procedure is sometimes called the method of snapshots [7].

15.3.3 Test Load Based Joint Interface Modes

In [11] it was suggested to define test loads within the contact interface and utilize the POD on the resulting displacement fields for computation of the joint interface modes. The respective steps are briefly recalled in this subsection.

First step is the decomposition of the discretized contact interface $\widehat{\Gamma}_{jc}$ into n subareas as suggested in [9]. Each of these subareas is loaded by a unit pressure distribution, according to Newton's 3rd law on both sides $^{\pm}$ of the contact interface. The resulting equivalent nodal force vectors are collected in the matrix of test loadcases

$$\boldsymbol{F}_{jc}^{(*)} = \left[f_{jc,1}^{(*)}, \ldots, f_{jc,n}^{(*)} \right]. \tag{15.23}$$

These test load cases are utilized within the static equilibrium

$$\boldsymbol{K}\widehat{U}^{(*)} = \boldsymbol{F}_{spc} + \boldsymbol{F}_{jc}^{(*)} \tag{15.24}$$

where the body is fixed at $\widehat{\Gamma}_d$ and $\widehat{\Gamma}_t$ as well. The constraints at $\widehat{\Gamma}_d$ are already considered within the principle of virtual work, the constraints at $\widehat{\Gamma}_t$ are enforced by the respective column vectors of the matrix $\boldsymbol{F}_{spc}$. The resulting displacement fields are collected in the matrix $\widehat{U}^{(*)} = \left[\widehat{u}_1^{(*)}, \ldots, \widehat{u}_n^{(*)} \right]$.

By setting $\boldsymbol{Y} = \widehat{U}^{(*)}$ and $\boldsymbol{W} = \boldsymbol{I}$ the application of the snapshot method Eq. (15.22) for computation of the joint interface modes $\boldsymbol{\Phi}_{jim}^{A} = \boldsymbol{V}$ is pretty straight forward.

Furthermore [11] suggests to use the stiffness matrix for weighting, $\boldsymbol{W} = \boldsymbol{K}$, which lead to promising results during own investigations of the authors. This requires a positive definite stiffness matrix, which is ensured in our case as $\boldsymbol{K}$ represents the constrained body. This method will be denoted as "method A" for the remainder of this contribution.

15.3.4 Contact Simulation Based Joint Interface Modes

To improve the convergence of the computed contact stresses with respect to the required number of joint interface modes the authors propose to utilize the contact forces resulting from selected contact simulations for the computation of joint interface modes. These contact simulations, which comprise static and dynamic loadcases, are presented in this subsection.

15.3.4.1 Static Loadcases

By means of the partitioning scheme introduced in Eq. (15.17) the static equilibrium equation of the body fixed at $\widehat{\Gamma}_d$ is given by

$$\boldsymbol{K}\widehat{\boldsymbol{u}}(s) = \begin{bmatrix} \boldsymbol{K}_{ii} & \boldsymbol{K}_{ib} \\ \boldsymbol{K}_{bi} & \boldsymbol{K}_{bb} \end{bmatrix} \begin{bmatrix} \widehat{\boldsymbol{u}}_i(s) \\ \widehat{\boldsymbol{u}}_b(s) \end{bmatrix} = \boldsymbol{f}_e^{(s)} + \boldsymbol{f}_{jc}^{(s)} = \begin{bmatrix} \boldsymbol{0}_{ie} \\ \boldsymbol{f}_{be}^{(s)} \end{bmatrix} + \begin{bmatrix} \boldsymbol{f}_{ijc}^{(s)} \\ \boldsymbol{0}_{bjc} \end{bmatrix} \tag{15.25}$$

The k nodal contact force $f_{jc,j}^{(s)}$ snapshots due to different external force vectors $\boldsymbol{f}_{e,j}^{(s)}$ are collected in the matrix

$$\boldsymbol{F}_{jc}^{(s)} = [\boldsymbol{f}_{jc,1}^{(s)}, \ldots, \boldsymbol{f}_{jc,k}^{(s)}] \tag{15.26}$$

for subsequent processing. It is worth to note that, due to the nonlinear relations within the contact interface, superposition is not possible any longer. For this reason the external force vectors $\boldsymbol{f}_{e,j}^{(s)}$ need to comprise meaningful load combinations covering each boundary degree of freedom.

15.3.4.2 Dynamic Loadcases

By means of the partitioning scheme introduced in Eq. (15.17) the initial value problem of the body fixed at $\widehat{\Gamma}_d$ is given by the equation of motion

$$\boldsymbol{M}\ddot{\widehat{\boldsymbol{u}}}^{(d)} + \boldsymbol{K}\widehat{\boldsymbol{u}}(d) = \begin{bmatrix} \boldsymbol{M}_{ii} & \boldsymbol{M}_{ib} \\ \boldsymbol{M}_{bi} & \boldsymbol{M}_{bb} \end{bmatrix} \begin{bmatrix} \ddot{\widehat{\boldsymbol{u}}}_i^{(d)} \\ \ddot{\widehat{\boldsymbol{u}}}_b^{(d)} \end{bmatrix} + \begin{bmatrix} \boldsymbol{K}_{ii} & \boldsymbol{K}_{ib} \\ \boldsymbol{K}_{bi} & \boldsymbol{K}_{bb} \end{bmatrix} \begin{bmatrix} \widehat{\boldsymbol{u}}_i(d) \\ \widehat{\boldsymbol{u}}_b(d) \end{bmatrix} = \boldsymbol{f}_e^{(d)} + \boldsymbol{f}_{jc}^{(d)} = \begin{bmatrix} \boldsymbol{0}_{ie} \\ \boldsymbol{f}_{be}^{(d)} \end{bmatrix} + \begin{bmatrix} \boldsymbol{f}_{ijc}^{(d)} \\ \boldsymbol{0}_{bjc} \end{bmatrix}. \tag{15.27a}$$

with the initial conditions

$$^0\widehat{\boldsymbol{u}}(d) = \boldsymbol{u}_0 \quad \forall \widehat{x} \in \widehat{\Omega} \qquad \text{and} \qquad ^0\dot{\widehat{\boldsymbol{u}}}^{(d)} = \boldsymbol{0} \quad \forall \widehat{x} \in \widehat{\Omega}. \tag{15.27b}$$

The initial displacement is either intended to consider the prestressed state due to pretension of the bolted connection or simply $\boldsymbol{u}_0 = \boldsymbol{0}$. To obtain stable responses, independently from the respective numerical time integration scheme, some damping should be introduced into the equation of motion. The authors suggest to utilize a Rayleigh damping approach where the damping matrix is a linear combination of mass and stiffness matrix. This is feasible as the main purpose of the simulations is to obtain meaningful snapshots due to the evolvement of the inner degrees of freedom resulting from the dynamic excitation of the boundary degrees of freedom.

The suggested k excitations are formally given by

$$\boldsymbol{f}_{e,j}^{(d)} = \begin{cases} \boldsymbol{0} & t \le 0 \\ \boldsymbol{f}_{e,j}^{(s)} & t > 0 \end{cases} \qquad \text{for} \quad j = 1, \ldots, k \tag{15.28}$$

where $\boldsymbol{f}_{e,j}^{(s)}$ corresponds to the static loadcases investigated in Sect. 15.3.4.1. The response to each of these k excitations, up to at least one period of the dominant vibration, is computed. To lower the computational burden a prior model reduction step, e.g. [5], is highly recommended.

The time discretized evolvement of the jth solution leads to n_j nodal contact force $f_{jc,i}^{(d,j)}$ snapshots which are collected in the matrix

$$F_{jc}^{(d,j)} = [f_{jc,1}^{(d,j)}, \ldots, f_{jc,n_j}^{(d,j)}] \tag{15.29}$$

for subsequent processing.

15.3.4.3 Trial Vectors

To compute the trial vectors all contact force snapshots Eqs. (15.26) and (15.29) are combined in a single matrix

$$F_{jc}^{(sd)} = [F_{jc}^{(s)}, F_{jc}^{(d,1)}, \ldots, F_{jc}^{(d,k)}]. \tag{15.30}$$

By setting $Y = F_{jc}^{(sd)}$ and $W = I$ the application of the snapshot method Eq. (15.22) again is pretty straight forward. Finally the resulting contact force vectors $F_{jim} = V$ are utilized within the static equilibrium Eq. (15.24). The resulting displacement shapes already represent the vector space of the suggested joint interface modes Φ_{jim}^{B}.

This method will be denoted as "method B" for the remainder of this contribution.

15.4 Numerical Example

15.4.1 Model Description

The reference method Sect. 15.3.3 and the proposed method Sect. 15.3.4 for computation of the joint interface modes are applied to the finite element model of a bolted cantilever beam depicted in Fig. 15.2.

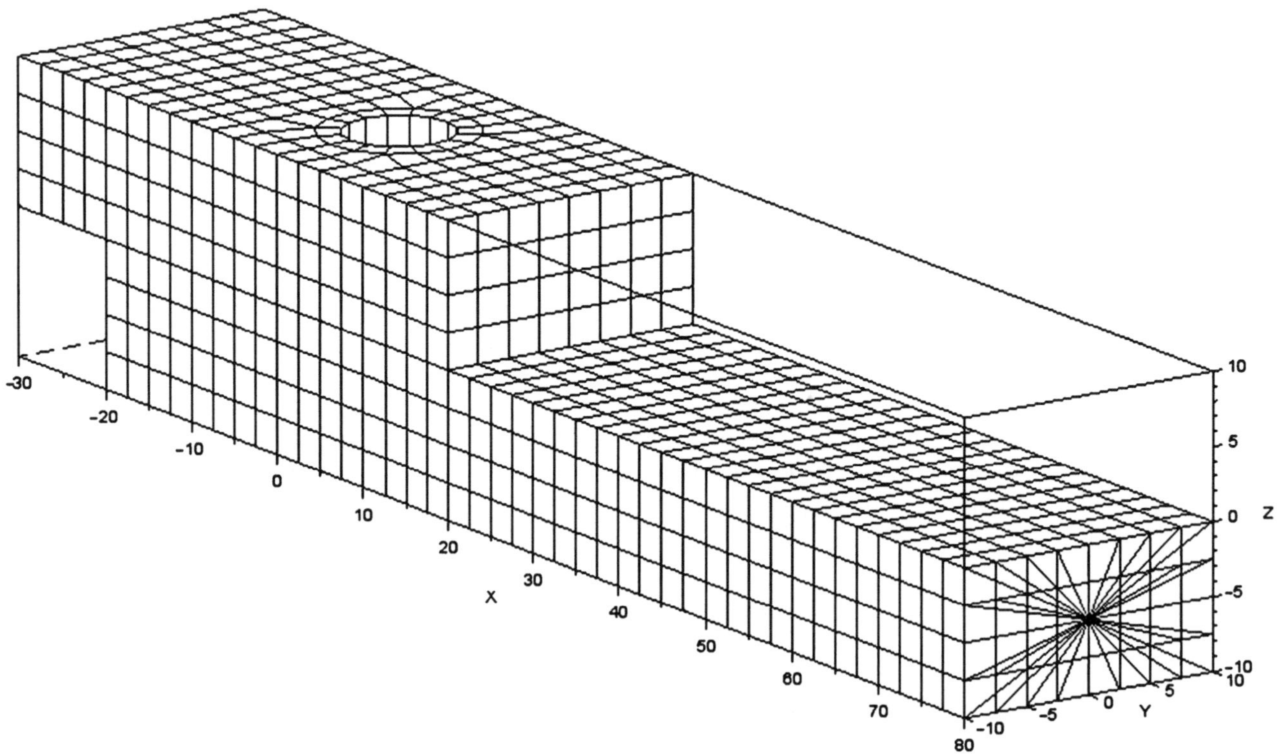

Fig. 15.2 Finite element model of the bolted cantilever (the beam element based bolt model is hidden)

The body consists of two substructures with rectangular shaped cross section ($a = 10$ mm, $b = 20$ mm) and an overall length of $l = 110$ mm. The substructures are connected by a contact interface and a single bolt. All structural components are modeled using a linear elastic material with Young's Modulus $E = 210{,}000$ N/mm^2, Poisson ratio $\nu = 0.29$ and density $\rho = 7820$ kg/m^3.

The discretization $\widehat{\Omega}$ of the full model comprises 1834 elements and 2780 nodes. The CHEXA, CBEAM, RBE2 and RBE3 element formulations of the commercial FEM software package MSC Nastran are used to generate the system matrizes within Eq. (15.10). The contact interface $\widehat{\Gamma}_{jc}$ is discretized with an 4 (8) node zero thickness element formulation which allows to obtain the contact stress equivalent nodal force vector within Eq. (15.13).

A convergence study regarding the meshsize of the full model is not within the scope of this contribution. It is assumed that the accuracy of the contact stresses obtained from this numeric model are sufficiently accurate for the purpose of this model. The purpose of the reduced order model is to reproduce these contact stresses with sufficiently low error but significantly higher computational efficiency.

15.4.2 Sticking Friction Definition

During generation of the system matrices using a commercial FEM software package there is no connection of the two contacting substructures except the beam element based bolt model. The resulting Eq. (15.13) is utilized for a contact simulation where the structure is constrained at $\widehat{\Gamma}_d$ and the bolt is incrementally pretensioned to a nominal value. The resulting contact stresses are depicted on the left hand side of Fig. 15.3.

All contact node pairs possessing a contact pressure over a certain threshold are assumed to stick together in tangential direction during all subsequent simulations. The node pairs selected this way are marked on the right hand side of Fig. 15.3. The sticking friction condition within this area is approximated utilizing a penalty approach. All subsequent steps consider these additional penalty stiffness related entries in the stiffness matrix.

15.4.3 Reduction Basis

The computation of static and dynamic loadcases to generate snapshots for obtaining a reduction basis was carried out according to Sect. 15.3.4. The static loadcases represent load combinations of bolt pretension, vertical tip load and torsional moment at the free end of the cantilever.

The dynamic loadcases, computed with a nominal pretension of the bolt, are limited to combinations of vertical load and torsional moment at the free end of the cantilever. For easier computation the equation of motion comprises an additional Rayleigh approach based damping matrix. The step responses are obtained from the equation of motion, which is transformed to a system of first order differential equations, utilizing the implizit Euler integration scheme.

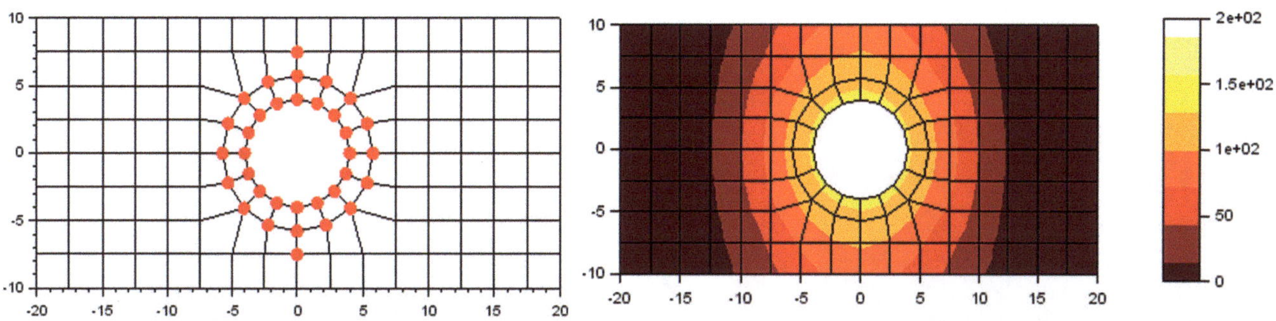

Fig. 15.3 Contact stresses in surface normal direction resulting from a nominal pretension of the bolted joint (*left*). Locations with sticking friction are identified based on a contact stress threshold. To suppress the tangential relative movement of the adjoining contact node pairs a penalty approach is utilized at the marked node pairs (*right*)

15.4.4 Contact Stress Accuracy Evaluation

15.4.4.1 Static Loadcases

Two exemplary loadcases are documented within this section. The first loadcase denoted as $LC1$ is a vertical tip load while the second loadcase denoted as $LC2$ additionally comprises a torsional moment. It is noted that both loadcases are not explicitly considered as certain load combination during computation of the snapshots for obtaining the reduction basis (Fig. 15.4).

The according contact pressure distributions in surface normal direction are depicted in Fig. 15.5.

The relative error of the contact stresses in surface normal direction is chosen as evaluation criteria. This quantity illustrates whether the computed contact stresses are useful for subsequent computations, e.g. for the computation of shear stresses due to friction. For a vector $\boldsymbol{y}$ the relative error vector is given by

$$e_{rel} = \frac{y - y_{ref}}{\left\| y_{ref} \right\|_{\infty}} \tag{15.31}$$

where $\|.\|_{\infty}$ denotes the maximum norm of a vector and $\boldsymbol{y}_{ref}$ denotes the reference values obtained from the full model. For computation of the contact pressure relative error the maximum contact pressure value of the reference solution is used as reference value.

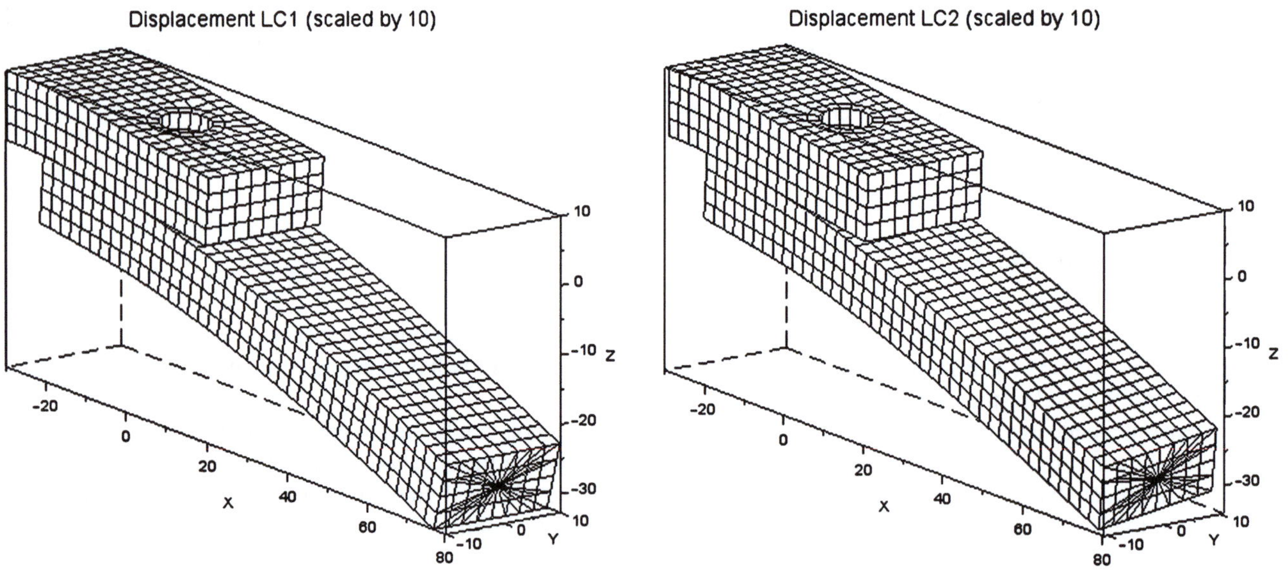

Fig. 15.4 Resulting displacement of the two substructures for $LC1$ (*left*) and $LC2$ (*right*)

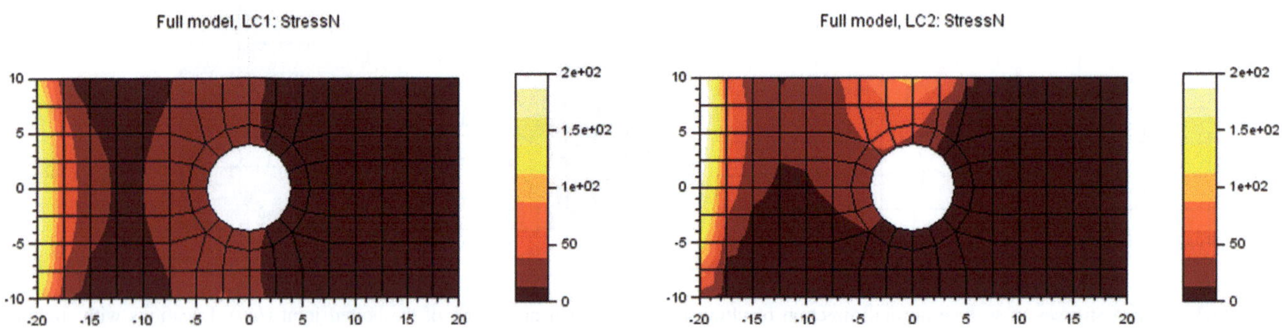

Fig. 15.5 Contact pressure obtained from the full model for loadcase LC1 (*left*) and loadcase LC2 (*right*)

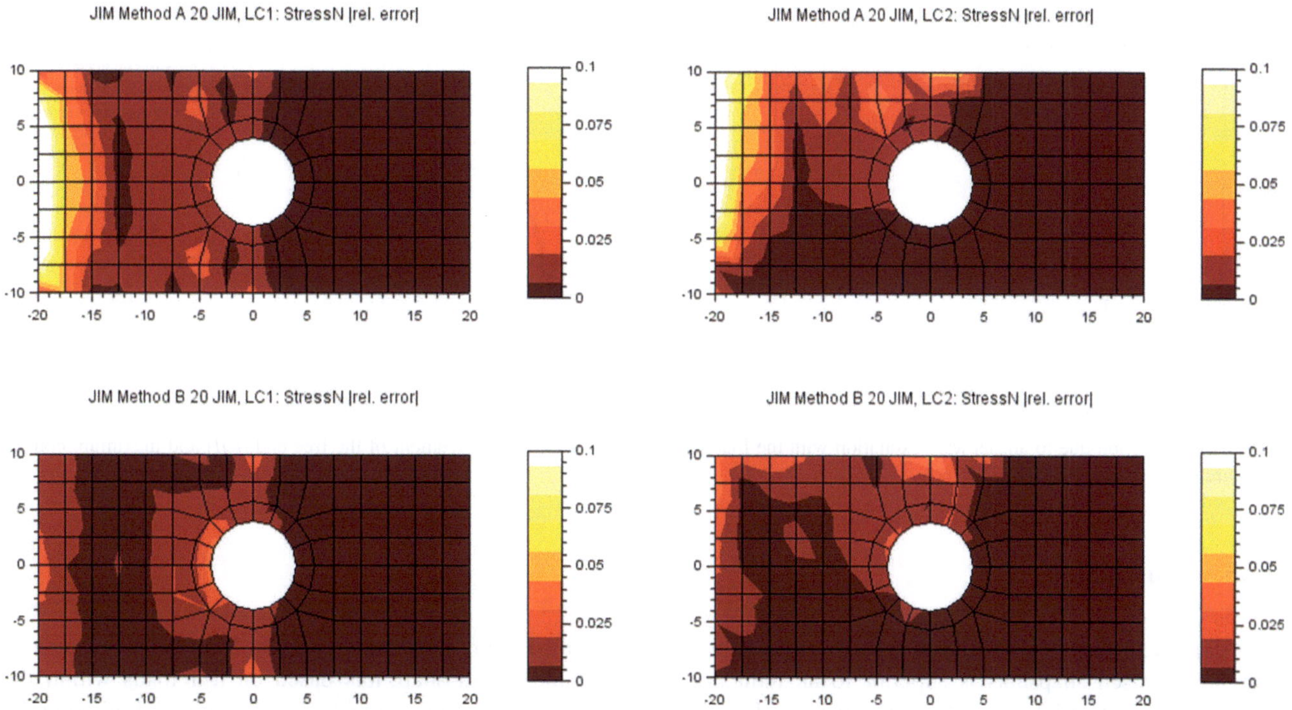

Fig. 15.6 Relative error of the contact pressure obtained from the reduced order model with method A based joint interface modes (*top row*) and method B based joint interface modes (*bottom row*)

The respective error distributions for the reduced order models are depicted in Fig. 15.6. The following conclusions can be drawn:

- Symmetric contact pressure distributions lead to symmetric error distributions.
- The maximum relative error is significantly larger than 10 % when utilizing 20 JIMs obtained by method A
- The maximum relative error is less than 5 % when utilizing 20 JIMs obtained by method B.

It can be noted that further investigations indicate a more rapid convergence rate of the joint interface modes computed by method B than those computed by method A.

15.4.4.2 Dynamic Loadcase

The response of the cantilevers free end due to LC2 applied as a step function is depicted in Fig. 15.7. Already five joint interface modes, either computed using method A or method B, are sufficient to resemble the cantilevers free end displacement of the reference solution. Evaluation of the maximum contact pressure leads to a different conclusion:

- Method A based JIMs lead to a significantly higher maximum contact pressure.
- Method B based JIMs almost resemble the maximum contact pressure of the reference solution.

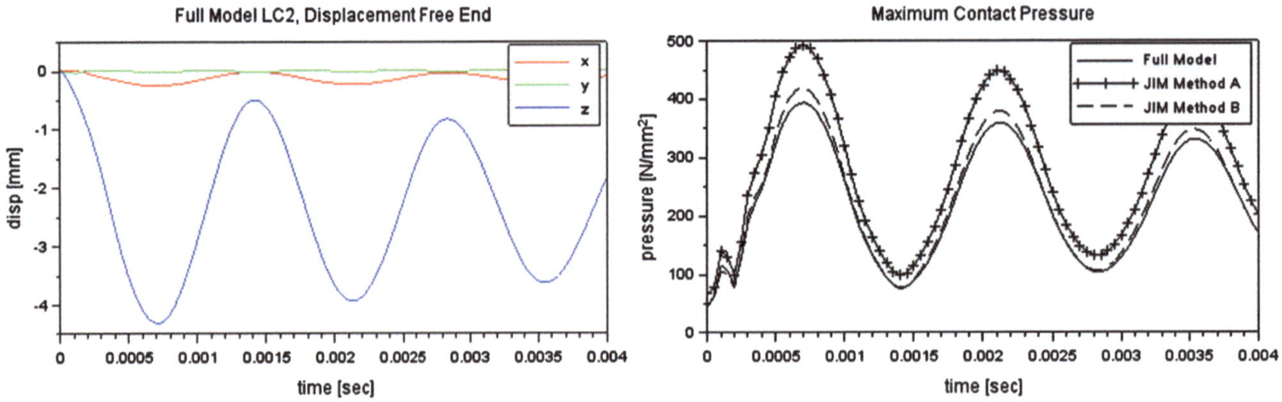

Fig. 15.7 Vibration due to an stepwise excitation with the load combination of LC2: displacement of the free end (*left*) and maximum contact pressure (*right*)

15.5 Conclusion

The numerical example reveals that an POD reduction basis obtained from a methodical defined collection of contact simulation based snapshots is superior to an reduction basis obtained from snapshots based on test loads within the investigated contact interface. This conclusion is based on the relative error of the contact stresses computed utilizing the according reduced order models with respect to the contact stresses obtained from the full model.

But it has to be noted that the test load based reduction basis can be obtained with reasonable effort while the computational burden involved to realize the contact simulations quickly gets tremendous, especially for complex structures comprising many boundary degrees of freedom. Therefore the results suggest the existence of an optimal basis of trial vectors for contact interfaces, but an efficient approach for obtaining such a basis still has to be found.

Acknowledgements Support of the authors by the K2 Austria Center of Competence in Mechatronics (ACCM) and the Engineering Center Steyr (MAGNA) is gratefully acknowledged.

References

1. Bathe, K.-L.: Finite Element Procedures. Prentice Hall, Upper Saddle River (1996)
2. Breitfuss, M., Irschik, H., Holl, H.J., Witteveen, W.: DEIM for the efficient computation of contact interface stresses. In: Allen, M., Mayes, R., Rixen, D. (eds.) Proceedings of the 32nd IMAC, A Conference and Exposition on Structural Dynamics, 2014. Conference Proceedings of the Society for Experimental Mechanics Series, vol. 1, pp. 435–445. Springer, New York (2014)
3. Craig, R.R., Bampton, M.C.C.: Coupling of substructures for dynamic analyses. AIAA J. **6**(7), 1313–1319 (1968)
4. Mayer, M.H., Gaul, L.: Segment-to-segment contact elements for modelling joint interfaces in finite element analysis. Mech. Syst. Signal Process. **21**(2), 724 – 734 (2007)
5. Qu, Z.-Q.: Model reduction for dynamical systems with local nonlinearities. AIAA J. **40**(2), 327–333 (2002)
6. Süß, D., Willner, K.: Investigation of a jointed friction oscillator using the multiharmonic balance method. Mech. Syst. Signal Process. **52–53**, 73 – 87 (2015)
7. Volkwein, S.: Model reduction using proper orthogonal decomposition. http://www.math.uni-konstanz.de/numerik/personen/volkwein/teaching/POD-Vorlesung.pdf (September 2014)
8. Witteveen, W., Irschik, H.: Efficient modal formulation for vibration analysis of solid structures with bolted joints. In: Conference Proceedings of IMAC-XXV: A Conference & Exposition on Structural Dynamics, Orlando, 19–22 February 2007
9. Witteveen, W., Irschik, H.: Efficient computation of joint interface modes. In: Proceedings of the IMAC-XXVII, Orlando, 9–12 February 2009
10. Witteveen, W., Irschik, H.: Efficient mode based computational approach for jointed structures: joint interface modes. AIAA J. **47**(1), 252–263 (2009)
11. Witteveen, W., Sherif, K.: Pod based computation of joint interface modes. In: Proulx, T. (ed.) Linking Models and Experiments. Conference Proceedings of the Society for Experimental Mechanics Series, vol. 2, pp. 19–28. Springer, New York (2011)
12. Zienkiewicz, O.C., Taylor, R.L., Zhu, J.Z.: The Finite Element Method: Its Basis and Fundamentals, 7th edn. Butterworth-Heinemann, Oxford (2013)

Chapter 16
Structural Design with Joints for Maximum Dissipation

M. Stender, A. Papangelo, M. Allen, M. Brake, C. Schwingshackl, and M. Tiedemann

Abstract Many engineered structures are assembled using different kinds of joints such as bolted, riveted and clamped joints. Even if joints are often a small part of the overall structure, they can have a massive impact on its dynamics due to the introduction of nonlinearities. Thus, joints are considered a design liability. Significant effort has been spent in joint characterization and modelling, but a predictive joint model is still non-existent. To overcome these uncertainties and ensure certain safety standards, joints are usually overdesigned according to static considerations and their stiffness. Especially damping and nonlinearity are not considered during the design process. This can lead to lower performance, lower payload, and as result of the joints structural dynamic models often do a poor job of predicting the dynamic response. However, it is well-known that, particularly for metal structures, joints represent the main source of energy dissipation. In this work a minimal model is used to show how structural performance can be improved using joints as a design variable. Common optimization tools are applied to a nonlinear joint model in order to damp undesired structural vibrations. Results illustrate how the intentional choice of joint parameters and locations can effectively reduce vibration level for a given operating point of a jointed structure.

Keywords Joints • Dissipation • Optimization • Design • Nonlinear dynamics

16.1 Introduction

Many structures from automotive, aerospace and civil engineering are assembled by joints such as bolted, riveted and clamped joints. These joints introduce nonlinear dynamical behavior and uncertainties [1–3] even if they are just a small part of the overall structure. Also, joints are one of the main sources of energy dissipation in metal structures [2, 3, 4]. During the design process of engineering structures, maximum stress levels and vibration amplitudes during operation are calculated using mathematical models. As joints can strongly affect the dynamic behavior, they have to be included in the mathematical model. A great effort has been spent in the characterization of joints, but reliable and predictive models are still non-existent. Therefore, joints are considered rather a design liability than a design parameter. Because joints are not considered in the design, non-optimal designs are achieved, which have large vibration amplitudes or are heavier than needed.

Various studies have considered the problem of finding the optimal linear damping and stiffness parameters which minimize the vibration level of an undamped linear structure with one [5] or more [6] degrees of freedom, while this work focuses on a nonlinear model of the joint, i.e. the Jenkins model. The ultimate goal of this paper is to demonstrate

M. Stender (✉)
Dynamics Group, Hamburg University of Technology, Schlossmuehlendamm 30, 21073 Hamburg, Germany
e-mail: merten.stender@tuhh.de

A. Papangelo
Department of Mechanics Mathematics and Management, Polytechnic of Bari, Viale Japigia 182, 70126 Bari, Italy

M. Allen
Department of Engineering Physics, University of Wisconsin, 1500 Engineering Drive, Madison, WI 53706, USA

M. Brake
Component Science and Mechanics, Sandia National Laboratories, MS 1070, Albuquerque, NM 87185-1070, USA

C. Schwingshackl
Imperial College London, Exhibition Road, SW7 2AZ London, UK

M. Tiedemann
Dynamics Group, Hamburg University of Technology, Schlossmuehlendamm 30, 21073 Hamburg, Germany

Audi AG, 85057 Ingolstadt, Germany

A. Brandt, R. Singhal (eds.), *Shock & Vibration, Aircraft/Aerospace, Energy Harvesting, Acoustics & Optics, Volume 9*,
Conference Proceedings of the Society for Experimental Mechanics Series, DOI 10.1007/978-3-319-30087-0_16

the performance improvement that might be achieved by exploiting joints, hopefully motivating the development of improved testing, modeling and design methods that consider the joints. A minimal model is studied and the joint model is considered to be predictive so that changes to the joint parameters (e.g. the clamping pressure) can be related to the response. While the simple model used here may not prove to be predictive in many real scenarios, it does provide a useful test case that captures some important physics. The structural response is numerically computed and optimization routines are adopted to show how it can be improved by the intentional choice of joint parameters. In particular the idea behind is to design not only the structure but also the joint to be able to damp vibrations in frequency ranges of interest. By minimizing the peak displacements in the simple model, the peak stresses in the structure would presumably decrease and the fatigue life would be improved, reducing wear and maintenance costs. We discuss different joint locations to show advantages and disadvantages of these choices which can help the designer during the design process.

16.2 The Model

This work employs a four degree of freedom (4-DOF) model including a nonlinear joint as depicted in Fig. 16.1. The structure consists of four masses m which are assembled by springs of stiffnesses k. The harmonic external forcing $F_{\mathrm{ext}} = A\cos(2\pi f_{\mathrm{ext}} t)$ acts on the first mass. In order to be able to compare Frequency Response Functions (FRF), constant modal damping ξ is applied to every mode [7]. The joint is represented by a Jenkins model. The joint's effect on the system response will be studied in two different positions: (a) joint located between the second and third mass, and (b) joint located between the third mass and the ground. These positions represent the fundamental locations of joints in engineering structures—either assembling two moving substructures or clamping a structure to a rigid body.

The LuGre formulation [5] is employed for the definition and implementation of the joint model, see Eq. (16.1). The tangential joint stiffness is denoted by k_{t} and the normal load by N. Therefore in the limit of full sliding and full stick, the sliding force is described by $\mu_{\mathrm{k}}N$, whereas the sticking force is f_{S}. The transition from the sticking to the sliding state is described by an exponential decay depending on the Stribeck velocity v_{s} and the relative joint velocity $\dot{u} = \frac{du}{dt}$, where u is the relative displacement experienced by the joint. During the numerical simulation the differential Eq. (16.1b) is solved and the joint force f_{LuGre} computed.

$$g(\dot{u}) = \mu_{\mathrm{k}}N = (f_s - \mu_{\mathrm{k}}N)e^{-\left(\dot{u}/v_s\right)^2} \tag{16.1a}$$

$$\dot{q} = \dot{u} - \frac{k_{\mathrm{t}}|\dot{u}|}{g(\dot{u})}q \tag{16.1b}$$

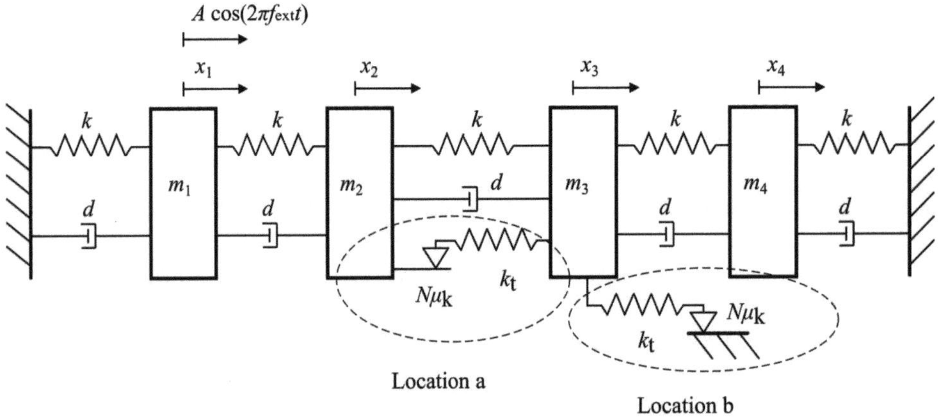

Fig. 16.1 4-DOF system with Jenkins element and joint locations discussed in this work

$$f_{\text{LuGre}}(\dot{u}) = k_t q \tag{16.1c}$$

The default model parameters read $m_{1,\ldots,4} = 1\,\text{kg}$, $k = 40{,}000\,\text{N/m}$, $\xi = 0.001$, $A = 0.1\,\text{N}$, $k_t = 40{,}000\,\text{N/m}$, $\mu_k = 1$, $N = 1\,\text{N}$, $f_s = 5\,\text{N} + N$ and $v_s = 5 \cdot 10^{-4}\text{m/s}$.

16.3 The Method

To compute the response of the structure, the nonlinear frequency response function of each mass has to be determined. Appropriate cost functions are defined and evaluated to obtain a scalar value ('cost value') which at each frequency is representative of the structural dynamic response. The resulting cost value is finally subject to the optimization routine. Hereafter, these key methods are introduced.

16.3.1 Numerical Continuation Method

The numerical continuation toolbox MATCONT [8] is utilized to compute frequency response functions of the system. It employs orthogonal collocation methods which discretize a (partial) differential equation at collocation points. Initialized with stable attractors of the system, such as equilibrium points or limit cycles, a solution curve can be approximated. The solution found previously can then be tracked in the direction of the continuation parameter using predictor-corrector steps. The continuation is initialized at a low value of the excitation frequency. In order to obtain the frequency response function, a continuation of the nonlinear solution is performed for the continuation parameter f_{ext}.

16.3.2 Merging of Frequency Response Functions

For the 4-DOF model the displacement x_i of every mass i is available for investigation. Dividing each displacement x_i by the excitation amplitude A yields four receptances R_i [m/N]. As the structural performance of the whole structure will be subject to an optimization, a scalar cost value is required to represent the vibration level. Thus, either the receptance of a single mass has to be selected as base for the scalar cost value, or the four receptances have to be merged into one combined receptance. In order to capture the overall system behavior a combined receptance $\overline{R}$ is used for this work. Therefore, the maximum receptance for every frequency point j is selected to compose a single data set $\overline{R}$ that represents the vibration level of the structure. The use of the maximum value ensures to consider the maximum displacement amplitude and thus the maximum stresses the structure undergoes.

$$\overline{R}(R_i) = \max_{\substack{i = 1, \cdots, 4 \\ j = 1, \cdots, n_f}} \left(R_i^j\right) \in [f_1, f_u] \tag{16.2}$$

In Eq. (16.2) $[f_l, f_u]$ is the frequency interval chosen and n_f the number of frequency points considered on the interval.

16.3.3 Cost Function

The objective of the optimization is to reduce vibration amplitudes in a certain frequency range. Therefore, an appropriate cost function based on the merged receptance $\overline{R}$ needs to be defined. Typically, frequency ranges in the vicinity of resonance frequencies are considered critical in practical applications. Hereafter, the peak value of the merged receptances within the frequency range $[f_l, f_u]$ is defined as the scalar cost value:

$$\Psi = R_{\text{peak}} \in [f_1, f_u] \tag{16.3}$$

16.3.4 Optimization Algorithm

The optimization is carried out by the MATLAB built-in method 'fmincon' which can perform bounded, nonlinearly constrained optimizations. This method is based on the sequential quadratic programming (sqp) algorithm [9], which iteratively solves a sequence of subproblems using a quadratic model of the objective. If no constraints are applied this method reduces to Newton's algorithm finding a point with zero gradient of the objective. The free joint parameters $\mathbf{p}$ serve as input values and the cost value Ψ as objective for the minimization giving the optimal parameter configuration $\hat{\mathbf{p}}$.

$$\hat{\mathbf{p}} = \min_{\mathbf{p}} (\Psi(\mathbf{p})) \in [\mathbf{p}_{\min}, \mathbf{p}_{\max}] \tag{16.4}$$

The termination tolerance on the parameter is chosen to be $tolX = 10^{-4}$ and the termination tolerance on the cost function value $tolFun = 10^{-6}$.

16.4 First Model Studies

The joint parameters k_t and N (see Fig. 16.1) are used as free parameters during the optimization process while all structure parameters are kept fixed. First, the qualitative influence of both joint parameters are discussed. The friction coefficient μ_k is kept constant so that the normal load N controls the sliding or sticking behavior of the joint for a given joint force. The limiting cases of very small and very large normal loads linearize the system. The joint becomes inactive for normal loads tending to zero as it always slides without dissipating. Large normal loads compared to the excitation amplitude result in a stuck joint adding the linear stiffness k_t to the system. The tangential stiffness affects the dominance of the joint and thus the impact the joint has on the structure: for small values of k_t compared to k, the joint becomes inactive and the system behaves linearly. Thus, the parallel connected stiffnesses k and k_t and the normal load N control the sliding state of the joint for a given excitation amplitude. This observation implies the introduction of two dimensionless parameters α and β.

$$\alpha = \frac{k_t}{k} \quad \beta = \frac{N}{A} \tag{16.5}$$

As the considered model is nonlinear, it is important to first verify that the two dimensionless parameters that were chosen remain valid. To check this, the frequency response functions were compared for different values of the dimensional parameters—once keeping the structure parameter (k, A, respectively) fixed, once keeping the joint parameter (k_t, N, respectively) fixed for the same dimensionless parameters. The results of this verification step are not shown here, but they did show perfect agreement of the frequency response functions: for a given value of the dimensionless parameter the system response is the same regardless of the choice of the dimensional parameters (either joint or a structure parameters). Hence, the system is completely defined by the two dimensionless parameters α and β which are therefore valid for further investigations. This result meets the experience from different fields of joint research: the force ratio β is a well-known parameter in the analysis of underplatform dampers [4, 6]. For the default configuration the parameters are chosen to be $\alpha = 1$ and $\beta = 1$.

Figures 16.2 and 16.3 depict the impact of the stiffness ratio α and the force ratio β on the system dynamics. Generally, stiffness is added to the structure when α is larger than 1. Consequently, the resonance frequencies rise and resonance amplitudes decrease as shown in Fig. 16.2. The study of different values of the force ratio β reveals the linearizing effect of the limiting cases $\beta \to 0$ (purely sliding joint) and $\beta \to \infty$ (completely stuck joint). The choice of a particular force ratio drives the system to a state within those linear limits. For this particular joint location (a), only the second and fourth modes are affected by the joint since only these modes exhibit out of phase movement of masses m_2 and m_3. If the joint is instead attached to the ground (joint location (b)) all modes are affected. Already at this stage it is clear that the intentional choice of the joint parameters has a strong influence on the system dynamics and the vibration levels that the structure experiences.

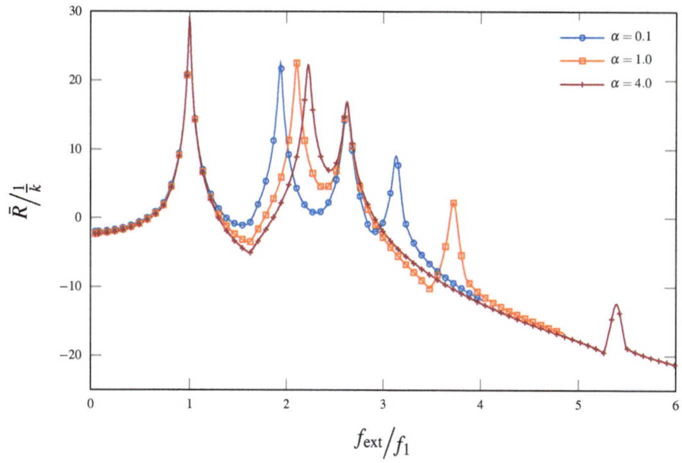

Fig. 16.2 Impact of stiffness ratio ($\alpha = 0.1$ [*circles*], $\alpha = 1.0$ [*squares*], $\alpha = 4.0$ [*crosses*]) on dynamic behavior of the default system configuration for fixed force ratio $\beta = 10$ and joint location (a)

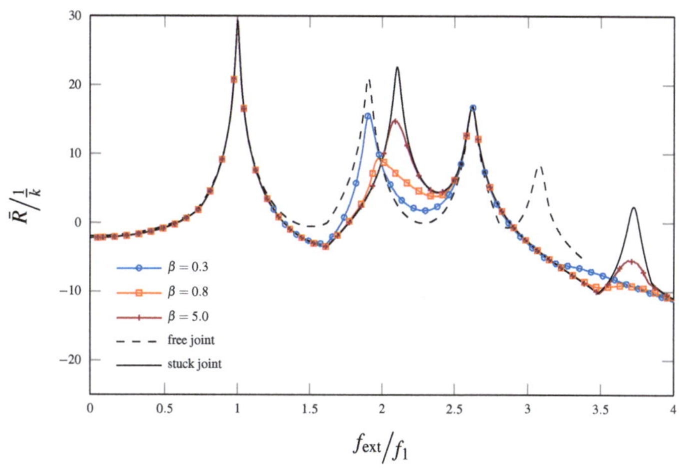

Fig. 16.3 Impact of force ratio ($\beta = 0.3$ [*circles*], $\beta = 0.8$ [*squares*], $\beta = 5.0$ [*crosses*]) on dynamic behavior of the default system configuration for fixed stiffness ratio $\alpha = 1$ and joint location (a). The limiting linear cases are labeled by *dashed* (free) and *solid* (stuck joint) *black lines*

16.5 Joint Included in Structure

The first optimization case focuses on the joint location (a) (joint interposed between second and third masses). Both dimensionless parameters α and β are considered as free parameters. All structure parameters are kept constant and the second mode is selected to be the objective of the optimization. Hence, the amplitude of the merged receptances at the second mode is employed as the cost value. Starting from the initial parameter configuration $\mathbf{p}_0 = [\alpha, \beta]^{\mathrm{T}} = [0.4, 4.5]^{\mathrm{T}}$ an unconstrained but bounded optimization is performed.

After $i = 14$ iterations the optimization algorithm provides the optimal parameter configuration $\hat{\mathbf{p}} = [2.0, 0.74]^{\mathrm{T}}$. Figure 16.4 depicts the contour of the cost value on the selected optimization regime and the optimization path of the unconstrained optimization, drawn with white triangles. The starting point corresponds to low joint stiffness and high normal load values. Generally, the single Jenkins element dissipates the maximum energy for a distinct relative displacement. As only the force and stiffness ratios are subject to the optimization, the cost value reveals one global minimum on the parameter space, compare the concave characteristic of the cost value in Fig. 16.4.

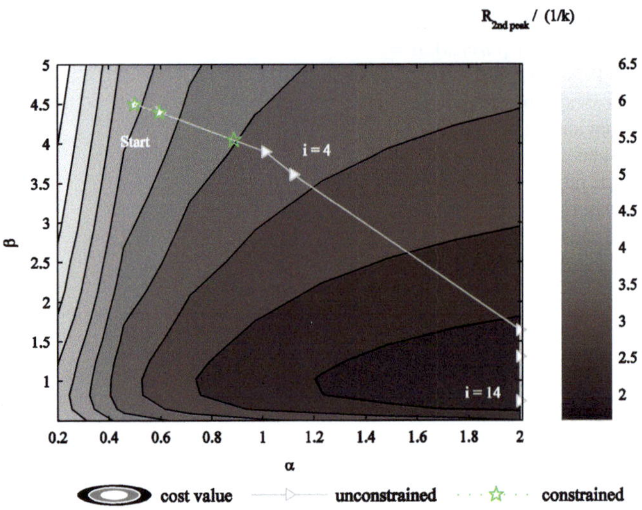

Fig. 16.4 Normalized vibration level of second mode and paths of unconstrained (*white, triangles*) and constrained (*green, stars*) optimization, for joint location (a)

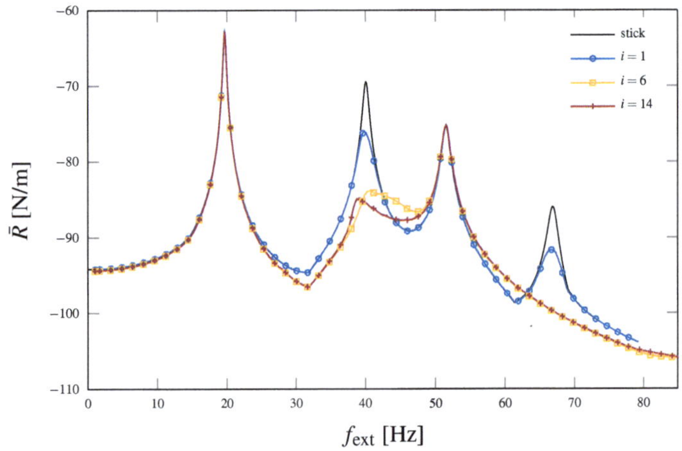

Fig. 16.5 Merged receptances $\overline{R}$ during optimization iterations $i = 1$ [*circles*], $i = 6$ [*squares*] and $i = 14$ [*crosses*] for joint location (a)

Figure 16.5 depicts the merged receptances for different optimization iterations i. Compared to the linear case of a stuck joint the vibration level of the second mode is reduced by 82.9 %. The reduction with respect to the nonlinear FRF of the initial configuration is 63.8 %. Only the second and fourth mode are affected by the joint since only these modes reveal out-of-phase motion with respect to mass m_2 and m_3. The optimization path from Fig. 16.4 can be divided into two main parts. First, the optimization algorithm mainly increases the joint stiffness value up to the maximum value allowed for the stiffness ratio ($\alpha = 2$) at iteration index $i = 6$. Then, the optimization heads in the direction of lower normal force on the Jenkins element to find the optimum. Physically, the deflection shapes are distorted by the increase of stiffness. Consequently, the energy feed-in is influenced (phase between vibration of mass m_1 and the forcing) as well as the energy dissipation (relative displacement of the joint). The first mechanism dominates the first part of the optimization path (increase of α) while the latter characterizes the second part (mainly decrease of β). Figures 16.6 and 16.7 emphasize these two generic concepts: During the first six optimization iterations the work put into the system is dramatically reduced while only a small decrease can be observed for iterations 7–14, compare Fig. 16.7. The corresponding hysteresis loops shrink during the first six iterations. Then, the maximization of dissipation takes place. This can be clearly seen from the growth of area enclosed by the hysteresis loops from iteration $i = 6$ to iteration $i = 14$, see Fig. 16.6.

Due to the added stiffness the fourth mode shifts by 10 Hz (8.8 %) during the optimization. In real application it typically is not possible to increase the stiffness of the joint so dramatically without incurring large weight penalties. Furthermore, increases to the frequencies of other modes may cause their response to increase leading to new vibration problems. Hence, constrained optimization was also pursued to seek a more realistic solution. In a second case study, the modes were not permitted to shift by more than 2 % and the vibration level of all modes must not increase during the optimization. The result

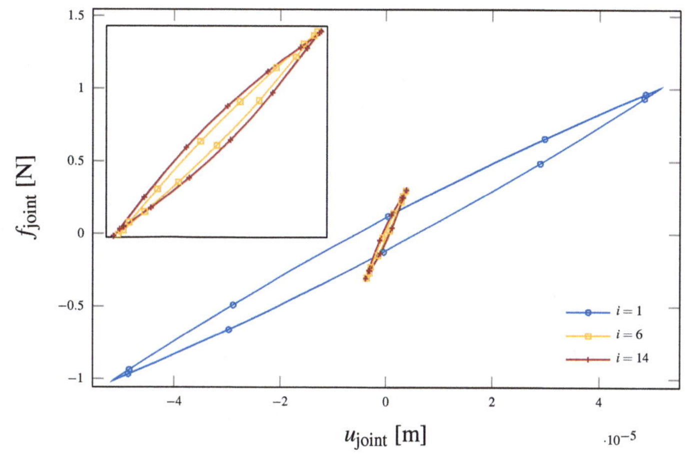

Fig. 16.6 Hysteresis loops and detailed view during optimization iterations $i = 1$ [*circles*], $i = 6$ [*squares*] and $i = 14$ [*crosses*] for the peak of 2nd mode, joint location (a)

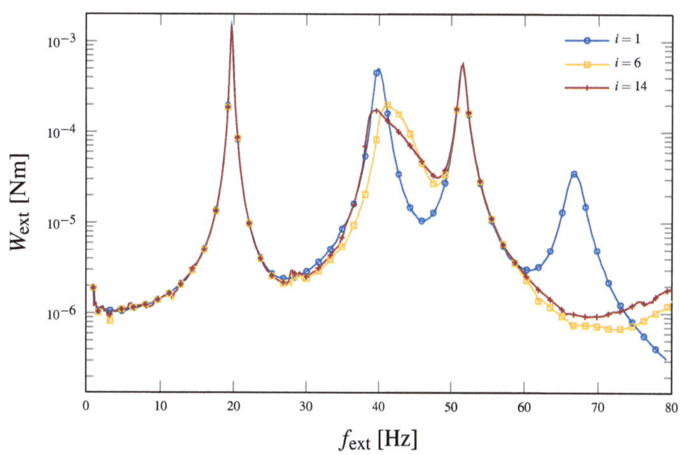

Fig. 16.7 Work put into the system by the external forcing during optimization iterations $i = 1$ [*circles*], $i = 6$ [*squares*] and $i = 14$ [*crosses*] for the peak of 2nd mode, joint location (a)

of the constrained optimization are depicted in Fig. 16.4 using green stars. Starting from the same point as the unconstrained optimization the direction of high joint stiffness is limited by the constraint on shifting modes. Therefore the optimization ends at a different optimal point with a 61.5 % reduction in the vibration level of the second mode, while simultaneously satisfying the constraints.

16.6 Joint Between Structure and Ground

The second case employs the joint location (b) (joint between third mass and ground). In this configuration the change of joint parameters affects every mode. Again, the optimization focuses on minimizing the response of the second mode which is started from the same point $\mathbf{p}_0$, in terms of joint parameters, as in the previous case for joint location (a). Figure 16.8 depicts the contour of the cost value for the 'ground' configuration (joint location (b)). In contrast to joint location (a) the

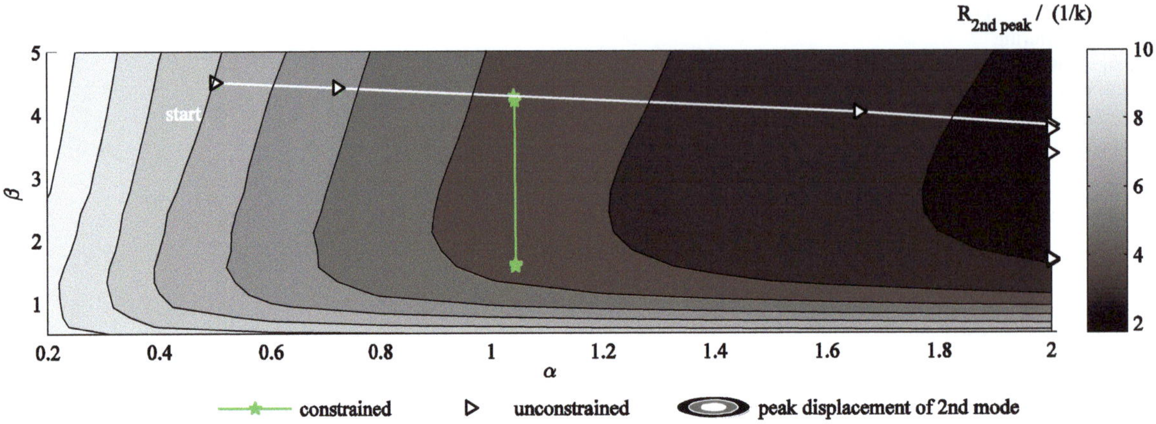

Fig. 16.8 Normalized vibration level of second mode and paths of unconstrained (*white*, *triangles*) and constrained (*green*, *stars*) optimization, joint location (b)

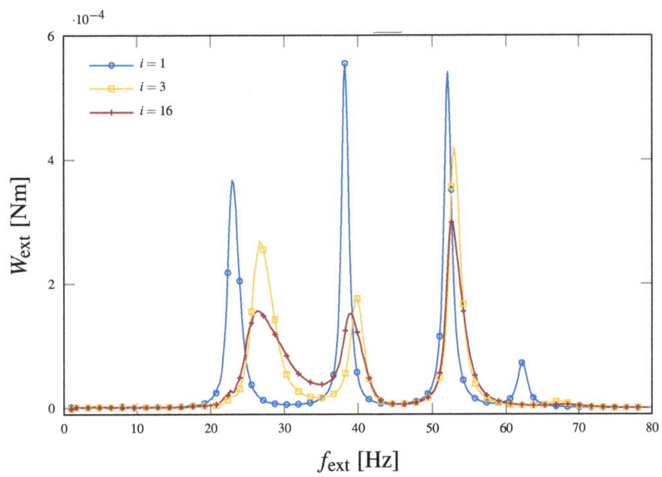

Fig. 16.9 Excitation work during optimization iterations $i = 1$ [*circles*], $i = 3$ [*squares*] and $i = 16$ [*crosses*], joint location (b)

cost value is much more sensitive to the stiffness ratio α than to the force ratio β. This observation is confirmed by the path of the unconstrained optimization. The optimization drives the system mainly to higher values of the joint stiffness. The reduction of the normal force does not affect the result as significantly. In this problem case, the optimization can generally take two different paths to minimize the vibration level of the structure: either maximize the dissipation in the joint or clamp the third mass to the ground via high values for the tangential stiffness. Here, the optimization pics the latter path. The introduction of a high stiffness between the structure and the ground physically represents a new constraint on the third mass which dramatically distorts the deflection shapes of the system. Thus, the minimization of vibration level is obtained by reduction of energy feed-in as already emphasized. Figure 16.9 illustrates the significant reduction of external work put into the system at the second mode. The reduction of vibration level of the second mode adds up to 77.3 % compared to the stick (linear) configuration, compare Fig. 16.10. In this case, a globally optimized system can be achieved since the vibration level of all modes decrease. This result illustrates the potential of intentional joint design in structural dynamics.

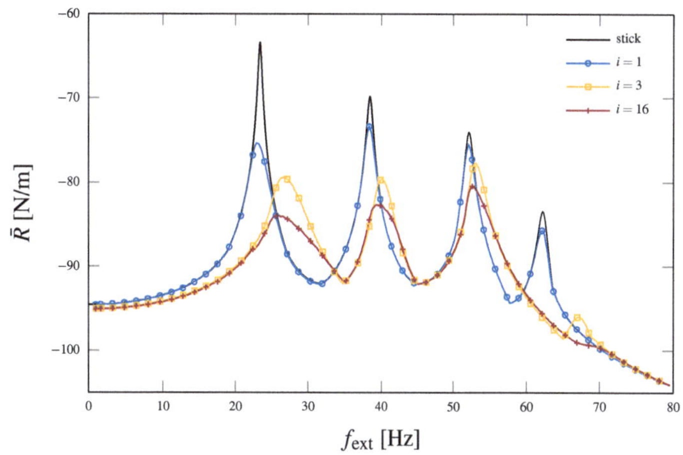

Fig. 16.10 Merged receptances during optimization iterations $i = 1$ [*circles*], $i = 3$ [*squares*], $i = 16$ [*crosses*] and the stick response [*solid black*], joint location (b)

16.7 Conclusion

This study on a minimal model shows how the structural performance of a given structure can be improved by joint design. For the studied model cases, the intentionally choice of joint parameters achieved vibration reductions of up to 82 % compared to the corresponding linearized joint stiffness model. In addition, different generic joint locations show different sensitivities to joint parameters. The main finding of this work is the twofold effect of joints on system dynamics: The optimization paths obtained can be clearly divided into two parts. Two different physical mechanisms, namely the reduction of energy feed-in and the increase of dissipation, are the root cause of this behavior. By varying the joint stiffness the deflection shapes are distorted. Therefore, the work put into the system is reduced and thus the vibration amplitudes decrease. In a second step, the optimization varies the normal load on the friction element to further decrease the vibration level by dissipation in the joint. In fact, the work dissipated in the joint is decreased in favor of higher joint stiffness to obtain the optimized solution. This result illustrates that when optimizing the response of a structure, it is important to consider both the damping and additional stiffness that joints introduce to the structure. The mutual reaction of the joint and the structure has to be accounted for. Despite the fact that a generic, oversimplified model is studied, these findings hopefully increase awareness of this interplay and are one step towards converting joints from design liabilities to design parameters which enable engineers to accomplish higher goals on reliability, wear and lightweight design than it is possible today. Further research is needed to develop accurate characterization and modeling for realistic joints, so this type of optimization can be performed on realistic structures.

References

1. Tiedemann, M., Kruse, S., Hoffmann, N.: Dominant damping effects in friction brake noise, vibration and harshness: the relevance of joints. Proc. Inst. Mech. Eng. D J. Automob. Eng. **229**(6), 728–734 (2015)
2. Padmanahban, K., Murty, A.: Damping in structural joints subjected to tangential loads. Proc. Inst. Mech. Eng. **205**, 121–129 (1991)
3. Padmanahban, K.: Prediction of damping in machined joints. Int. J. Mach. Tools Manuf. **32**, 305–312 (1992)
4. Sanliturk, K.Y., Ewins, D.J., Stanbridge, A.B.: Underplatform dampers for turbine blades: theoretical modelling, analysis and comparison with experimental data. Journal of Engineering for Gas Turbines and Power 123(4), 919–929 (2001)
5. Bograd, S., Reuss, P., Schmidt, A., Gaul, L., Mayer, M.: Modeling the dynamics of mechanical joints. Mech. Syst. Signal Process. **25**(8), 2801–2826 (2011)
6. Griffin, J.H.: Friction damping of resonant stresses in gas turbine engine airfoils. J. Eng. Gas Turbines Power **102**(2), 329–333 (1980)
7. Ewins, D.: Modal Testing: Theory, Practice, and Application, Mechanical Engineering Research Studies: Engineering Dynamics Series. Research Study Press (2000)
8. Dhooge, A., Govaerts, W., Kuznetsov, Y. A., Mestrom, W., Riet, A. M., & Sautois, B. : MATCONT and CL MATCONT: Continuation toolboxes in matlab. Universiteit Gent, Belgium and Utrecht University, The Netherlands (2006)
9. Nocedal, J., Wright, S.: Numerical Optimization, Springer Science & Business Media (2006)

Chapter 17
A Hybrid Piezoelectric and Electrostatic Vibration Energy Harvester

H. Madinei, H. Haddad Khodaparast, S. Adhikari, and M.I. Friswell

Abstract Micro Electro Mechanical Systems for vibration energy harvesting have become popular over recent years. At these small length scales electrostatic forces become significant, and this paper proposes a hybrid cantilever beam harvester with piezoelectric and electrostatic transducers for narrow band base excitation. One approach would be to just combine the output from the different transducers; however, this would require accurate tuning of the mechanical system to the excitation frequency to ensure the beam is resonant. In contrast, this paper uses the applied DC voltage to the electrostatic electrodes as a control parameter to change the resonant frequency of the harvester to ensure resonance as the excitation frequency varies. The electrostatic forces are highly non-linear, leading to multiple solutions and jump phenomena. Hence, this paper analyses the non-linear response and proposes control solutions to ensure the response remains on the higher amplitude solution. The approach is demonstrated by simulating the response of a typical device using Euler Bernoulli beam theory and a Galerkin solution procedure.

Keywords MEMS • Energy harvesting • Electrostatic forces • Nonlinear

17.1 Introduction

Energy harvesting from ambient vibration has been a popular research topic in recent years [1, 2]. The most common types of transduction methods are electromagnetic [3], piezoelectric [4] and electrostatic [5] and each has its own advantages and disadvantages. At the scale of Micro Electro Mechanical Systems (MEMS) devices electrostatic forces become significant. The use of piezoelectric effects is also feasible, whereas electromagnetic harvesters become less practical. Hence this paper considers a hybrid harvester that includes piezoelectric and electrostatic transducers.

Jeon et al. [6, 7] employed a piezoelectric thin film in a MEMS-scale energy harvester using a resonant system at the excitation frequency of 13.9 kHz. Renaud et al. [8] and Shen et al. [9] also considered piezoelectric MEMS harvesters for low excitation frequencies of 1.8 kHz and at 461 Hz respectively. Most MEMS scale energy harvesters are designed to work at resonance in order to obtain maximum output power, and they are usually manufactured to have natural frequencies that match the frequencies of excitation. In many cases, there will be a mismatch between the natural and excitation frequencies due to manufacturing errors or changes in the working conditions. To overcome this problem, harvesters with adjustable natural frequencies may be designed [10]. A range of concepts have been suggested for this tuning, such as mechanical nonlinear strain stiffening [11], permanent magnets [12, 13], or a sliding proof mass [14].

This paper uses an electrostatic device to adjust the resonant frequency of a piezoelectric MEMS harvester. The equations of motion are briefly outlined and a single mode approximation used to simulate the response of the beam and the power harvested. The nonlinear response to the system is simulated and the possibility of a system to control the response to maximize the harvested power is discussed.

H. Madinei • H.H. Khodaparast • S. Adhikari • M.I. Friswell (✉)
College of Engineering, Swansea University, Bay Campus, Fabian Way, Crymlyn Burrows, Swansea SA1 8EN, UK
e-mail: m.i.friswell@swansea.ac.uk

A. Brandt, R. Singhal (eds.), *Shock & Vibration, Aircraft/Aerospace, Energy Harvesting, Acoustics & Optics, Volume 9*,
Conference Proceedings of the Society for Experimental Mechanics Series, DOI 10.1007/978-3-319-30087-0_17

17.2 The Hybrid Harvester and Its Equations of Motion

The proposed harvester consists of an isotropic micro-beam of length L, width a, thickness h, density ρ and Young's modulus E, shown in Fig. 17.1. Piezoceramic layers of thickness h_0, Young's modulus E_0 and density ρ_0 are located on each side of the micro-beam along its entire length. Electrostatic electrodes (1 and 2) are located at the free end of the beam. A tip mass, M_t, is attached to the cantilever beam in order to decrease its natural frequency. When the tip mass is much larger than the mass of the cantilever beam, a simple SDOF model can be used to model the harvester. Madinei et al. [15] considered the full model for the continuous beam harvester and discussed the approximation of the equations of model using the Galerkin method for the first mode of the cantilever beam without piezoelectric or electrostatic effects. Only a summary will be provided here, with some discussion of the physical original of the terms.

The beam is modeled as an Euler-Bernoulli beam, where the flexural rigidity and mass density are equivalent properties that include the effect of the piezoelectric material. The output of the piezoelectric transducers is proportional to the local axial strain, which is proportional to the beam curvature. For the uniform beam and piezoelectric patch considered here the output can be integrated so that the total output is proportional to the difference in beam slopes at the ends of the patches. Furthermore, the piezoelectric coupling also gives moments on the beam at the ends of the patches proportional to the voltage. The device is base excited at a frequency Ω and amplitude z_0.

The electrostatic force per unit length is proportional to

$$\frac{\varepsilon_0 a}{2}\left(\frac{V_{DC}^2}{(g_0-w)^2}-\frac{V_{DC}^2}{(g_0+w)^2}\right) \tag{17.1}$$

where V_{DC} is the voltage applied to the electrostatic electrodes, $w(x)$ is the beam displacement at position x along the beam, g_0 is the clearance between the beam and the electrodes, and ε_0 is the permittivity of free space. Note that both the clearance and the DC voltage is equal for both electrodes so that the system is symmetrical. This force is only applied to that part of the beam where the electrode is located.

A single degree of freedom approximation to the equations of motion, based on the first mode of the linear undamped system, is obtained as [15]

$$m\ddot{U}+c\dot{U}+kU-2\theta_p v_p=F_e+F_b\sin(\Omega t) \tag{17.2}$$

$$2C_p\dot{v}_p+\frac{1}{R}v_p=-2\theta_p\dot{U}\cdot \tag{17.3}$$

where U is the mode shape amplitude, v_p is the piezoelectric voltage, m, c and k are the equivalent mass, damping and stiffness properties of the beam, including the mechanical effects of the piezoelectric patches, θ_p is the electromechanical

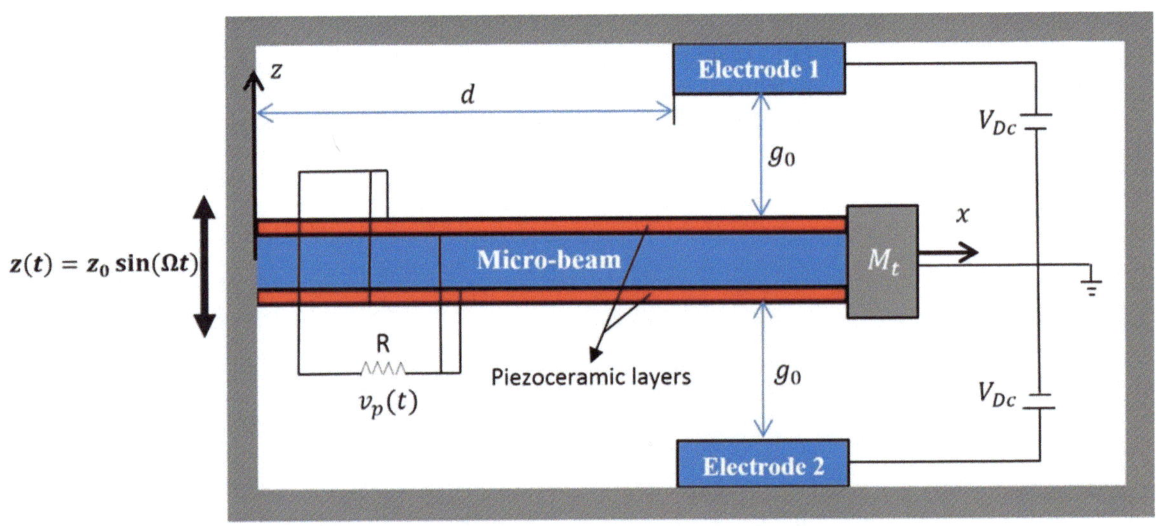

Fig. 17.1 Schematic of the proposed energy harvester

coupling constant for one of the piezoelectric patches and F_b is the base excitation force amplitude. C_p is the capacitance of one of the piezoelectric patches and R is the load resistance. Note that the piezoelectric patches are assumed to be connected in parallel. The electrostatic force is

$$F_e = \frac{\varepsilon_0 a\, V_{DC}^2}{2} \int_d^L \left(\frac{1}{(g_0 - \overline{w})^2} - \frac{1}{(g_0 + \overline{w})^2} \right) \phi(x)\, \mathrm{d}x \tag{17.4}$$

where $\phi(x)$ is the first mode shape. For a given time step the beam displacement, denoted $\overline{w}$, is assumed to be constant, and is calculated using the modal displacement, U, from the previous time step [15]; the electrostatic force is then calculated numerically from Eq. (17.4). The contribution to the linear stiffness from the electrostatic transducers is obtained by linearizing the electrostatic force about the equilibrium position, which is zero for the symmetric case. For a given clearance, this stiffness will be negative and proportional to V_{DC}^2. Hence there will be a DC voltage where the total stiffness becomes negative, leading to the well-known pull-in instability.

The power produced by the harvested is given by the power in the load resistor, given by $P = v_p^2/R$.

17.3 An Example

A clamped-free micro-beam is considered with the characteristics given in Table 17.1. The variation of the linear open circuit natural frequency with the applied DC electrostatic voltage is shown in Fig. 17.2 when $d = L/2$. The natural frequency of the system decreases with increasing DC voltage and becomes zero at the pull-in voltage. This phenomenon motivates the adjustment of the natural frequency of the system to match the frequency of the base excitation to increase the output power from the piezoelectric patches. The load resistance may also be optimised to increase the harvested power for a given excitation frequency.

The electrostatic force is nonlinear and hence affects the response and performance of the harvester. The nonlinear effect depends on the applied DC voltage and the air gap between the electrodes. Figure 17.3 show the effect of the applied DC voltage between electrodes on the performance of the system for an air gap of 25 μm. The system response is given in terms of peak power, which is closely related to peak beam displacement. It is clear from Fig. 17.3 that the electrostatic force is softening, and that multiple solutions and jumps occur in the response. Furthermore, the resonance frequency decreases with applied DC voltage, which corresponds to the effect shown in Fig. 17.2, and thus the resonant frequency can be adjusted to match the excitation frequency. In a similar way, the resonance frequency decreases with increasing air gap, and this may also be used to control the system, although varying the air gap is not as convenient and practical as varying the applied DC voltage.

For efficient energy harvesting the beam response should always occur at the higher of the two solutions and close to resonance (but not too close to risk jumping down to the low amplitude solution). However the solution actually obtained

Table 17.1 Geometrical and material properties of the micro-beam and piezoelectric layers

Design variable	
Beam length (L)	3000 μm
Beam width (a)	1000 μm
Beam thickness (h)	4 μm
Piezoelectric thickness (h_p)	2 μm
Beam Young's modulus (E)	169.6 GPa
Piezoelectric Young's modulus (E_p)	65 GPa
Viscous air damping coefficient (c_a)	0.0002 N.s/m
Poisson's ratio (υ)	0.06
Beam mass density (ρ)	2330 kg/m^3
Piezoelectric mass density (ρ_p)	7800 kg/m^3
Equivalent piezoelectric coefficient (e_{31})	-11.18 Cm^{-2}
Piezoelectric permittivity component ($\varepsilon_{33}{}^s$)	13.48 nF/m
Tip mass (M_t)	9.724 $\times 10^{-8}$ kg
Length of the tip mass (L_m)	20 μm
Thickness of the tip mass (h_m)	10 μm

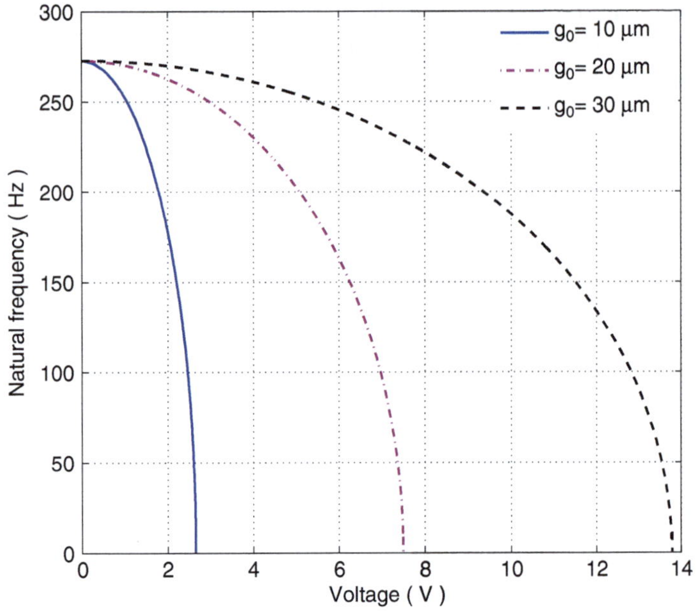

Fig. 17.2 The variation of the natural frequency of the linear system with electrostatic voltage for different air gaps

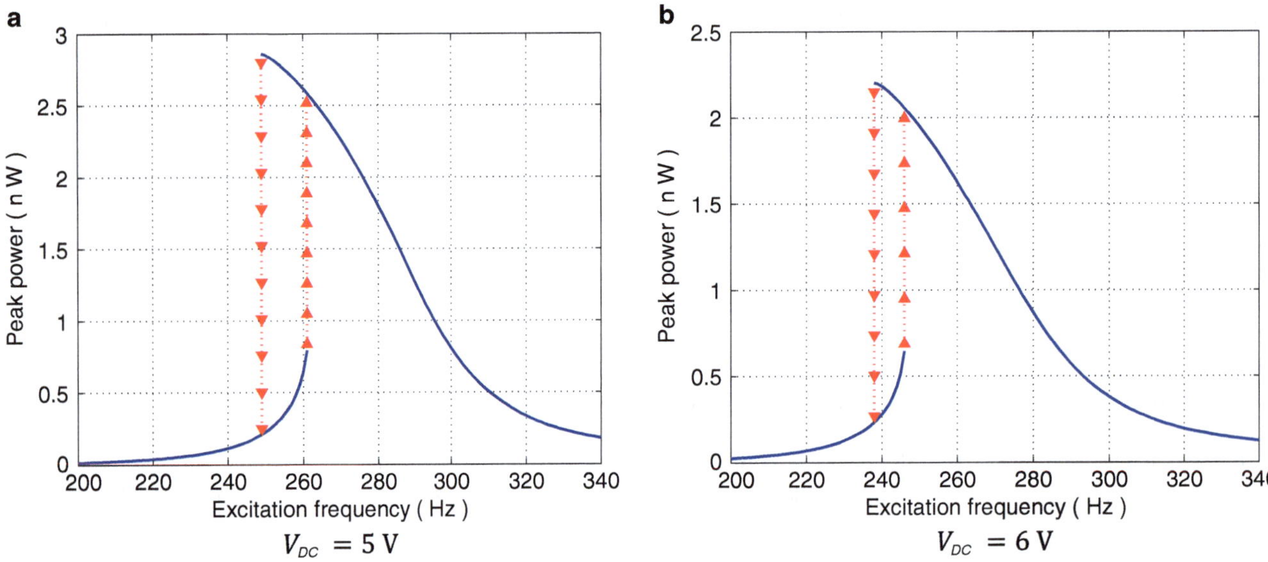

Fig. 17.3 Multiple solutions and jumps in the piezoelectric peak power with the frequency of base excitation ($z_0 = 0.3$ μm, $g_0 = 25$ μm, $R = 80$ kΩ)

will depend on the initial conditions and hence the response at the high amplitude solution cannot be guaranteed. The control system using the applied DC voltage can be used to ensure the harvester always responds in the higher amplitude solution. For a given excitation frequency if the harvester response happens to be in the lower amplitude solution the DC voltage is increased until a region is reached where the harvester only has a single solution. The DC voltage is then slowly reduced and the harvester follows high amplitude solution until the resonance is obtained.

Based on the design considerations outlined previously the optimal applied DC voltage to harvest maximum power for a base excitation with amplitude 0.3μm and an air gap of 35 μm is shown in Fig. 17.4 for different frequencies of excitation. Without any DC voltage, significant power can be harvested only at the resonant frequency (see point A). However, when the frequency of excitation is changed, the harvested power will be reduced significantly. By increasing voltage from zero at point A to 12 V at point E the optimal applied DC voltage is found for a frequency range between 232 and 312 Hz. Thus, a variable voltage source can be used to increase the operational frequency band of the proposed harvester by simply matching the resonance frequency of the system to the frequency of the base excitation.

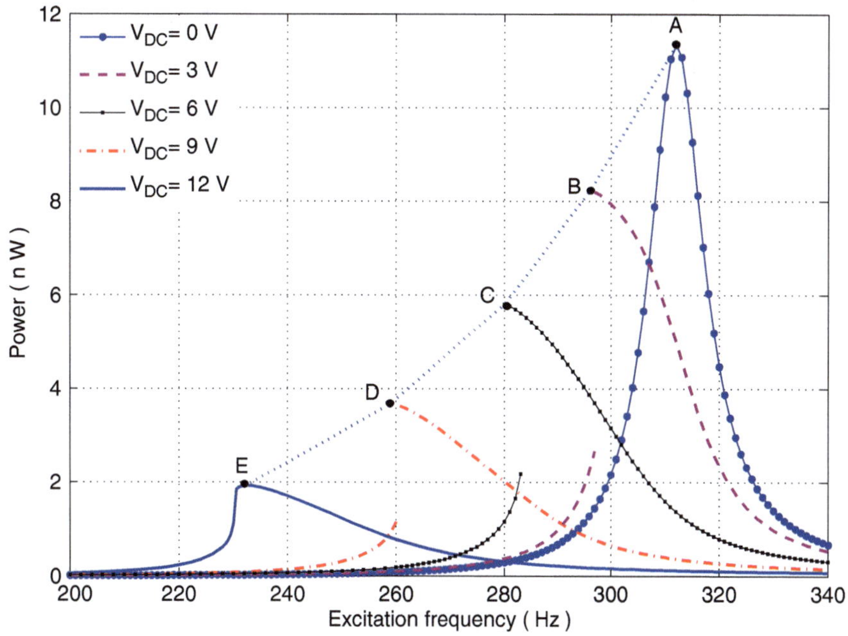

Fig. 17.4 Optimizing the harvested power by varying the DC voltage for a given frequency of base excitation ($z_0 = 0.3$ µm, $g_0 = 35$ µm, $R = 80$ kΩ)

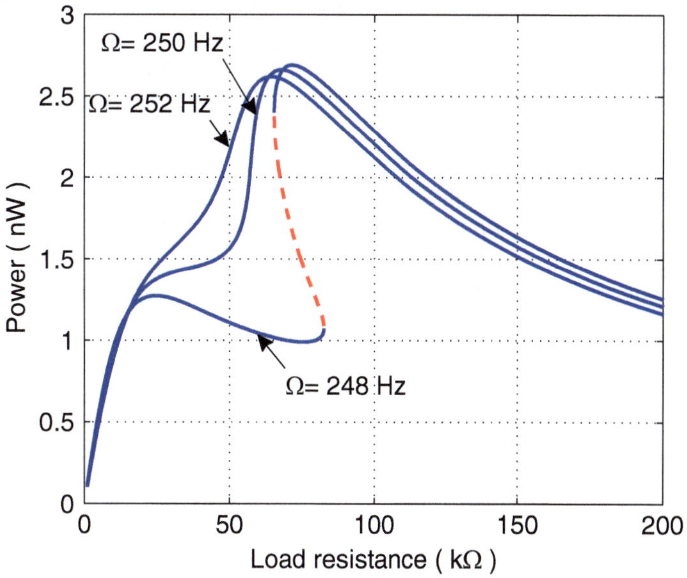

Fig. 17.5 The variation of harvested power with load resistance for different frequencies of base excitation ($z_0 = 0.3$ µm, $g_0 = 30$ µm, $V_{DC} = 8\ V$)

The load resistance also affects the dynamics of the systems and therefore also the harvested power. Figure 17.5 shows that the optimal value of the resistance depends on the frequency of the base excitation for an air gap of 30 µm. Notice that the dynamic behavior of the system is also significantly affected by the resistance, and multiple solutions and jumps are clearly present. Similarly the amplitude of base excitation also affects the character of the solutions and the harvested power, as shown in Fig. 17.6.

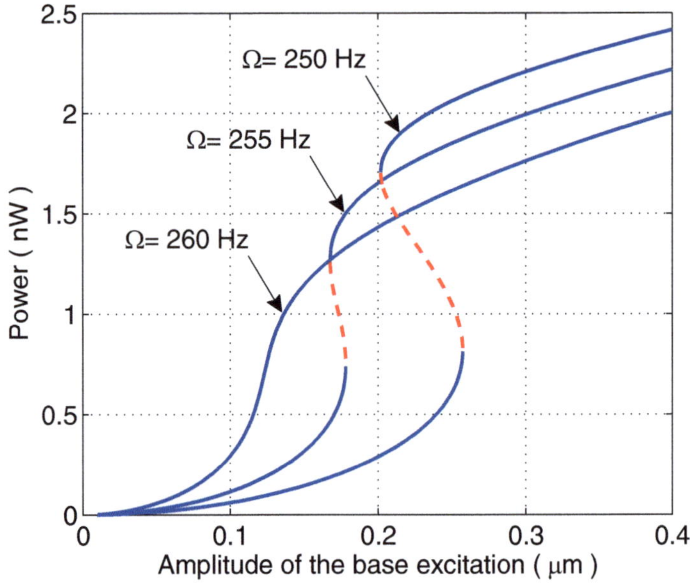

Fig. 17.6 The variation of harvested power with excitation amplitude for different frequencies ($g_0 = 30$ μm, $V_{DC} = 8\ V$, $R = 100$ kΩ)

17.4 Conclusions

A hybrid piezoelectric electrostatic MEMS harvester which is capable of adjusting its resonance frequency to the excitation frequency is proposed in this paper. The main advantage of the proposed system is the use of an electrostatic device for the adaptive control of the natural frequency of the system. The numerical results showed that the natural frequency of the hybrid system is extremely sensitive to the applied DC voltage and therefore can be tuned by a variable voltage source in order to increase the operating frequency bandwidth of the harvester system. For a certain design, not necessarily the optimal design, it was shown the harvested system can cover a wide range of excitation frequencies, i.e. 232–312 Hz. The disadvantage of the proposed system is the effect of the softening nonlinearity of the electrostatic part of the harvester which results in a lower level of harvested energy.

Acknowledgement Hadi Madinei acknowledges the financial support from the Swansea University through the award of the Zienkiewicz scholarship.

References

1. Borowiec, M., Litak, G., Friswell, M.I., Adhikari, S.: Energy harvesting in a nonlinear cantilever piezoelastic beam system excited by random vertical vibrations. Int. J. Struct. Stab. Dyn. **14**(8), 1440018 (2014)
2. Vijayan, K., Friswell, M.I., Haddad Khodaparast, H., Adhikari, S.: Non-linear energy harvesting from coupled impacting beams. Int. J. Mech. Sci. **96–97**, 101–109 (2015)
3. Zorlu, Ö., Topal, E.T., Külah, H.: A vibration-based electromagnetic energy harvester using mechanical frequency up-conversion method. IEEE Sensors J. **11**(2), 481–488 (2011)
4. Galchev, T., Aktakka, E.E., Najafi, K.: A piezoelectric parametric frequency increased generator for harvesting low-frequency vibrations. J. Microelectromech. Syst. **21**(6), 1311–1320 (2012)
5. Hoffmann, D., Folkmer, B., Manoli, Y.: Fabrication, characterization and modelling of electrostatic micro-generators. J. Micromech. Microeng. **19**(9), 094001 (2009)
6. Jeon, Y.B., Sood, R., Jeong, J.H., Kim, S.G.: MEMS power generator with transverse mode thin film PZT. Sensors Actuators A Phys. **122**, 16–22 (2005)
7. Choi, W.J., Jeon, Y., Jeong, J.H., Sood, R., Kim, S.G.: Energy harvesting MEMS device based on thin film piezoelectric cantilevers. J. Electroceram. **17**, 543–548 (2006)
8. Renaud, M., Karakaya, K., Sterken, T., Fiorini, P., van Hoof, C., Puers, R.: Fabrication, modelling and characterization of MEMS piezoelectric vibration harvesters. Sensors Actuators A Phys. **145–146**, 380–386 (2008)
9. Shen, D., Park, J., Ajitsaria, J., Choe, S., Wikle, H., Kim, D.: The design, fabrication and evaluation of a MEMS PZT cantilever with an integrated Si proof mass for vibration energy harvesting. J. Micromech. Microeng. **18**, 055017 (2008)

10. Youngsman, J.M., Luedeman, T., Morris, D.J., Anderson, M.J., Bahr, D.F.: A model for an extensional mode resonator used as a frequency-adjustable vibration energy harvester. J. Sound Vib. **329**(3), 277–288 (2010)
11. Marzencki, M., Defosseux, M., Basrour, S.: MEMS vibration energy harvesting devices with passive resonance frequency adaptation capability. J. Microelectromech. Syst. **18**(6), 1444–1453 (2009)
12. Zhu, D., Roberts, S., Tudor, M.J., Beeby, S.P.: Design and experimental characterization of a tunable vibration-based electromagnetic micro-generator. Sensors Actuators A **158**, 284–293 (2010)
13. Challa, V.R., Prasad, M.G., Fisher, F.T.: Towards an autonomous self-tuning vibration energy harvesting device for wireless sensor network applications. Smart Mater. Struct. **20**, 025004 (2011)
14. Miller, L.M., Pillatsch, P., Halvorsen, E., Wright, P.K., Yeatman, E.M., Holmes, A.S.: Experimental passive self-tuning behavior of a beam resonator with sliding proof mass. J. Sound Vib. **332**(26), 7142–7152 (2013)
15. Madinei, H., Haddad Khodaparast, H., Adhikari, S., Friswell, M.I., Fazeli, M.: Adaptive tuned piezoelectric MEMS vibration energy harvester using an electrostatic device. Eur. Phys. J. Special Topics. **224**(14), 2703–2717 (2015)

Chapter 18
Design of Scaled-Down Composite I-Beams for Dynamic Characterization in Subcomponent Testing of a Wind Turbine Blade

Mohamad Eydani Asl, Christopher Niezrecki, James Sherwood, and Peter Avitabile

Abstract Blade certification in the wind industry starts with coupon testing of materials and eventually culminates with full-scale blade testing. Coupon testing is not always representative of the materials' performance and full-scale testing is expensive and time consuming. Subcomponent testing can bridge this gap and increase the assurance of blade manufacturers for introducing new materials and designs into wind industry. In this study, similitude theory is applied to the I-beam structure of a utility-scale wind turbine blade to design scaled down models that emulate the dynamic characteristics of the full-scale I-beam. The governing equations of motion for vibration of a thin walled laminated I-beam are analyzed to derive the scaling laws. Derived scaling laws are used as a design criterion to develop models that can accurately predict the fundamental frequency of the full-scale I-beam. Both complete and partial similarity cases are investigated. The distorted layup scaling technique is introduced as a novel approach to design scaled down composite models with totally different layups than the full-scale component. According to the results, depending on the desired size of the scaled models and ply scheme of the full-scale I-beam, models could be found with very good accuracy in predicting the fundamental frequency of the full-scale I-beam using derived scaling laws.

Keywords Wind turbine blade • Similitude • Sub-component • Composite I-beam • Distorted layup scaling

18.1 Introduction

The certification procedure of utility scale wind turbine blades starts with coupon testing of the materials and eventually is finalized with a full-scale test. Coupon testing is not fully representative of the structural performance of the blade and full-scale testing is very expensive (e.g. hundreds thousands of dollars) and time consuming. As the size of wind turbine blades increase, full-scale testing gets more challenging because of the logistical difficulties to accommodate the blade in testing facility and also increased amount of time to apply a certain number of fatigue cycles to the blade. Subcomponent testing can bridge the gap between coupon and full-scale testing and increase the assurance of the manufacturers to introduce new materials to the blade industry. If meaningful scaled-down subcomponent tests can be designed which are representative of their parent components in the utility scale blade, certification procedure can be expedited and facilitated.

Subcomponent testing of wind turbine blades has gained momentum in recent years. Different parts of utility-scale wind turbine blades have been case studies for researchers. Mandell et al. [1] tested composite I-beams with flanges and shear webs out of components which are used in the cross section of wind turbine blades. Stiffness and strain measurements of the beams resulted from four-point bending test were in agreement with predictions from simple beam theory and finite element analysis. Cairns et al. [2] studied the root section of the blade in which the root specimens represented a single insert of a blade root to hub joint. The primary focus of the study was manufacturing, but a significant amount of static and fatigue strength data were generated by performing pull-out tests. A study with the focus on skin-stiffener intersections and sandwich panel closeout was conducted by Mandell et al. [3]. The goal was to predict skin-stiffener fracture loads and evaluate performance at locations where the sandwich panel transitions into the normal laminate.

A few studies investigated the performance of adhesive joints and bond lines of a wind turbine blade using subcomponent testing. The idea was to test the static and fatigue properties of the shear web to spar cap bond under stress states that are representative of those seen during service for a wind turbine blade. Sayer et al. [4] proposed an asymmetric three-point bending test called a Henkel beam. It was meant to give a comparable combination of bending moment and shear forces as a three-point bending test while reducing the stress concentrations at the clamped end. The specimen was used for a parametric

M.E. Asl (✉) • C. Niezrecki • J. Sherwood • P. Avitabile
Department of Mechanical Engineering, University of Massachusetts Lowell, Lowell, MA, USA
e-mail: Mohamad_eydaniasl@student.uml.edu

A. Brandt, R. Singhal (eds.), *Shock & Vibration, Aircraft/Aerospace, Energy Harvesting, Acoustics & Optics, Volume 9*, Conference Proceedings of the Society for Experimental Mechanics Series, DOI 10.1007/978-3-319-30087-0_18

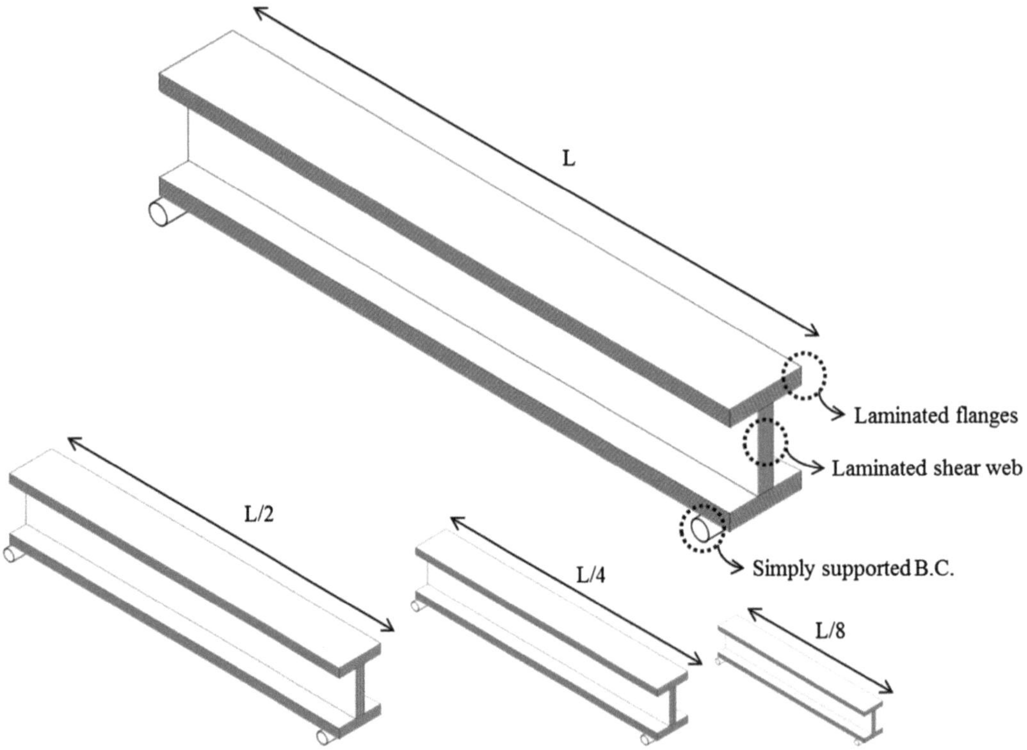

Fig. 18.1 Prototype (*top*) and Models with different scales and layups (*bottom*) showing the assumed boundary condition and applied loading

study, investigating the influence of the design and manufacturing variables on shear web to spar caps adhesive joints [5]. Zarouchas et al. [6] performed a static four-point bending test on two symmetric I-beams. The tested I-beams represented the spar cap and shear web structure inside a wind turbine blade.

Although subcomponent testing is usually categorized as laboratory-scale test rigs, there are a few mid-level blade test approaches that fall into the subcomponent category. A study of the structural static strength of a box girder of a 34 m wind turbine blade, loaded in flap-wise direction, was conducted by Jensen [7]. A combination of experimental and numerical work was used to address the critical failure mechanisms of the box girder of wind turbine blade. White et al. [8] developed a dual-axis test setup on a truncated 37 m wind turbine blade which combined resonance excitation with forced hydraulic loading to reduce the total test time required for evaluation.

The designed subcomponent (referred to as "model") regardless of the size and complexity needs to be correlated with the full-scale component (referred to as "prototype", see Fig. 18.1). The connection between scaled model and the prototype must be based on the structural parameters that predict the behavior of the system under study. Similitude theory can extract scaling laws from the governing equations of the system to connect the response of the scaled-down model to the prototype. In order to take advantage of the scaling laws, a proper scaled model should be designed to work well with the derived scaling law. In other words, the designed scaled model should be able to predict the response the prototype accurately by using the derived scaling laws. Otherwise, the experimental data of the scaled model cannot be correlated with the prototype and therefore the designed model cannot be representative of its corresponding prototype. As the model is not always an exact scaled-down replica of the prototype, a deliberate methodology should be implemented to design scaled models that can be used with scaling laws to predict the response of the prototype.

Similitude theory deals with determining the necessary and sufficient conditions of similarity between two systems. These similarity conditions may be derived directly from the governing equations of the system which lead to more specific similarity conditions than dimensional analysis. Simitses and Rezaeepazhand [9] established a technique that could be applied directly to the governing equation of a system to derive the scaling laws. The derived scaling laws are then used to predict or estimate the response of a prototype by the response of its associated model. The buckling response of an orthotropic and symmetric cross-ply laminated plate was investigated as a benchmark in that study. In later studies, they analyzed the vibration of scaled laminated rectangular plates [10]. Additionally, they studied the effect of axial and shear load on stability of scaled laminated rectangular plates [11, 12]. According to their results, the scaling laws that are obtained directly from the governing equations can be used with perfect accuracy for cross-ply laminates, while for the angle-ply laminates the scaling laws did not show good accuracy. Later, this method was extensively used in their works regarding the vibration response of laminated shells [13, 14].

The design of scaled models for a structure made of composite materials is more challenging than same structure made of isotropic materials. Laminated structures cannot be scaled down to any arbitrary size because of the practical difficulties in scaling the thickness of the plies of a laminate. Design of scaled-down composite models with same layup as the prototype will be limited by manufacturing constraints because only fabrics with specific thicknesses are available in industry. Therefore, making scaled-down composite models with a completely similar lamination scheme as the prototype is hard to implement. Therefore, use of partially similar scaled models can be considered as an alternative. Although ply-level scaling [9] within the scope of complete similarity has been implemented successfully, the design of partially similar models is still lacking a systematic methodology. Within this study, which is an extension of Authors' previous work on design of scaled composite models [15–17], similitude theory is applied to the governing equations of motion for vibration of a thin-walled composite I-beam [18] to design scaled-down composite I-beams which are representations of the spar caps and shear web of a utility scale wind turbine blade [19]. This paper presents the first work to design partially similar laminated models with totally different layups than their prototype using a systematic approach to predict the vibration response of the prototype. The models designed with the proposed approach are shown to have a very good accuracy in predicting the fundamental frequency of their corresponding prototype.

The spar caps and shear web are the backbone of wind turbine blades which make an interesting case study for subcomponent test design. In this study, similitude theory is applied to the subcomponent test concept, to develop scaled-down models that are representative of the dynamic characteristics of the I-beam structure inside a utility-scale blade. Governing equations for vibration of a simply-supported thin-walled shear deformable composite I-beam are considered to derive the scaling laws. Derived scaling laws are used as design criteria to develop scaled models that can accurately predict the fundamental frequency of the prototype. Within this analysis, the geometry and ply scheme of the prototype were based on a portion of Sandia National Laboratories 100-m long wind turbine blade near the maximum chord. Both complete and partial similarity cases are studied based on the considered layup for the prototype. Models with different sizes and ply schemes are designed that can accurately predict the fundamental frequency of their corresponding prototype.

18.2 Description of the Mathematical Model

This section presents the governing equations of motion for flexural vibration of a shear deformable thin-walled composite I-beam shown in Fig. 18.2. The closed form solution is derived for natural frequencies of the I-beam with a simply supported boundary condition. Then similarity transformation is applied to derive the scaling laws. The objective is to design scaled down beams that can accurately predict the fundamental flexural frequency of the prototype using scaling laws. For this study, free vibration in y-direction, in the absence of thermal effects for a symmetric I-beam is considered.

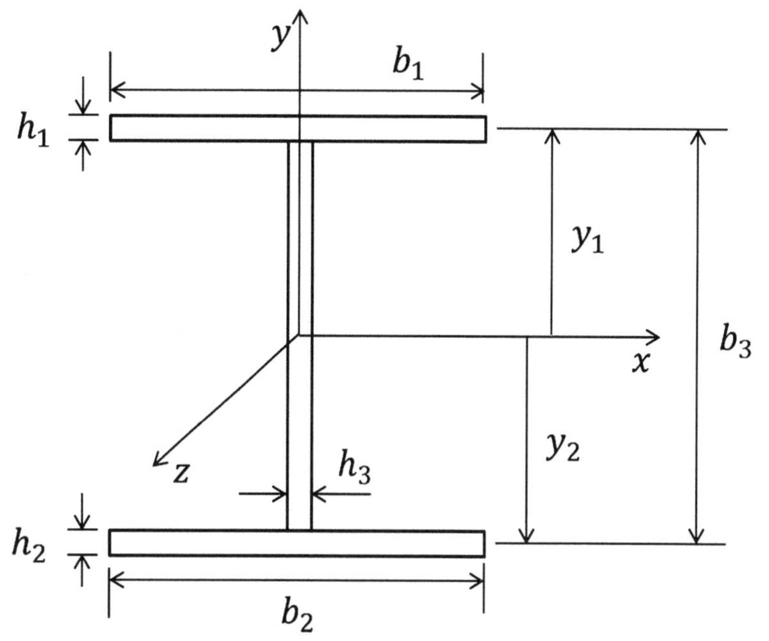

Fig. 18.2 Geometry and coordinate of the I-beam [18]

Neglecting all coupling effects due to the symmetric geometry, the equations governing for free flexural vibration in y-direction are given by [18] which are well known Timoshenko beam equations:

$$(GA_x)_{com}(V'' + \Psi'_x) = \rho A\ddot{V} \tag{18.1a}$$

$$(EI_x)_{com}\Psi''_x - (GA_x)_{com}(V' + \Psi_x) = \rho I_x\ddot{\Psi}_x \tag{18.1b}$$

where Ψ_x denotes the rotation of the cross section with respect to x axis shown in Fig. 18.2, V the displacement in y direction, q the distributed load, the prime (′) is used to indicate differentiation with respect to z and (˙) is time differentiation. Density and area of cross section are expressed by ρ and A respectively and I_x is moment of inertia with respect to x-axes. $(GA_x)_{com}$ and $(EI_x)_{com}$ are shear and flexural rigidity of thin walled composite with respect to x, respectively which could be expressed as:

$$(EI_x)_{com} = \left[A_{11}^{\alpha}y_{\alpha}^{(2)} - 2B_{11}^{\alpha}y_{\alpha} + D_{11}^{\alpha}\right]b_{\alpha} + \frac{b_3^{(3)}}{12}A_{11}^{3} \tag{18.1c}$$

$$(GA_x)_{com} = A_{55}^{\alpha}b_{\alpha} + A_{66}^{3}b_3 \tag{18.1d}$$

where A_{11}, A_{66}, A_{55}, B_{11} and D_{11} are elements of extensional, coupling and bending stiffness matrices for a composite layup [20]. The superscript in the parenthesis () denotes the power of the exponent, and the repeated index denotes summation. Index α varies from 1–3 where the indices 1 and 2 represent the top and bottom flanges, and 3 is for the web, respectively as shown in Fig. 18.2 and b_{α} denotes width of the flanges and web. The closed-form solution for flexural natural frequencies in the y-direction may be directly calculated for the simple support boundary condition as [18]:

$$\omega_{y_n} = \sqrt{\left[\frac{\rho A}{(EI_x)_{com}}\frac{L^4}{n^4\pi^4} + \frac{\rho A}{(GA_x)_{com}}\frac{L^2}{n^2\pi^2}\right]^{-1}} \tag{18.2}$$

18.3 Scaling Laws for Vibration of Composite I-Beams

Natural frequencies for vibration of a simply supported shear deformable composite I-beam are described by Eq. (18.2). To derive the scaling laws, it is assumed that all the variables of the governing equations for the prototype (x_p) can be connected to their corresponding variables in a scaled model (x_m) by a one to one mapping. Then, the scale factor for each variable can be defined as $\lambda_x = x_p/x_m$ which is ratio of each variable of the prototype to that of the scaled model. Rewriting Eq. (18.2) for the model and prototype and applying similarity transformation, the scaling laws can be extracted as follows based on the standard similitude procedure [9]:

$$\lambda_l^2 = \frac{\lambda_{EI}}{\lambda_{GA}} \tag{18.3}$$

$$\lambda_{\omega} = \sqrt{\frac{\lambda_n^4\lambda_{EI}}{\lambda_{\rho}\lambda_A\lambda_l^4}} \tag{18.4a}$$

$$\lambda_{\omega} = \sqrt{\frac{\lambda_n^2\lambda_{GA}}{\lambda_{\rho}\lambda_A\lambda_l^2}} \tag{18.4b}$$

Equation (18.3) which is referred to as the design scaling law is a prerequisite for deriving constitutive response scaling laws Eqs. (18.4a–b). Design scaling law Eq. (18.3) denotes that for having complete similarity between two shear deformable beams, ratio of flexural to shear rigidity must be equal to the square of the length for the two scales. Having Eq. (18.3) satisfied, the ratio of natural frequencies between two scales can be obtained using constitutive scaling laws Eqs. (18.4a–b). Expanding Eq. (18.3) using the definition of flexural and shear rigidities Eqs. (18.1c–1b) yields more specific design scaling laws. Assuming that the flanges and shear web of the I-beam are identical, Eqs. (18.1b) and (18.1c) can be simplified as follows:

$$(EI_x)_{com} = 2A_{11}y^2b - 4B_{11}yb + 3D_{11}b + \frac{A_{11}b^3}{12} \tag{18.5}$$

$$(GA_x)_{com} = 3A_{55}b + A_{66}b \tag{18.6}$$

The second terms of Eq. (18.5) can be dropped by assuming a symmetric layup for the flanges and the web. Also the third term in Eq. (18.5) is negligible for a thin-walled beam. This study focuses on the fundamental frequency (n = 1), but same methodology can be utilized for the higher modes. Applying the similarity transformation to Eqs. (18.5) and (18.6) yields the following scaling laws:

$$\lambda_{EI} = \lambda_b \lambda_{A_{11}} \lambda_y^2 = \lambda_{A_{11}} \lambda_b^3 \tag{18.7}$$

$$\lambda_{GA} = \lambda_b \lambda_{A_{55}} = \lambda_b \lambda_{A_{66}} \tag{18.8}$$

Substitution of Eqs. (18.7) and (18.8) in Eq. (18.3), the following scaling laws are derived for a symmetric thin-walled laminated I-beam:

$$\lambda_b = \lambda_y = \lambda_l \tag{18.9}$$

$$\lambda_{A_{11}} = \lambda_{A_{66}} = \lambda_{A_{55}} \tag{18.10}$$

Equations (18.9) and (18.10) are specific versions of Eq. (18.3) which are valid for a thin walled composite I-beam with a symmetric layup. In order to use the constitutive response scaling laws Eqs. (18.4a–b) to predict the vibration frequencies of the prototype using a scaled model, Eqs. (18.9) and (18.10) must be valid between prototype and the scaled model. Equation (18.9) can be satisfied assuming that model and the prototype have a same geometrical aspect ratio. However, validation of Eq. (18.10) depends on the layup schemes of the prototype and the model.

In the next sections of this study, validation of Eq. (18.10) is investigated by considering different layups for both prototype and scaled models. For simplicity, all models are assumed to have a same geometrical aspect ratio as their corresponding prototype in order to have Eq. (18.9) satisfied for all the case studies. Equation (18.10) is then used as a design criterion to search through all possible model layups for a fixed prototype layup, and select the layups that completely or partially satisfy the terms of the equation. Finally, fundamental vibration frequency of the prototype is predicted using the frequency of the models and response scaling laws Eqs. (18.4a–b).

18.4 Complete Similarity

To achieve the complete similarity between the prototype and the model, all the scaling laws must be satisfied simultaneously. In this case, the resulting model is a scaled replica of the prototype and theoretically can predict the response of the prototype exactly using the constitutive response scaling laws. Based on the derived scaling laws, Eqs. (18.9) and (18.10) must be satisfied to have complete similarity between the prototype and its corresponding scaled-down composite I-beam model. Equation (18.9) is satisfied by design of a model that has the same geometrical aspect ratio as its corresponding prototype. However, having Eq. (18.10) satisfied depends on the layup scheme of the model and prototype.

Ply-level scaling method $\left[(+\Theta_n/-\Theta_n)_s, \mathrm{n} = 1, 2, 3, \cdots\right]$ might be used to achieve complete similarity for laminated plate models in which the angle of plies in the layup scheme of the laminate is denoted by Θ [14]. In this approach, the total number of the plies with the same fiber angle n, is divided by an integer number resulting in models with the same extension, coupling and bending ratios as the prototype [15, 16]. Ply-level scaling if applicable, yields models which automatically satisfy Eq. (18.10).

To apply ply-level scaling and achieve complete similarity, the prototype layup must have multiple plies for each fiber direction. The prototype geometry and lay-up scheme considered in this study emulate the spar-cap flanges of the Sandia 100 m wind turbine blade [19] near the max chord. The ply scheme used for the prototype in this analysis is based upon the

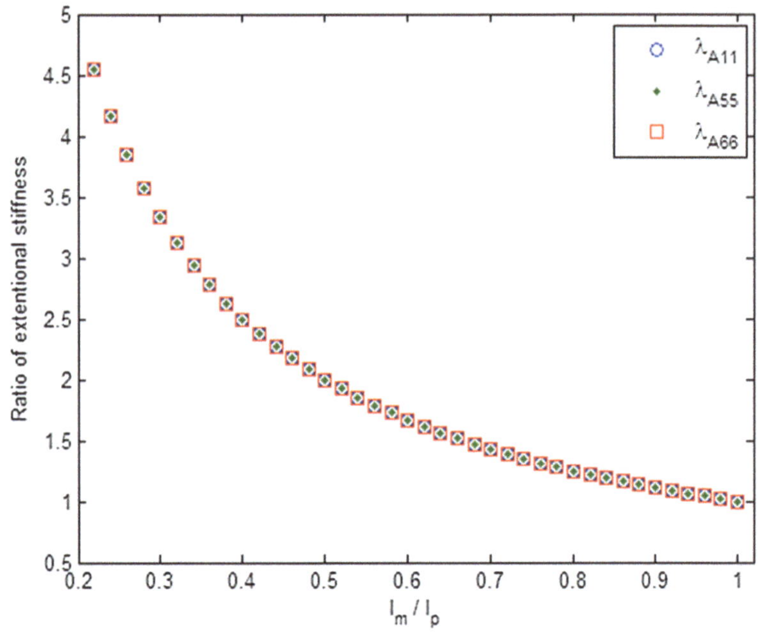

Fig. 18.3 Comparison of the terms in Eq. (18.10) as a function of the size of the model which demonstrates complete similarity

ply scheme of the Sandia blade near maximum chord. This area was chosen for analysis because many failures in blades occur near maximum the chord, and the area of interest is responsible for carrying most of the aerodynamic loads. The prototype is assumed to be an I-beam consisting of three identical laminated plates having an overall dimension of 29.3 m × 2.446 m × 0.136 m with layup $[0]_{100}$ and ply thickness of t = 1.36 mm. Because the blade is tapered throughout its length and the geometry of the structure changes significantly, the geometry of the full-scale component (i.e. the first 1/3 of the Sandia blade from its root) is assumed to have a uniform cross section whose height along its length was averaged to simplify the analysis. All computations are implemented for the glass/epoxy materials with following material characteristics [19]: $E_1 = 41.8$ GPa, $E_2 = E_3 = 14$ GPa, $G_{12} = G_{13} = 2.63$ GPa, $G_{23} = 1.83$ GPa, $\nu_{12} = \nu_{13} = 0.28$, $\nu_{23} = 0.47$. Within this analysis, it was assumed that the geometry and layup scheme of the prototype was kept fixed while comparing to the developed models with various dimensions and ply stack up.

Assuming a unidirectional layup for the prototype, scaled-down models were developed with a same geometrical aspect ratio as the prototype and unidirectional layup. Thickness of the plies for the models was considered fixed and same as the prototype. Sizes of the developed models range from 1 to 100 plies in which geometry and layup of the largest model is identical to those of the prototype. Figure 18.3 demonstrates that Eq. (18.10) is valid, meaning that all three extensional stiffness ratios between the different models and the prototype are equivalent. All the three terms in Eq. (18.10) have been plotted in Fig. 18.3 for models whose sizes range from 1 to 100 plies assuming a prototype with the specified dimensions and unidirectional layup. Validation of Eq. (18.3) for the developed models and the prototype as shown in Fig. 18.3 allows using response scaling laws Eqs. (18.4a–b) to predict the frequency of the prototype. Because all models satisfy all the necessary scaling laws for complete similarity which are Eqs. (18.9) and (18.10), they can predict the fundamental vibration frequency of the prototype exactly using the constitutive response scaling laws Eqs. (18.4a–b).

Next, the developed models are utilized to theoretically predict the fundamental frequency of the prototype. Constitutive response scaling laws Eqs. (18.4a–b) are applied to the frequency of the developed models to predict the vibration frequency of the prototype. Figure 18.4 shows the overlapping of the predicted frequencies by each model to that of the prototype, which demonstrates excellent accuracies in prediction. The first bending mode resonant frequency is 8.23 Hz and the models are able to exactly predict this frequency by using constitutive response scaling laws. Both of the response scaling laws Eqs. (18.4a–b) yield the same results because all the necessary conditions for complete similarity have been met for the developed models.

Although complete similarity holds between the developed models and the considered prototype, use of ply-level scaling may not be applicable to design scaled models for a prototype with non-unidirectional layup. Feasibility of ply-level scaling is limited to layups with multiple plies of a same angle. Also the smallest scale model that can be designed through this approach is dictated by the ratio of the stacks of the angle plies in a given layup for a laminated plate. For example, consider the prototype with the ply stack up $[\pm 45_2/0_{46}]_s$ which is more representative of a utility-scale blade layup for the spar caps. It is not possible to scale down the thickness to half of the prototype thickness using ply-level scaling, because it will result in a

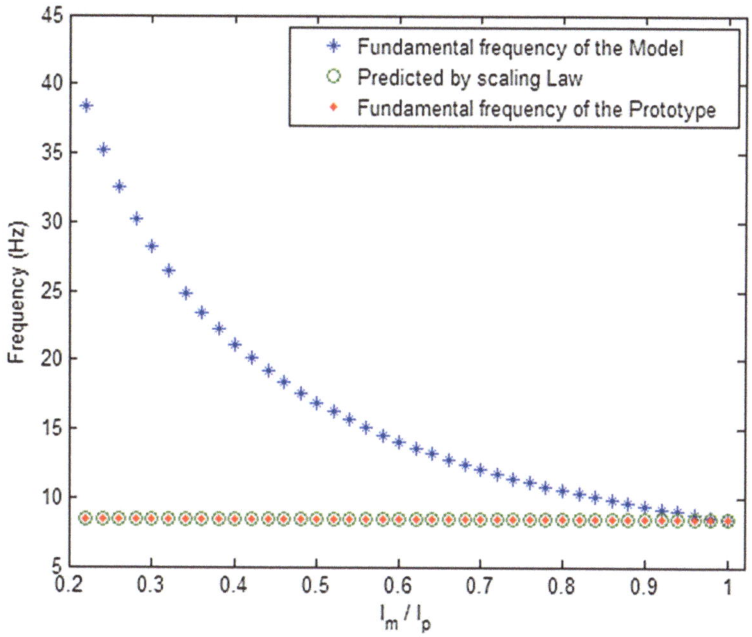

Fig. 18.4 Comparison of the predicted fundamental frequency of the prototype (*green circles*) to the actual frequency (*red diamond*) by using the fundamental frequency of the model (*blue star*) via the response scaling law for complete similarity

non-integer number of plies. Design of partially similar models is considered an option when complete similarity is not feasible. In the next section, a systematic methodology is developed to design scaled down composite I-beams that are partially similar to their corresponding prototype.

18.5 Partial Similarity

When satisfying all the scaling laws between model and prototype is not applicable and thus complete similarity is not the case, partial similarity is investigated. In partial similarity, some of the design scaling laws may be relaxed to allow more freedom in design of scaled model. Although the partially similar model is not a scaled replica of its prototype, a designed model can be accurate enough to replicate the desired aspects of the prototype.

In the previous section complete similarity was demonstrated for a composite I-beam with unidirectional layup representative of the spar caps and shear web structure inside a utility scaled wind turbine blade. To extend the applicability of the scaled models, two non-unidirectional layups with same geometry are considered for the prototype. Then scaled models with same geometrical aspect ratio as the prototype but distorted layup, are developed to capture the dynamic behavior of the prototype. In this case, scaling laws regarding the geometry of the models Eq. (18.9) will be still valid because all models have a same geometrical aspect ratio as the prototype. However, scaling laws regarding the layup of the models in Eq. (18.10) are not satisfied completely for the developed models in this section. Discrepancy in the design scaling laws will result in error in prediction of the frequency using the constitutive response scaling laws Eqs. (18.4a–b). The goal is to find lamination layups for the model which lead to minimum error in prediction of the frequency of the prototype using appropriate response scaling laws.

18.5.1 Distorted Layup Scaling Approach

To design a scaled laminated structure that is partially similar to its prototype, a common practice is to examine layups for the model that are similar to the layup of the prototype [12, 13]. However, the only condition that needs to be met to yield similarity between the model and prototype is satisfying the design scaling law. Any layup that fully or even partially satisfies the design scaling laws between the model and prototype, can be considered as a candidate layup for the model as long as selected layup obeys the governing equations of the motion (e.g. balanced and symmetric in this case). This terminology makes a broad range of models available to designers and increases the probability of finding acceptable models with smaller scales. To find acceptable layups that satisfy the design scaling laws and eventually yield accurate partially similar models, design scaling laws can be rearranged and used as quantitative criteria to search through all possible layups for the model and select the most accurate ones to predict the response of the prototype.

To design a partially similar scaled down model with a distorted layup compared to the prototype, Eq. (18.10) is the only condition that needs to be approximately satisfied. Equation (18.9) is already satisfied by assuming a geometrical aspect ratio for the model being the same as the prototype. Based on the prototype layup scheme, a permutation algorithm was implemented to search for the potential ply schemes for the model with overall laminate thickness less than that of the prototype which approximately satisfies Eq. (18.10). Any ply scheme for the model that approximately satisfies Eq. (18.10) may be considered a solution.

To evaluate the feasibility of the proposed approach, a non-unidirectional layup $[\pm45_2/0_{46}]_s$ was considered for the prototype, which is more representative of a layup in a utility scale wind turbine blade. The geometry of the prototype was kept same as the previous section. The error criteria to find the desired layups based on satisfying design scaling laws in Eq. (18.10) then were defined as:

$$e_1 = [(\lambda_{A_{11}} - \lambda_{A_{66}})/\lambda_{A_{11}}] * 100 \tag{18.11a}$$

$$e_2 = [(\lambda_{A_{11}} - \lambda_{A_{55}})/\lambda_{A_{11}}] * 100 \tag{18.11b}$$

18.5.2 Analytical Observations

Layup $[\pm45_2/0_{46}]_s$ is considered for the prototype and three different sets of models with different length scales are developed. Equations (18.11a–b) were used as accuracy metric to select the most accurate layup for models with a given scale to predict the fundamental frequency of the prototype. For all the case studies, error in satisfying the design scaling law Eqs. (18.11a–b) is examined primarily, and then the fundamental frequency of the prototype is predicted using the constitutive response scaling laws Eqs. (18.4a–b).

As the first benchmark for the distorted layup scaling technique, models with a total of 12 layers for each of the flanges and the web and $\lambda_l = 8.33$ were developed. The error defined by Eqs. (18.11a–b) for all possible ply schemes for the symmetric and balanced models with a total of 12 layers that only include 0, 90 and ±45 plies is shown in Fig. 18.5. The 0, 90 and ±45 plies were selected intentionally to make the designed model easy to manufacture. Among all 34 possible layups, the cases with permutation number 4, 5, 7, 10 and 16 were found with less than 30 % error in both error criteria. Figure 18.6

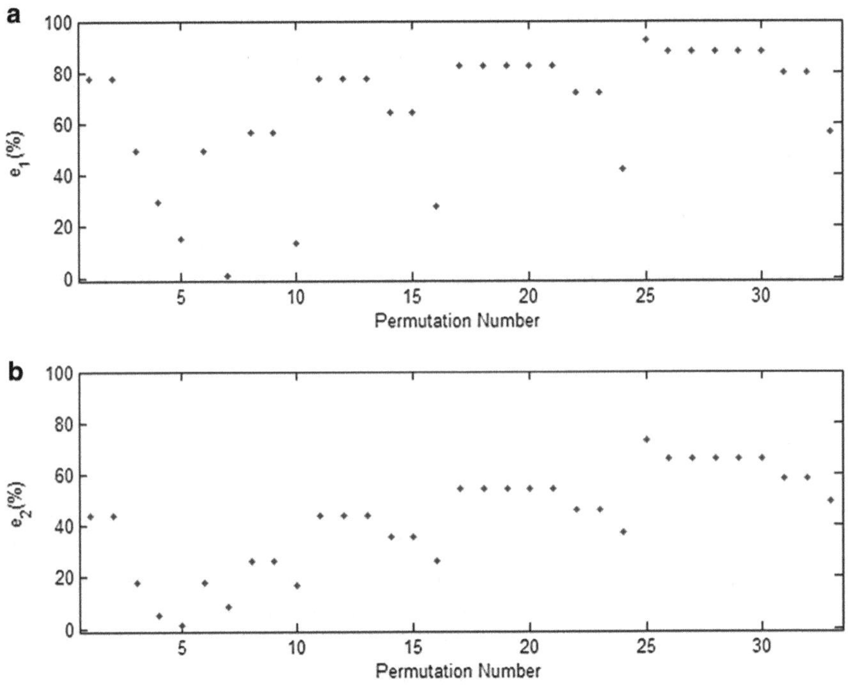

Fig. 18.5 Error percentage (defined by Eqs. 11a–b) for all possible ply schemes for the symmetric and balanced models with a total of 12 layers that only include 0, 90 and ±45 plies for the prototype layup $[\pm45_2/0_{46}]_s$

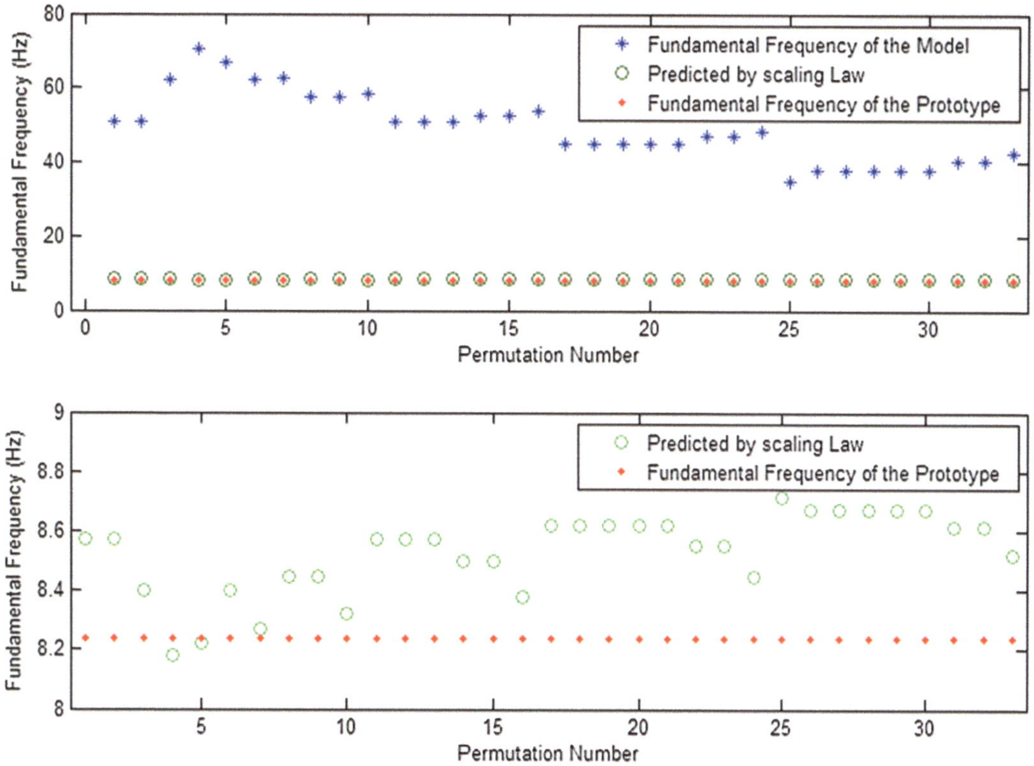

Fig. 18.6 *Top*: comparison of the predicted fundamental frequency of the prototype (*green circles*) to the actual frequency (*red diamond*) by using the fundamental frequency of the model (*blue star*) via the response scaling law for partial similarity (models with 12 layers) for the prototype layup $[\pm45_2/0_{46}]_s$; Bottom: zoom in for the region that has predicted values

Table 18.1 Selected ply schemes for the symmetric models with a total of 12 layers that only include 0, 90 and ±45 plies for the prototype layup $[\pm45_2/0_{46}]_s$

Layup	Permutation number	e_1(%)	e_2(%)	Error percentage in predicting fundamental frequency
$[0_5/90]_s$	5	15.0	1.3	0.18
$[0_4/90_2]_s$	7	0.7	8.7	0.38
$[0_{12}]$	4	29.3	5.3	0.71
$[0_3/90_3]_s$	10	13.6	17.1	1.01
$[0_2/90_4]_s$	16	27.0	18	1.70

shows the accuracy of the developed models using the distorted layup scaling technique in predicting the fundamental frequency of the prototype along with response scaling law Eq. (18.4a). The selected cases 4, 5, 7, 10 and 16 also show the best predictions among all possible permutation cases. Although the prototype layup includes ±45 plies, none of the selected models have ±45 plies in the layup but still show a very good accuracy in prediction of the response of the prototype.

Table 18.1 lists the selected layups for the models with a total of 12 layers organized based on their corresponding error percentage in predicting the fundamental frequency of the prototype. The amount of error in the scaling law as shown in Fig. 18.5 for all 35 layups is related to the accuracy in predicting the fundamental frequency by each layup shown in Fig. 18.6. Layups that have less error in satisfying the design scaling laws, show less discrepancy in prediction of the fundamental frequency of the prototype as shown in Fig. 18.6.

Extending the analysis to a smaller size, models with a total of 8 layers and $\lambda_l = 12.5$ were investigated. The error defined by Eqs. (18.11a–b) for all possible ply schemes for the symmetric and balanced models with a total of eight layers that only include 0, 90 and ±45 plies is shown in Fig. 18.7. Among the 12 possible cases, the cases with permutation number 3, 2 and 5 were found with less than 30 % error in both error criteria for design scaling laws. Figure 18.8 shows the accuracy of all the potential models using the distorted layup scaling technique in predicting the fundamental frequency of the prototype along with response scaling law Eq. (18.4a). The selected cases with permutation number 3, 2 and 5 also show the best predictions among all other potential models. Table 18.2 shows a list of the selected layups for the models with a total of eight layers. As shown in Table 18.2, layups with less amount of error in satisfying design scaling laws predict the fundamental frequency of

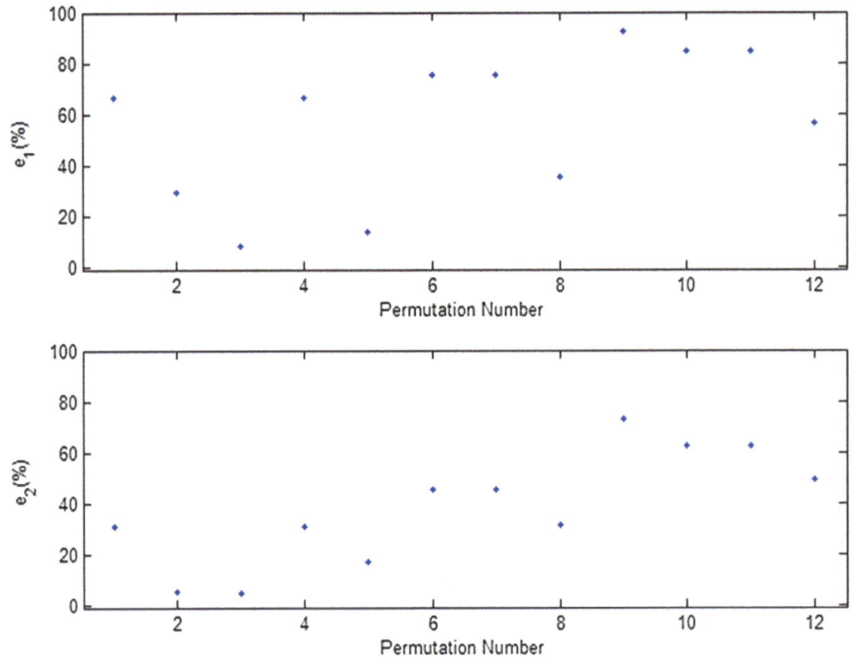

Fig. 18.7 Error percentage (defined by Eqs. 11a–b) for all possible ply schemes for the symmetric models with a total of eight layers that only include 0, 90 and ±45 plies for the prototype layup $[\pm45_2/0_{46}]_s$

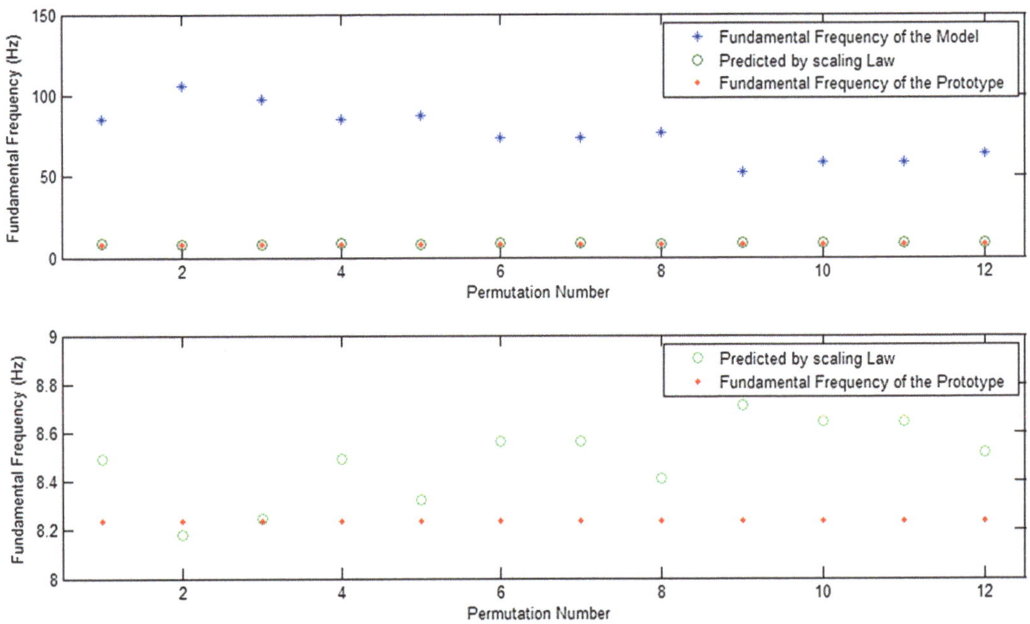

Fig. 18.8 *Top*: comparison of the predicted fundamental frequency of the prototype (*green circles*) to the actual frequency (*red diamond*) by using the fundamental frequency of the model (*blue star*) via the response scaling law for partial similarity (models with eight layers) for the prototype layup $[\pm45_2/0_{46}]_s$; Bottom: zoom in for the region that has predicted values

Table 18.2 Selected ply schemes for the symmetric models with a total of 8 layers that only include 0, 90 and ±45 plies for the prototype layup $[\pm45_2/0_{46}]_s$

Layup	Permutation number	$e_1(\%)$	$e_2(\%)$	Error percentage in predicting fundamental frequency
$[0_3/90]_s$	3	7.8	4.9	0.09
$[0_8]$	2	29.3	5.3	0.7
$[0_2/90_2]_s$	5	13.6	17.1	1.0

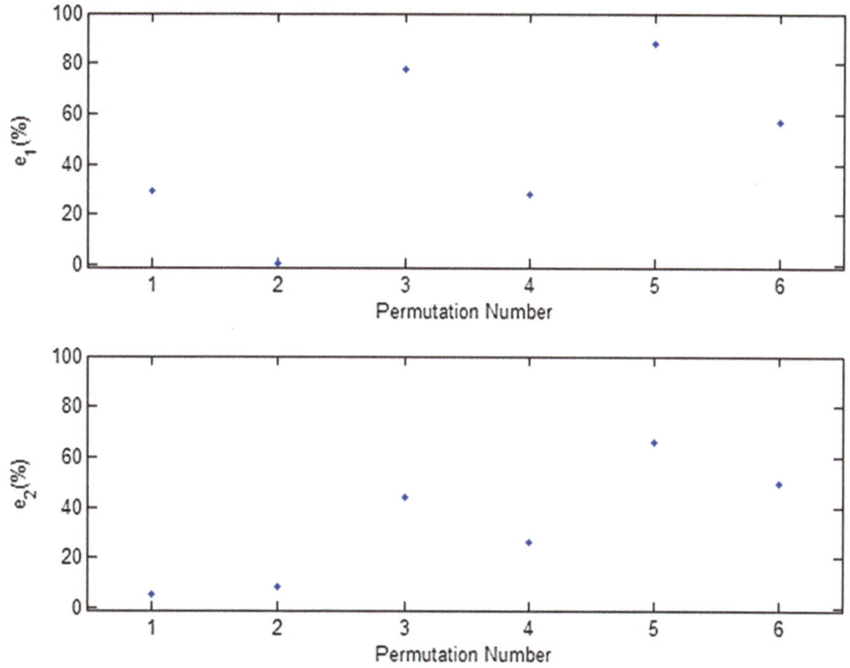

Fig. 18.9 Error percentage (defined by Eqs. 11a–b) for all possible ply schemes for the symmetric models with a total of six layers that only include 0, 90 and ±45 plies for the prototype layup $[\pm45_2/0_{46}]_s$

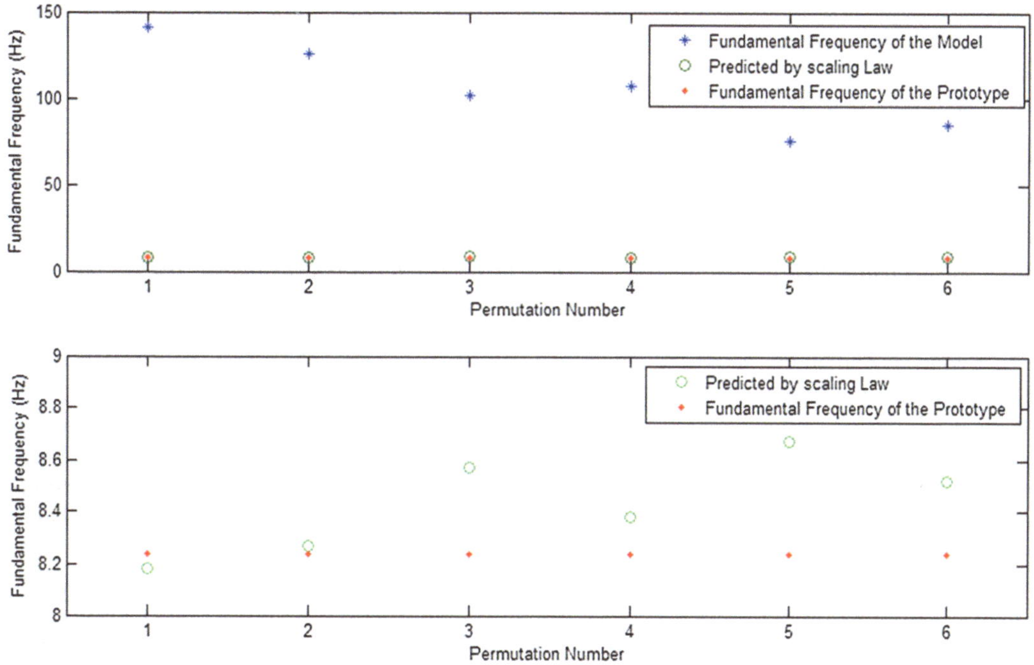

Fig. 18.10 *Top*: comparison of the predicted fundamental frequency of the prototype (*green circles*) to the actual frequency (*red diamond*) by using the fundamental frequency of the model (*blue star*) via the response scaling law for partial similarity (models with six layers) for the prototype layup $[\pm45_2/0_{46}]_s$; Bottom: zoom in for the region that has predicted values

the prototype more accurately. Also, the total number of possible cases is less than the previous case study because the total number of layers considered for the models in this case is fewer (eight layers) which results in fewer permutations.

Extending the analysis to an even smaller size having a total of six layers and $\lambda_l = 16.66$ were investigated. Figure 18.9 shows the value of the error criteria defined by Eqs. (18.11a–b) for all possible symmetric and balanced layup for models with a total of six layers that only include 0, 90 and ±45 plies. Among six possible cases, the cases with permutation number 2 and 1 were found with less than 30 % error in both design scaling laws. Figure 18.10 shows the accuracy of all the potential models using the distorted layup scaling technique in predicting the fundamental frequency of the prototype along with response scaling law Eq. (18.4a). The selected cases with permutation numbers 2 and 1 also show the best predictions among all other potential models. Table 18.3 shows a list of selected layups for the models with total of six layers.

Table 18.3 Selected ply schemes for the symmetric models with a total of six layers that only include 0, 90 and ±45 plies for the prototype layup $[\pm45_2/0_{46}]_s$

Layup	Permutation number	$e_1(\%)$	$e_2(\%)$	Error percentage in predicting fundamental frequency
$[0_2/90]_s$	2	7.0	5.3	0.38
$[0_6]$	1	29.3	8.7	0.71

Table 18.4 Different case studies were conducted using complete and partial similarity for the I-beam

Category	Type	Layup	Total number of the plies	$b_1 = b_2 = b_3$ (m)	Length (m)
Complete similarity	Prototype	$[0\ ]_{100}$	100	2.446	29.3
	Models	$[0\ ]_n$; n = 1−100	1–100	0.02446–2.446	0.293–29.3
Partial similarity (Case 1)	Prototype	$[\pm45_2/0_{46}]_s$	100	2.446	29.3
	Models	$[0_5/90]_s$, $[0_4/90_2]_s$	12	0.29352	3.516
	Models	$[0_3/90]_s$, $[0_8]$	8	0.19568	2.344
	Models	$[0_2/90]_s$, $[0_6]$	6	0.14676	1.758

18.6 Discussion

The design scaling laws that need to be satisfied for complete similarity are Eqs. (18.9) and (18.10). Equation (18.9) can be simply satisfied by designing scaled models with same geometrical aspect ratios as the prototype. However, satisfying Eq. (18.10) depends on the proportionality of the extensional stiffness elements between model and the prototype. As extensional stiffness of a laminate is not dependent on the position of its plies, changing the position of the plies in a developed model does not result in a different model. This fact limits the total number of the possible permutations because shuffling the plies in the flanges does not generate a new model in this study. Based on this observation, the number of the 0, 90 and ±45 plies in each flange of the I-beam was varied to develop different models instead of changing the position of the plies in flanges.

Assuming a unidirectional layup for the prototype, ply level scaling technique was applied which resulted in models with also a unidirectional layup that were completely similar to the prototype and showed an excellent accuracy in predicting the frequency using the derived scaling laws. For the prototypes with a non-unidirectional layup $[\pm45_2/0_{46}]_s$, the Distorted Layup Scaling technique resulted in partially similar models in different size scales with a very good accuracies in predicting the fundamental frequency of their prototypes. Based on the results presented in Tables 18.1, 18.2 and 18.3, models with less error in satisfying the design scaling laws show more accurate prediction of the prototype frequency. This observation indicates that the proposed methodology for use of scaling laws as a quantitative criterion to design scaled laminated structures, works properly.

By decreasing the desired total number of the layers for a model, the probability to find a model which works well with the response scaling laws Eqs. (18.4a–b) to predict the frequency of the prototype will decrease. This decrease implies that there is a practical limit to the amount of scaling that can be achieved using the proposed Distorted Layup Scaling technique for laminated structures. Although the proposed approach was investigated for specific laminated I-beams as the prototype, its applicability may be expanded for prototypes with other geometries and layups. As long as accurate governing equations can be found for the laminated prototype, the Distorted Layup Scaling technique can develop partially similar models that predict the static and dynamic characteristics of the prototype. A summary of the results is shown in Table 18.4 that includes the layups and overall dimensions for the most accurate developed models and their corresponding prototypes.

18.7 Conclusion

A meaningful scaled-down model which is representative of its prototype can expedite and facilitate the certification for use of new materials in a wind turbine blade. A theoretical study was carried out to design scaled-down I-beams replicating the dynamic characteristic of the I-beam structure inside a utility scale wind turbine blade. Similitude theory was applied to the governing equations for vibration of a shear deformable thin-walled composite I-beam to derive the scaling laws.

Derived scaling laws were used to measure accuracy of the designed scaled-down models that can predict the fundamental frequency of the full-scale I-beam.

The ply level scaling approach was applied to a prototype with a unidirectional layup to develop models that are completely similar to their prototype. For a prototype with non-unidirectional layup, the Distorted Layup Scaling approach was developed to design partially similar scaled-down I-beams which have layups different than their prototype. The accuracy of the designed models in replicating the dynamic characteristic of the prototype was investigated by predicting the fundamental frequency of the prototype using derived scaling laws. According to the results, depending on the layup of the prototype and desired size of the model, scaled-down I-beams were designed which could predict the fundamental frequency of their corresponding prototype with a very good accuracy.

Acknowledgments This material is based upon work supported by the National Science Foundation under Grant Number 1230884 (Achieving a Sustainable Energy Pathway for Wind Turbine Blade Manufacturing). Any opinions, findings, and conclusions or recommendations expressed in this material are those of the author(s) and do not necessarily reflect the views of the National Science Foundation.

References

1. Mandell, J.F., Combs, D., Samborsky, D.D.: Fatigue of fiberglass beam substructures. Wind Energy **16**, 99 (1995)
2. Cairns, D.S., Skramstad, J.D., Mandell, J.F. Evaluation of hand lay-up and resin transfer molding in composite wind turbine blade structures. In: 20th 2001 ASME Wind Energy Symposium, 2001
3. Mandell, J.F., Creed Jr., R., Pan, Q., Combs, D.W., Shrinivas, M.: Fatigue of fiberglass generic materials and substructures. Wind Energy **15**, 207 (1994)
4. Sayer, F., Post, N., Van Wingerde, A., Busmann, H., Kleiner, F., Fleischmann, W., Gansow, M.: Testing of adhesive joints in the wind industry. In: European Wind Energy Conference and Exhibition 2009, EWEC 2009, pp. 288–315 (2009)
5. Sayer, F., Antoniou, A., Van Wingerde, A.: Investigation of structural bond lines in wind turbine blades by sub-component tests. Int. J. Adhes. Adhes. **37**, 129–135 (2012)
6. Zarouchas, D.S., Makris, A.A., Sayer, F., Van Hemelrijck, D., Van Wingerde, A.M.: Investigations on the mechanical behavior of a wind rotor blade subcomponent. Compos. B Eng. **43**, 647–654 (2012)
7. Jensen, F.M., Falzon, B.G., Ankersen, J., Stang, H.: Structural testing and numerical simulation of a 34 m composite wind turbine blade. Compos. Struct. **76**, 52–61 (2006)
8. White, D., Musial, W., Engberg, S.: Evaluation of the B-REX fatigue testing system for multi-megawatt wind turbine blades. In Collection of the 2005 ASME Wind Energy Symposium Technical Papers at the 43rd AIAA Aerospace Sciences Meeting and Exhibit, pp. 52–65 (2005)
9. Simitses, G.J., Rezaeepazhand, J.: Structural similitude for laminated structures. Compos. Eng. **3**, 751–765 (1993)
10. Rezaeepazhand, J., Simitses, G.J.: Use of scaled-down models for predicting vibration response of laminated plates. Compos. Struct. **30**, 419–426 (1995)
11. Rezaeepazhand, J., Simitses, G.J., Starnes Jr., J.H.: Design of scaled down models for stability of laminated plates. Aiaa j. **33**, 515–519 (1995)
12. Simitses, G.J., Rezaeepazhand, J.: Structural similitude and scaling laws for buckling of cross-ply laminated plates. J. Thermoplast. Compos. Mater. **8**, 240–251 (1995)
13. Rezaeepazhand, J., Simitses, G.J., Starnes Jr., J.H.: Design of scaled down models for predicting shell vibration response. J. Sound Vib. **195**, 301–311 (1996)
14. Rezaeepazhand, J., Simitses, G.J.: Structural similitude for vibration response of laminated cylindrical shells with double curvature. Compos. B Eng. **28**, 195–200 (1997)
15. Eydani Asl, M., Niezrecki, C., Sherwood, J., Avitabile, P.: Predicting the vibration response in subcomponent testing of wind turbine blades. In: Proceedings of the IMAC-XXXIII, Orlando, 2–5 February 2015
16. Eydani Asl, M., Niezrecki, C., Sherwood, J., Avitabile, P.: Application of structural similitude theory in subcomponent testing of wind turbine blades. In: Proceedings of the American Society for Composites 2014, La Jolla, 8–10 September 2014
17. Eydani Asl, M., Niezrecki, C., Sherwood J., Avitabile, P.: Similitude analysis of composite I-beams with application to subcomponent testing of wind turbine blades. In: SEM 2015 Annual Conference and Exposition on Experimental and Applied Mechanics, Costa Mesa, 8–11 June 2015
18. Vo, T.P., Lee, J.: Flexural–torsional coupled vibration and buckling of thin-walled open section composite beams using shear-deformable beam theory. Int. J. Mech. Sci. **51**, 631–641 (2009)
19. Griffith, D.T., Ashwill, T.D.: The Sandia 100-meter all-glass baseline wind turbine blade: SNL100-00, Sandia National Laboratories, Albuquerque, Report no.SAND2011-3779, 2011
20. Reddy, J.N.: Mechanics of Laminated Composite Plates and Shells: Theory and Analysis. CRC press, Boca Raton (2004)

MIX
Papier aus verantwortungsvollen Quellen
Paper from responsible sources
FSC® C105338

If you have any concerns about our products,
you can contact us on
ProductSafety@springernature.com

In case Publisher is established outside the EU,
the EU authorized representative is:
Springer Nature Customer Service Center GmbH
Europaplatz 3, 69115 Heidelberg, Germany

Printed by Libri Plureos GmbH
in Hamburg, Germany